Springer-TUP Physics Series

The Springer-TUP Physics series is a collaborative effort between Springer and Tsinghua University Press (TUP). It publishes monographs and edited volumes authored/edited by leading researchers in China and globally. The series is dedicated to presenting the latest advancements across the entire spectrum of physics, including theoretical research, experiments, technology, and applications. Under the leadership of Series Editors Yichun ZHOU, Mingbo PU, and Chen WANG, the series maintains a high standard of academic excellence. The managing series editor from Tsinghua University Press is Yongfang LU.

Yanbiao Liao · Min Li · Wu Kuang

Fiber Optic Interferometer

Fundamentals and Applications

清華大學出版社
Tsinghua University Press

Springer

Yanbiao Liao
Tsinghua University
Beijing, China

Min Li
Wuhan University of Technology
Wuhan, Hubei, China

Wu Kuang
Realtimin Technology Co., Ltd.
Beijing, China

ISSN 3091-3012 ISSN 3091-3020 (electronic)
Springer-TUP Physics Series
ISBN 978-981-95-2557-7 ISBN 978-981-95-2558-4 (eBook)
https://doi.org/10.1007/978-981-95-2558-4

Translation from the Chinese Simplified language edition: "Fiber Optic Interferometer: Fundamentals and Applications" by Yanbiao Liao et al., © Tsinghua University Press 2024. Published by Tsinghua University Press. All Rights Reserved.

Jointly published with Tsinghua University Press, Beijing, China

This Springer imprint is published by the registered company Springer Nature Singapore Pte Ltd.
The registered company address is: 152 Beach Road, #21-01/04 Gateway East, Singapore 189721, Singapore

Preface

Modern sensors are evolving with increased sensitivity, precision, compactness, and intelligence. Fiber-optic sensors are increasingly favored for their inherent slimness, sensitivity, resistance to electromagnetic interference, and the integration of sensing and transmission capabilities. Fiber-optic interferometers have been widely used in aerospace, national defense, marine, and power industries. They are typical representatives of the first fiber-optic sensors applied in engineering and possess extremely high sensitivity. Nevertheless, numerous technical obstacles persist in its engineering application, including the resolution of temperature drift and stability issues.

Since the inception of optical fiber sensing technology in the late 1970s, there have been only a handful of professional books published, such as Eric Udd's *Fiber Optic Sensors*, Herve C. Lefevre's *Fiber Optic Gyroscope*, Wei Jin's *Novel Fiber Optic Sensors*, Chuan Li's *Fiber Gratings: Principle, Technology and Sensing Application*, and Libo Yuan's *Optical Fiber White Light Interferometric Sensing Technology*, among a few others. The content provides either a comprehensive overview of optical fiber sensors or an in-depth discussion on a specific type of sensor. There is no comprehensive monograph on optical fiber interferometers, which are based on phase modulation and demodulation, feature complex sensing structures, employ mature signal processing technology, offer versatile applications, and still possess broad potential for future development. Thus, practitioners require a professional tutorial that systematically explains the fundamental theory of interferometers and key technologies, such as signal demodulation and processing.

Beginning with the wave theory and coherence of light, this book systematically explores the signal processing technology of optical fiber interferometers. It builds upon an explanation of the transmission and control of polarized light within optical fibers, the optical characteristics of fiber interferometers, and provides a targeted analysis of optical fiber interferometers in typical application cases.

Building upon the coverage of the primary types of optical fiber interferometers, the short-cavity FPI structure, broadband light source interferometers, and long-range interferometers have been added. The signal processing methods for double-beam, multi-beam, and long-range interferometers are systematically classified and explained. The application technologies of typical optical fiber interferometers, such as fiber-optic underwater acoustic sensor, fiber-optic vector/acceleration sensor, fiber-optic gyroscope, fiber-optic hydrogen sensor and LPG-MZ composite refractive index sensor, which are widely used in commercial applications, are analyzed.

The book presents a clear theoretical framework, detailed technical schemes, and vivid application examples that have been extensively used by practitioners to explain the concepts and technologies behind fiber-optic interferometers. It has been compiled based on extensive research experience conducted by the author and his research team in the field of fiber-optic interferometer development. This book aims to provide valuable insights for researchers and professionals engaged in photoelectric information, automatic control, intelligent sensing, and related fields.

Beijing, China — Yanbiao Liao
Wuhan, China — Min Li
Beijing, China — Wu Kuang

Contents

1 **Optical Interference** 1
1.1 Superposition of Light Waves 1
1.1.1 Overview 1
1.1.2 Method for Obtaining Coherent Light 2
1.1.3 Superposition of Two Plane Light Waves 5
1.2 Amplitude-Splitting Two-Beam Interferometry 6
1.2.1 Equal-Inclination Interference 6
1.2.2 Multi-beam Interference of PPP 13
1.3 Low-Coherence Light Source Interferometry 19
1.4 Coherence of Light 26
1.4.1 Overview 26
1.4.2 Coherence Time and Coherence Length 27
1.4.3 Coherence Time and Spectral Linewidth 29
1.4.4 Mutual Coherence Function and Complex Correlation Coefficient 32
1.4.5 Coherence Time and Coherence Length of a Light Source 35
1.4.6 Spatial Coherence 41
1.4.7 Interference Characteristics of Partially Coherent Light 49
1.4.8 Laser Coherence 50
1.5 Typical Multi-beam Interferometer: Fabry–Perot Interferometer 51
1.5.1 Structure of a Fabry–Perot Interferometer 51
1.5.2 Performance Parameters of Interferometer 52
References 58

2 **Transmission and Control of Polarized Light in Optical Fiber** 59
2.1 Introduction 59
2.2 Wave Theory of Uniform Refractive Index Fiber 61
2.2.1 Maxwell's Equations and Wave Equations 61
2.2.2 Vector Mode 65

2.2.3 Linear Polarization Mode and Scalar Method 74
2.2.4 Two-Layer Uniform Optical Fiber 76
2.3 Transmission and Control of Polarized Light in Optical Fiber 87
2.3.1 Polarization Characteristics of Single-Mode Optical Fibers 87
2.3.2 Perturbation Coupled Mode Theory of Weakly Guided Fiber 91
2.3.3 Perturbation Coupled Mode Theory of Birefringence Analysis in Optical Fiber 93
2.3.4 Description of Polarization of Single-Mode Fiber Using the Poincaré Sphere 98
2.3.5 Polarization Characteristics of Twisted Fiber and Spun Fiber 103
2.4 Jones Matrix of Optical Fiber and Polarization Perturbation 108
2.4.1 Transmission Matrix of Polarized Light in Optical Fibers 108
2.4.2 Variation in Polarization Characteristics Caused by Extrusion and Twist 110
2.4.3 Thermal Effect on Polarization State 111
2.5 Polarization Control in Optical Fiber and Devices 111
2.5.1 Polarization-Maintaining Fiber, Polarizing Fiber and Spun Fiber 112
2.5.2 Optical Fiber Polarization Devices 115
References 129

3 Types and Optical Characteristics of Optical Fiber Interferometers 131
3.1 Optical Fiber Interferometer: Phase Modulation Optical Fiber Sensor 131
3.2 Double Beam Interferometer 132
3.2.1 Mach-Zehnder Interferometer 132
3.2.2 Michelson Fiber Interferometer 134
3.2.3 Sagnac Fiber Interferometer 137
3.3 Fiber FP Interferometer-Double Beam and Multi-beam Interferometry 147
3.3.1 Introduction 147
3.3.2 Fundamentals 149
3.3.3 Classification and Characteristics 150
3.3.4 Resonance Modes and Enhancement of Short Cavity FP Structures 153
3.3.5 Signal Demodulation of Fiber FP Sensor 156
3.4 Multi-beam Interferometer: Fiber Ring Resonator 158
3.5 Fiber Optic White Light Interferometer 160
3.5.1 Principles and Characteristics 160
3.5.2 Advantages and Difficulties 163

3.5.3 Strain and Temperature Measurement Technology Based on Optical Fiber White Light Interferometry 164
3.6 Optical Fiber Polarization Interferometer 167
3.7 Optical Fiber Long-Range Interferometer 168
3.7.1 Principle of Event Monitoring with MZ Interferometer 168
3.7.2 The Principle of Event Location Using Time Delay Estimation 169
3.8 Phase Modulation of Optical Fiber by External Stress 174
3.9 Temperature Induced Phase Modulation of Fiber 178
References 180

4 Signal Processing of Fiber Interferometer 181
4.1 Overview 181
4.2 Signal Processing of Dual-Beam Fiber Interferometer 182
4.2.1 Laser-Based Interferometric Relative Measurement 182
4.2.2 Fiber-Optic Interferometers Utilize Broad-Spectrum Light Sources for Absolute Measurements 195
4.3 Multi-beam Fiber Interferometer Signal Processing 210
4.3.1 Absolute Measurement 212
4.3.2 Relative Measurement 212
4.4 Long-Range Fiber Interferometer Signal Processing 214
4.4.1 ϕ-OTDR 215
4.4.2 OFDR 223
4.5 Influence of Light Source Spectrum 227
4.5.1 Laser Light Source 228
4.5.2 Broadband Light Sources 230
4.6 Photoelectric Conversion in Fiber-Optic Sensors 232
4.6.1 Single-End Detection Structures 232
4.6.2 Differential Balanced Detection 235
References 236

5 Applications of Fiber-Optic Interferometers 239
5.1 Fiber Optic Underwater Acoustic Sensor 239
5.1.1 Design and Preparation of Fiber Optic Hydrophone Sensing Unit 241
5.1.2 Phase Shift Sensitivity of Composite Structured Fiber Hydrophones 242
5.1.3 Lightsource and Noise in the Fiber Underwater Acoustic Sensor 249
5.2 Fiber-Optic Vector/Acceleration Sensor 250
5.2.1 Membrane-Type Acceleration Sensor 253
5.2.2 Interference Fiber Acceleration Sensor Based on Compliant Cylinder 257
5.2.3 The Fundamental Architecture of a Fiber Optic Vector Hydrophone (Acoustic Pressure Gradient Sensor) 262
5.3 Fiber-Optic Rotation Sensor (Gyro) 264

5.3.1 Interferometric Fiber Optic Gyroscope 264
5.3.2 Resonant Fiber Optic Gyro 267
5.4 Fiber-Optic Hydrogen Sensor 270
5.4.1 Fiber Fizeau Broadband Interferometric Hydrogen Sensor 271
5.4.2 Cavity Length Demodulation and Hydrogen Sensing Based on Correlation Principles 272
5.5 Long-Period Grating MZ Interferometric Refractive Index Sensor 276
5.5.1 Principle of the Long-Period Grating MZ Interferometer 276
5.5.2 LPG-MZ Interferometric Refractive Index Sensor 278
References 280

Chapter 1
Optical Interference

Optical fiber interference technology is a subset of optical interference technology that utilizes optical fibers. Consequently, it adheres to the fundamental principles of optical interference. The unique waveguide properties of optical fibers have led to the emergence of numerous distinctive phenomena, theories, and application techniques. This chapter primarily explores the interference process of light, encompassing the requirements for a light source to produce interference, the laws governing interference phenomena, and their practical applications. Firstly, we discuss the fundamental laws of two-beam interference. For instance, how to determine the distribution of the interference field and the characteristics of interference fringes through the optical path difference (OPD), as well as two typical interferences due to the division of amplitude: equal-thickness interference and equal-inclination interference. We also explore the light intensity distributions of the interference field and related applications; the coherence of light; and the characteristics of double-beam interference (such as the Michelson interferometer, MI) and multi-beam interference (such as the Fabry–Perot interferometer, FPI). The focus of this chapter is on the calculation of optical path, the coherence of the optical field, and the characteristics of multi-beam interference.

1.1 Superposition of Light Waves

1.1.1 Overview

The superposition of light waves adheres to specific rules: in areas where superposition does not occur, light waves propagate according to their inherent laws, as illustrated by the shaded area in Fig. 1.1; within the superposition zone, the propagation of light waves varies depending on the difference in the characteristic parameters of the two waves.

Y. Liao et al., *Fiber Optic Interferometer*, Springer-TUP Physics Series,
https://doi.org/10.1007/978-981-95-2558-4_1

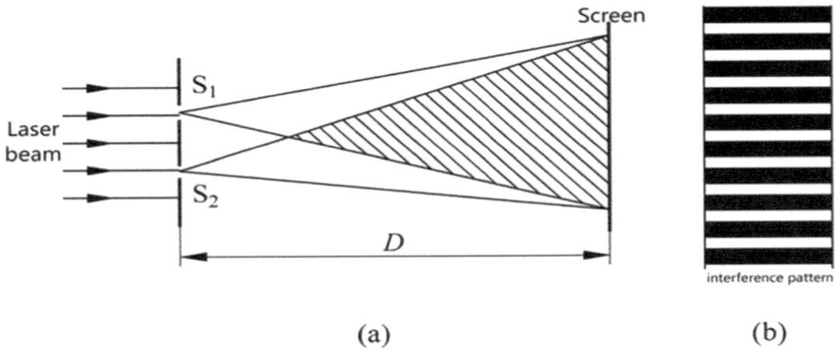

Fig. 1.1 Double-slit interference of light

Light waves have four characteristic parameters: amplitude, phase, polarization, and wavelength. Depending on the differences in these parameters between two light waves, various outcomes can result from their superposition.

1. Superposition of light vectors occurs when they have the same direction and wavelength (though they may differ in amplitude and phase). Interference and diffraction of monochromatic light are examples of this type of superposition.
2. Superposition of light vectors with the same direction but different wavelengths, amplitudes, and phases includes phenomena such as white light interference, light pulses, and light modulation.
3. Superposition occurs when waves have the same wavelength but differ in light vector direction, amplitude, and phase. When the light vectors' directions are similar, they can be approximated as interference and diffraction. When the light vectors' directions are perpendicular to each other, it pertains to polarized light interference.

1.1.2 Method for Obtaining Coherent Light

To demonstrate the interference phenomenon of a double-beam in the laboratory, the simplest method involves allowing a laser beam to pass through two narrow slits—commonly referred to as slits or two small circular holes—resulting in a typical double-slit interference pattern (typically known as Young's double slit experiment) on the white screen behind them: alternating light and dark stripes with equal spacing. As depicted in Fig. 1.1, the left side of the figure depicts the experimental device, while the right side illustrates the interference fringes. In the figure, S_1 and S_2 represent two long, narrow slits. These can be created by carving two grooves on a piece of waste film using a blade. There is no strict requirement for the distance between the two grooves; approximately 1 mm is sufficient. The purpose of the double slit is to generate two coherent beams of light, and the superposition of these two coherent light beams results in interference fringes.

However, when light waves from two independent sources overlap, they simply add to each other's intensity, resulting in no light or dark fringes. For instance, when two fluorescent lamps illuminate the same area simultaneously, the intensity increases (yielding a uniform light), but no bright or dark fringes emerge. The reason is that two independent fluorescent lamps are not coherent light sources. To comprehend why they are incoherent, we must explore some characteristics of light sources.

Double-beam interference is a phenomenon that occurs when two beams of light intersect: in the overlapping region of the two light waves, the light intensity is redistributed. Stable interference patterns, alternating between light and dark, often emerge. The stability of these patterns refers to the spatial distribution of light intensity within a specific time frame, which typically far exceeds the response time of the photodetector. For instance, the visual persistence time of the human eye, the exposure time of film, and the response time of a photocell do not change over time. Whether this intensity distribution remains stable is the primary indicator for distinguishing coherence from incoherence at present.

Upon examining the nature of light emission, we discover that a light source comprises numerous luminous atoms, molecules, or electrons, with each atom or molecule serving as a luminous center (essentially a tiny point light source). Every beam of light we see is emitted and collected from a multitude of atoms, which serve as luminous centers. On the other hand, the luminescence of individual atoms is not perpetual. The luminous action can only last for a brief period each time, which is very short (experimentally proven to be less than 10^{-8} s), so atomic luminescence can only produce a limited spatial wave train at a time. The luminescence mode of ordinary light sources is primarily spontaneous radiation. Each atom is an independent luminous center, and the luminescence action of each atom is chaotic and has nothing to do with each other. Therefore, there is no constant phase relationship between the various wave trains generated by the same atom successively and between the wave trains generated by different atoms. The superposition of such light waves can certainly not observe interference. Because at a certain moment, the superposition of light waves may be constructive; at another moment, due to the change of the initial phase, its superposition may be destructive. Within one second, this change in strength may occur more than 10^8 times. The overall effect is that the time average of the interference term $\boldsymbol{J}_{12}$ is zero:

$$\frac{1}{\tau}\int_0^\tau \sqrt{I_1 I_2}\cos\delta dt = 0$$

It can be seen that not only does the light emitted from two ordinary light sources not interfere, but even the light emitted from two different parts of the same light source is incoherent. Therefore, the ordinary light sources are incoherent light sources.

A new type of light source, the laser, was invented in the 1960s. Its primary luminous characteristics include: the main mechanism of atomic luminescence is stimulated emission, meaning that the movements and timing of each luminescent atom are orderly, regular, and coordinated with one another. Consequently, there is a consistent phase relationship between the wave trains emitted by the same atom successively, as well as between wave trains emitted by different atoms in space. As a result, lasers possess excellent coherence and are considered coherent light sources.

From the above discussion, it can be seen that in order to obtain stable interference fringes, the following requirements are imposed on the light source: ① The two light waves have the same frequency; ② The two light waves vibrate in the same direction at the point where they meet; ③ The two light waves have a fixed phase difference between at the point where they meet. For the light interference, due to the characteristics of light source, the most critical thing is to meet the requirement ③. Therefore, the only way to obtain a coherent light source in optics is to divide a wave train into two or several waves, and then make them overlap to produce a stable interference effect. There is no other better way.

The double-slit interference device will be a good help to understand that the initial phase difference of the two light waves can be kept constant by splitting one into two. In Fig. 1.1, it is assumed that two slits S_1 and S_2 are illuminated by a light wave from point S_0. This light wave is divided into two parts by the double slit, so the two waves emitted from S_1 and S_2 are from the same wave train, and their initial phase difference is determined by the path difference $\Delta_0 = S_0S_2 - S_0S_1$, which has nothing to do with the initial phase of the wave train itself. Therefore, no matter how the initial phase of this wave train changes with time, The initial difference between the two light waves emitted from S_1 and S_2 remains unchanged. Therefore, the problem of unstable initial phase of the light source can solved by using this "one in two" method. The above three points are the necessary conditions for the occurrence of interference phenomenon, and in order to obtain a certain quality of interference fringes (that is, the contrast of the fringes can be visually observed or measured), the following two sufficient conditions need to be met: ① The amplitudes (or light intensity) of the light waves participating in the interference are similar at the superposition of the light waves; ② The time difference between the two light waves participating in the interference reaching the superposition is much less than the coherence time of the light source, or the optical path difference, i.e., the product of refractive index and geometric path difference, is much less than the coherence length of the light source. This will be explained in Sect. 1.4 of coherence of light.

There are generally two ways to obtain coherent light [1]: amplitude-splitting interference and wavefront-splitting interference. The amplitude of the incident light is decomposed into several parts by the sequential reflection of the incident light waves from the first and the second surface of a transparent thin plate. The interference caused by the encounter of these parts of light waves is called amplitude-splitting interference, which is a common method to generate interference. The color of oil film and soap is the result of this kind of interference. The so-called wavefront-splitting method is to split the wavefront of a wave train into two parts or several parts. When the waves emitted by each part meet again, they must be coherent. The double-slit interference introduced above is an early wavefront-splitting interference. The wavefront of a wave train is split into two parts by two slits S_1 and S_2, these two parts become two coherent light sources.

1.1.3 Superposition of Two Plane Light Waves

Assume that two light waves $u_1 = a_1 e^{i\varphi_1}$ and $u_2 = a_2 e^{i\varphi_2}$ with the same frequency and the same vibration direction are superimposed in space, then the light vibration at any point P is the result of the superposition of the vibrations caused by the two waves at that point. The light intensity after the superposition of the two waves is

$$I = u \times u^* = \left(a_1 e^{i\phi_1} + a_2 e^{i\phi_2}\right)\left(a_1 e^{-i\phi_1} + a_2 e^{-i\phi_2}\right)$$
$$= a_1^2 + a_2^2 + 2a_1 a_2 \cos(\phi_1 - \phi_2)$$

or

$$I = I_1 + I_2 + 2\sqrt{I_1 I_2}\cos\delta \tag{1.1}$$

In formula (1.1), $I_1 = a_1^2$, $I_2 = a_2^2$ are the light intensities of the two light waves at point P (omitting a constant factor), and $\delta = \varphi_1 - \varphi_2$ is the phase difference of the two light waves at point P. The result shows that the total intensity of the superimposed two beams is not equal to the sum of the intensities of the two waves, that is $I \neq I_1 + I_2$, there is an extra cross term. It reflects the interference effect of two beams of light, usually called the interference term, J_{12} is represented by

$$J_{12} = 2\sqrt{I_1 I_2}\cos(\varphi_1 - \varphi_2) \tag{1.2}$$

Therefore, in order to observe interference in the superposition of two light waves, $J_{12} \neq 0$ must be satisfied. The interference term J_{12} is related to the phase difference δ, so its value varies depending on the position of point P and can be positive or negative. When $J_{12} > 0$, $I > I_1 + I_2$, is constructive interference; When $J_{12} < 0$, $I < I_1 + I_2$, is destructive interference. Obviously, when $\delta = 0, \pm 2\pi, \pm 4\pi, \ldots$, the interference intensity is maximum (interference maximum)

$$I_{max} = I_1 + I_2 + 2\sqrt{I_1 I_2}$$

When $\delta = \pm\pi, \pm 3\pi, \pm 5\pi, \ldots$, the interference intensity is minimum (interference minimum)

$$I_{min} = I_1 + I_2 - 2\sqrt{I_1 I_2}$$

where, the formula $I_1 = I_2$ Eq. (1.1) becomes

$$I = 4I_1\cos^2(\delta/2) \tag{1.3}$$

At this time, $I_{max} = 4I_1$, $I_{min} = 0$. Figure 1.2 shows the function curve of formula (1.3).

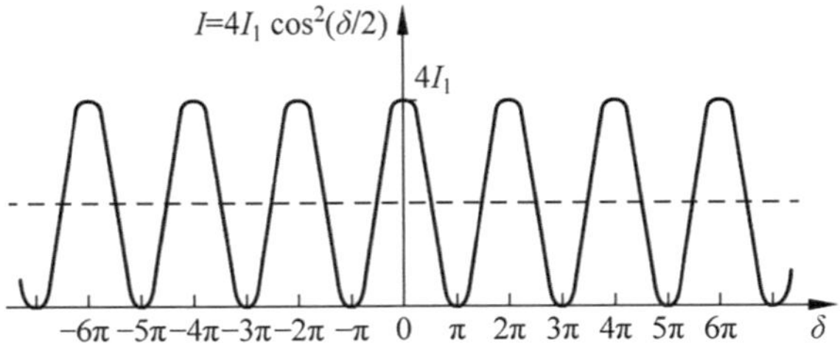

Fig. 1.2 Light intensity distribution of equal-intensity double-beam interference

1.2 Amplitude-Splitting Two-Beam Interferometry

1.2.1 Equal-Inclination Interference

Calculation of Optical Path Difference

To study the variation of optical path difference (OPD) in amplitude-splitting interference, it is essential to first discuss the interference phenomenon of plane parallel plates (PPP).

Suppose there is a uniform transparent PPP with a thickness of h and a refractive index of n. Monochromatic light with a wavelength of λ incidents on the upper surface of the plate with an incident angle of θ_0. After being reflected by the upper and lower surfaces, the reflected beams $1'$ and $2'$ are obtained, as illustrating in Fig. 1.3. Obviously, the two reflected light beams $1'$ and $2'$ are coherent light because they are separated from the same incident light beam. Since the upper and lower surfaces of the PPP are assumed to be parallel to each other, the reflected light beams $1'$ and $2'$ are also parallel to each other, so these two coherent light beams can only intersect at infinity, that is, the interference occurs at infinity. For this reason, a convex lens L is necessary to move the interference fringes closer (to the focal plane of L) for observation and measurement. Therefore, the interference fringes of PPP appear at infinity, which is a kind of localized interference.

To find out the light intensity at point P, it is necessary to calculate the phase difference between the two reflected lights $1'$ and $2'$ when they reach point P. There is no phase difference when the two beams are separated from point A. In the distance from DC to P, there will be no phase difference between the two beams (because the ideal lens does not produce additional path difference for these two beams). Such as in Fig. 1.3, $CD \perp AD$. Therefore, the phase difference between the two light beams $1'$ and $2'$ only comes from the OPD between the two paths of AD and ABC. If the flat plate is placed in a medium with a refractive index of n_0, the OPD between the two beams is

$$\Delta = 2nAB - n_0AD = \frac{2h}{\cos\theta}(n - n_0 \sin\theta \sin\theta_0)$$

Then using the law of refraction and simplifying, we can get

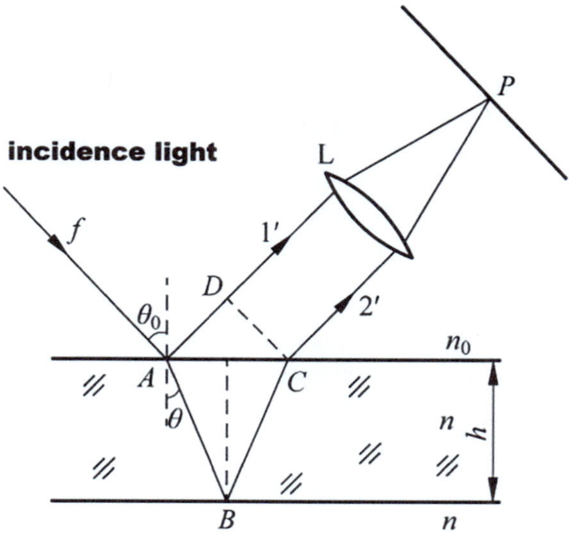

Fig. 1.3 The optical path diagram of parallel plate interference

$$\Delta = 2nh\cos\theta = 2h\sqrt{n^2 - n_0^2\sin^2\theta_0} \tag{1.4}$$

Formula (1.4) is the basic formula to discuss PPP. Where, h and n are the thickness and refractive index of the plate, n_0 is the refractive index of the surrounding medium, θ_0 and θ are the incident angle and refractive angle.

From the above results, it can be drawn that:

1. The interference effect of PPP is mainly determined by the OPD of the two light waves, and its value $\boldsymbol{\Delta}$ is calculated by Formula (1.4).
2. The interference fringes of PPP are mainly determined by three factors: the refractive index n and the thickness h of the plate, the incident angle θ_0 of the light beam. If the refractive index n of the plate is constant (i.e. uniform plate), the change in the fringe is only controlled by two quantities h and θ_0. Therefore, for ease of discussion, plate interference is classified into two categories: when θ_0 is *fixed* (the incident angle is constant) and h *changes* (the plate thickness at each point is different), each interference fringe corresponds to the trajectory of the points of the same thickness, so this type of interference is called equal-thickness interference. The other type is that the plate thickness remains unchanged (h is the same at each point on the plate) and the incident angle θ_0 is changed. At this time, each interference fringe corresponds to the result of interference by incident light with the same inclination angle (incident angle), so it is called equal-inclination interference.

In addition, in the interference of layered media (multilayer film), refractive index is also an important factor controlling the interference effect.

3. In this type of interference, we should pay attention to the problem of additional OPD. Although never occurs between the refracted light and the incident

light, a phase mutation often exists between the reflected light and the incident light. Therefore, the additional optical path caused by the phase mutation must be considered for the reflected light. Specific problems should be discussed by Fresnel formula. However, for the common interference problems of small angle incidence or grazing incidence, the following rules can be used: When the reflected light is reflected at the interface from a medium with a small refractive index (optically sparse medium) to a medium with a large refractive index (optically dense medium), there is an additional optical path of $\frac{\lambda}{2}$. On the contrary, when it is incident from a medium with a large refractive index to a medium with a small refractive index, there is no additional optical path. Alternatively, if the plate is immersed in the same uniform medium, such as air or water, there will always be an additional optical path of $\frac{\lambda}{2}$ (independent of the angle of incidence) between the two beams reflected from the upper and lower surfaces of the plate. In a specific device, if there is an additional OPD between the two light beams, the OPD formula (1.4) should be changed to

$$\Delta' = 2nh\cos\theta \pm \frac{\lambda}{2} \tag{1.5}$$

4. The distribution of fringe intensity is determined by the following formula

$$I = I_1 + I_2 + 2\sqrt{I_1 I_2}\cos k\Delta$$

Therefore, the fringe intensity distribution is uniquely determined by the OPD Δ. When there is no additional OPD

$$\Delta = 2hn\cos\theta = m\lambda (m = 0, 1, 2 \ldots) \tag{1.6}$$

When the interference light intensity is maximum,

$$\Delta = 2nh\cos\theta = m\lambda \left(m = \frac{1}{2}, \frac{3}{2}, \frac{5}{2} \cdots \right) \tag{1.7}$$

where the interference light intensity is minimum.

When there is an additional OPD, the interference intensity will be maximum at

$$\Delta = 2nh\cos\theta \pm \frac{\lambda}{2} = m\lambda (m = 0, 1, 2, \ldots)$$

and minimum at,

$$\Delta = 2nh\cos\theta \pm \frac{\lambda}{2} = m\lambda \left(m = \frac{1}{2}, \frac{3}{2}, \frac{5}{2} \cdots \right)$$

Experimental Apparatus for Observing Equal-Inclination Interference

Equal-inclination refers to a set of interference fringes generated when the thickness of a thin film (generally, when the thickness of a plate is very small, it is called a thin film) is the same everywhere, and the light beam is incident at various angles. Therefore, in order to obtain iso-inclined fringes, there are two key points: one is to have a uniform film with constant thickness (or an air layer with constant thickness), and the other is that the light beam incidents on the film at different angles. The specific device for observing equal-inclination interference is shown in Fig. 1.4. The light emitted by an extended light source S pass through a half-reflecting mirror M and is incident on the upper and lower surfaces M_1 and M_2 of the thin film at various angles. The thickness of the film is h. The refractive index inside and outside the film are n and n_0, respectively. The incident light is reflected by the upper and lower surfaces of the thin film and focuses on its focal plane by lens L, where equal-inclination interference fringes are formed. The equal-inclination interference fringes are circular (observed in the direction perpendicular to the film), and the formation process of these circular fringes is as follows. Suppose a certain light beam S_0 emitted by the light source incidents on the surface of the film at an certain angle of θ_1. After being reflected by the upper and lower surfaces of the film, two beams of the reflected light R_1 and R_2 parallel to each other are generated. The two light beams intersect at point P on the focal plane of L to produce interference, and the effect of interference is determined by their OPD. If a light beam from the light source S_0' is also incident on the surface of the film at the same angle θ_1, then its two reflected lights R_1' and R_2' must be parallel to R_1 and R_2, and intersect at the same point P to interference after passing through the lens L. Due to the OPD between R_1' and R_2' is the same as that between R_1 and R_2, the interference effect at point P is also the same. Therefore, as long as the incident angle is equal, the two reflected lights have the same OPDs, and their meeting points on the focal plane belongs to the same interference order. On the other hand, it is obvious from the experimental device that all light beams on the cone generatrix with the optical axis *OO'* of lens L as the axis and the incident angle θ_1 as the semi-apex angle of the cone, after being reflected by the flat plate, will be concentrated on the focal plane with O as the center of the circle and OP as a circumference of the radius, that is, the fringes of the same interference order are a circle.

Because the light source is an extended light source, it emits light in a certain area, so a set of interference rings are formed on the focal plane. From the experimental device, it can be seen that the formation of interference fringes is that (incident angle) changes, and h remains unchanged (film thickness). So the same fringe represents the same inclination angle (angle of incidence or angle of refraction). Equal-inclination fringes are formed at infinity, that is, the fringes are localized at infinity.

Fig. 1.4 Optical path diagram of equal-inclination interference

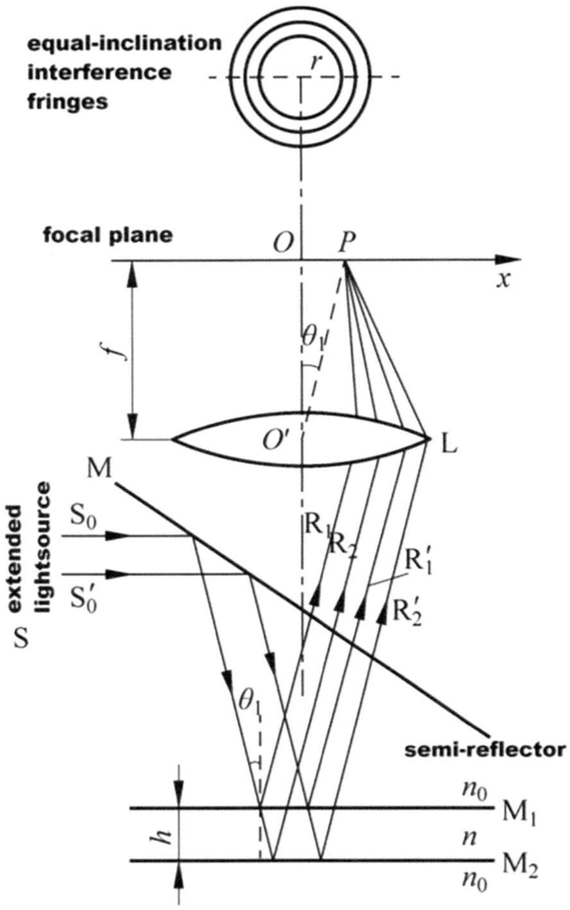

Properties of Equal-Inclination Fringes

1. Central fringe

Because the central fringe is the center point O of the interference fringe, it is formed by the interference of light with an incident angle $\theta_1 = 0$ (of course, there is $\theta = 0$). Therefore, according to the OPD equation, m_0 has an expression as,

$$m_0 = \frac{\Delta_0}{\lambda} = \frac{2nh}{\lambda} + \frac{1}{2} \tag{1.8}$$

m_0 is the interference order of the central fringe. It is often a large number in practical interference devices. For example, when an equal-inclination circular ring is produced by a flat glass sheet with a thickness of 1 mm ($n = 1.50$), the order of its central bright ring m_0 for green light of $\lambda = 0.5\ \mu\text{m}$ is

$$m_0 = \frac{2 \times 1.50 \times 1\,\text{mm}}{0.5 \times 10^{-3}\,\text{mm}} + \frac{1}{2} \approx 6000$$

From the OPD Eq. (1.6) of equal-inclination interference, cos θ decreases with the incident angle θ_1 increases (the refraction angle θ also increases), so the OPD also decreases accordingly. Therefore, when the light is perpendicularly incident, the OPD is the largest, and the interference order is the highest. Therefore, the closer the equal-inclination interference ring is to the edge, the lower the interference order is, which is a feature of equal-inclination interference. For ease of measurement, the ring closest to the center is often recorded as the first ring, and the second ring, the third ring,..., and the Nth ring are recorded in sequence towards the edge. The corresponding relationship between this artificially marked ring number N and the interference order m of the ring itself is as follows:

Interference order $m_0, m_1, m_1 - 1, m_1 - 2, \ldots, m_N$,

fringe numbers 1, 2, 3, ..., N

The relationship between them is

$$m_N + N = m_1 + 1$$

m_1 is the interference order of the first ring closest to the center. Then substitute the above equation into the OPD formula (1.5) and simplify, the relationship between N and the parameters can be found. Because

$$\Delta_N = 2nh\cos\theta + \frac{\lambda}{2} = m_N\lambda = (m_1 + 1 - N)\lambda \tag{1.9}$$

It is also generally found in interference devices, so $\theta \approx 0$

$$\cos\theta \approx 1 - \frac{\theta^2}{2}$$

Reuse

$$2nh + \frac{\lambda}{2} = m_0\lambda = (m_1 + \varepsilon)\lambda$$

where, $m_1 = m_0 - \varepsilon$, ($\varepsilon < 1$), ε is the fractional part of the interference order of the central fringe, because the order of the central fringe is not necessarily an integer. By substituting the above two expressions into Δ_N and simplifying, we can get:

$$\Delta_N = \sqrt{\frac{\lambda}{nh}} \cdot \sqrt{(N-1) + \varepsilon} \tag{1.10}$$

Then use the law of refraction at small angles: $n_0\theta_0 = n\theta$

$$\theta_{0N} = \frac{1}{n_0}\sqrt{\frac{n\lambda}{h}} \cdot \sqrt{(N-1)+\varepsilon} \tag{1.11}$$

where, θ_{0N} is the incident angle of the light beam corresponding to the N-th ring, λ is the wavelength of the incident light, n and h are the refractive index and thickness of the film, n_0 is the refractive index of the medium surrounding the film.

2. Radius and spacing of interference fringes

It can be seen from the experimental setup of equal-inclination interference in Fig. 1.4 that the radius r_{N} of the interference ring is determined by the focal length f of the lens L and the opening angle α_N of the interference ring to the center of the lens:

$$r_N = f \tan \alpha_N$$

α_N also known as the anglular radius of fringes, its value is generally very small, so there are

$$r_N \approx f \cdot \alpha_N$$

Obviously: $\alpha_N = \theta_0$. Substituted formula (1.11) in to the equation, the fringe angle radius is:

$$\alpha_N = \frac{1}{n_0}\sqrt{\frac{n\lambda}{h}} \cdot \sqrt{(N-1)+\varepsilon} \tag{1.12}$$

The fringe radius is:

$$r_N = f \cdot \frac{1}{n_0}\sqrt{\frac{n\lambda}{h}} \cdot \sqrt{(N-1)+\varepsilon} \tag{1.13}$$

Then $r_{N+1}^2 - r_N^2 = \frac{f^2}{n_0^2} \cdot \frac{n\lambda}{h}$, the expression of fringe spacing e_N can be obtained. Because e_N

$$e_N = r_{N+1} - r_N = \frac{f^2}{n_0^2} \cdot \frac{n\lambda}{h} \cdot \frac{1}{r_{N+1} + r_N}$$

and,

$$r_{N+1} + r_N \approx 2r_N = \frac{2f}{n_0}\sqrt{\frac{n\lambda}{h}} \cdot \sqrt{(N-1)+\varepsilon}$$

So

$$e_N = \frac{f}{2n_0} \cdot \sqrt{\frac{n\lambda}{h(N - 1 + \varepsilon)}} \tag{1.14}$$

It can be seen that the closer towards the edge (N Increase), the denser the fringes are.

1.2.2 *Multi-beam Interference of PPP*

In the interference of PPP (Sect. 1.2.1), we only discuss the dual-beam interference phenomenon produced by the superposition of two light beams reflected from the upper and lower surfaces of the plate. In fact, due to multiple reflections and refractions on the upper and lower surfaces, the number of reflected and transmitted light is far from two beams. Because it is a parallel plate, all reflected waves are parallel to each other, as well as all transmitted waves are. When the reflectivity is very low, for example, a glass plate ($n = 1.52$) placed in the air, the reflectivity of its upper and lower interfaces is 4%. At this time, only the first two reflected light beams with similar intensity can produce interference fringes with good visibility. And after that, the intensity of each reflected light drops sharply, and so does the transmitted light, so its interference effect can be ignored. On the contrary, when the reflectivity is high, this neglect is not allowed. For example, when using plate glass with high reflectivity on both sides, such as 90%, the intensity of each transmitted light beam is very similar. At this time, each light beam contributes to the interference effect and cannot be ignored. This effect caused by the superposition of multiple beams of coherent light is called **multi-beam interference**.

Intensity Distribution of Multi-beam Interference [2]

As shown in Fig. 1.5, assume that $E^{(i)}$ is the complex amplitude of the electrical vector of the incident plane wave, the incident wave is a linearly polarized wave, and the electrical vector is parallel to the incident plane or perpendicular to the incident plane. It can be seen from Fig. 1.5 that the OPD between two adjacent reflected light or transmitted light is.

$$\Delta = 2nh\cos\theta_t \tag{1.15}$$

The phase difference caused by the OPD is

$$\delta = k\Delta = \frac{4\pi}{\lambda_0} nh\cos\theta_t \tag{1.16}$$

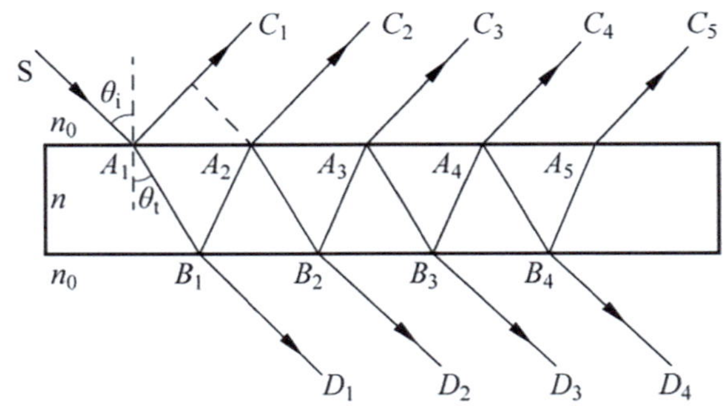

Fig. 1.5 Acquisition of multiple beams of coherent light

where, h and n are the thickness and the refractive index of the plate, λ_0 is the wavelength in vacuum, and θ_t is the refraction angle of light. The transmission amplitude ratio (the ratio of the amplitude of the transmitted light to that of the incident light) and reflection amplitude ratio (the ratio of the amplitude of the reflected light to that of the incident light) of the upper and lower surfaces are respectively: when the beam enters the flat plate from the medium, the transmission amplitude ratio is t and the reflection amplitude ratio is r; When the beam enters the medium from the flat plate, it is $t'$and r', as shown in Fig. 1.6. With these coefficients, the complex amplitude of each reflected light when it reaches the rendezvous point can be immediately calculated.

$$\left.\begin{array}{l} E_1^{(r)} = rE^{(i)} \\ E_2^{(r)} = r'tt'E^{(i)}e^{-i\delta} \\ E_3^{(r)} = r'^3tt'E^{(i)}e^{-i2\delta} \\ E_4^{(r)} = r'^5tt'E^{(i)}e^{-i3\delta} \\ \cdots\cdots\cdots\cdots\cdots\cdots \\ E_{P+2}^{(r)} = r'^{(2P+1)}tt'E^{(i)}e^{-i(P+1)\delta} \end{array}\right\} \tag{1.17}$$

The complex amplitude of each beam of transmitted light when it reaches the rendezvous point is

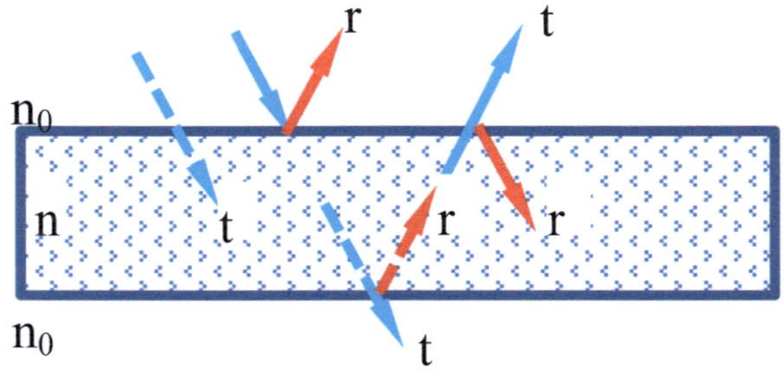

Fig. 1.6 Reflection and transmission of light on a PPP

$$\left.\begin{aligned} E_1^{(t)} &= tt'E^{(i)} \\ E_2^{(t)} &= r'^2 tt'E^{(i)}e^{-i\delta} \\ E_3^{(t)} &= r'^4 tt'E^{(i)}e^{-i2\delta} \\ &\ldots\ldots \\ E_P^{(t)} &= r'^{2(P-1)} tt'E^{(i)}e^{-i(P-1)\delta} \end{aligned}\right\} \tag{1.18}$$

Taking advantage of

$$\sum_{p=0}^{\infty} x^{P} = \frac{1}{1-x}, (\mathrm{p} = 1, 2, 3, \ldots)$$

The complex amplitude of the superimposed reflected light is

$$E^{(r)} = E_1^{(r)} + E_2^{(r)} + E_3^{(r)} + \cdots E_P^{(r)} + \cdots = \left[r + \frac{r'tt'e^{-i\delta}}{1 - r'^2 e^{-i\delta}}\right]E^{(i)} \tag{1.19}$$

From the Fresnel formula, it is known that the amplitude ratios have the following relationships:

$$r' = -r;\ tt' + r^2 = 1$$

Substitute to Eq. (1.19), and simplify to get

$$E^{(r)} = \frac{r\left(1 - e^{-i\delta}\right)}{1 - r^2 e^{-i\delta}}E^{(i)} \tag{1.20}$$

Thus obtained that the reflected light intensity after superposition is

$$I^{(r)} = I^{(r)} \propto E^{(r)}E^{(r)*} = \frac{r^2\left[2 - \left(\mathrm{e}^{\mathrm{i}\delta} + \mathrm{e}^{\text{-}\mathrm{i}\delta}\right)\right]}{1 + r^4 - r^2\left(\mathrm{e}^{\mathrm{i}\delta} + \mathrm{e}^{\text{-}\mathrm{i}\delta}\right)}I_0$$

Then the above equation can be simplified using equation $e^{i\delta} + e^{-i\delta} = 2\cos\delta$ and $1 - \cos\delta = 2\sin^2\frac{\delta}{2}$.

$$I^{(r)} = \frac{4R\sin^2\frac{\delta}{2}}{4(1-R)^2 + R\sin^2\frac{\delta}{2}}I_0 \tag{1.21}$$

where, $R = r^2$ is the reflectivity of the plate surface, and $I_0 \propto E^{(i)}E^{(i)*}$ is the incident light intensity.

$$\delta = \frac{2\pi}{\lambda_0}\Delta = \frac{2\pi}{\lambda_0}2nh\cos\theta_t = \frac{4\pi}{\lambda_0}nh\sqrt{1 - \sin^2\theta_t} \tag{1.22}$$

is the phase difference between two adjacent light beams. Where n and h are the refractive index and the thickness of the plate, and θ_t is the angle of refraction.

Similarly, for transmitted light,

$$I^{(t)} = \frac{(1-R)^2}{(1-R)^2 + 4R\sin^2\frac{\delta}{2}} I_0 \tag{1.23}$$

Set $F = \frac{4R}{(1-R)^2}$, Eqs. (1.21) and (1.23) can be rewritten as

$$\frac{I^{(r)}}{I_0} = \frac{F\sin^2\frac{\delta}{2}}{1 + F\sin^2\frac{\delta}{2}} \tag{1.24}$$

$$\frac{I^{(t)}}{I_0} = \frac{1}{1 + F\sin^2\frac{\delta}{2}} \tag{1.25}$$

The above equations are the intensity distribution of multi-beam interference. Obviously, from Eqs. (1.24) and (1.25), there is

$$I^{(\mathrm{t})} + I^{(r)} = I_0$$

It can be seen from the above formula that since the sum of the reflected and transmitted light intensity is a constant, when the reflected light is constructive due to interference, its transmitted light must be destructive, and vice verso. This is also commonly referred to as the complementarity of reflected light and transmitted light. Moreover, this is also an inevitable result of energy conservation. Therefore, usually, as long as $I^{(R)}$ (or $I^{(\mathrm{t})}$) has been obtained, $I^{(\mathrm{t})}$ (or $I^{(R)}$) can be directly calculated from the above formula.

The light intensity distribution of the fringes can be determined by Eq. (1.24) or (1.25). Only the relative intensity distribution of the transmitted light is discussed below. Taking the phase difference δ between two adjacent beams as the abscissa and the relative light intensity as the ordinate. The distribution curve of transmitted light intensity of multi-beam interference can be drawn by Eq. (1.25), shown in Fig. 1.7. The extreme light intensity and its conditions can be obtained as follows. It is obviously from Eq. (1.25),

when $\sin(\frac{\delta}{2}) = 0$, i.e., $\delta = 2m\pi$ or

$$\Delta = 2nh\cos\theta_t = m\lambda\ (m = 0, 1, 2, 3, \cdots) \tag{1.26}$$

there is a maximum values $I^{(\mathrm{t})}_{max} = I_0$: That is, the maximum transmitted light intensity $I^{(t)}$ is equal to the incident light intensity. Obviously, the corresponding reflected light intensity is zero.

When $\sin\frac{\delta}{2} = 1$, i.e., $\delta = (2m+1)\pi$, or

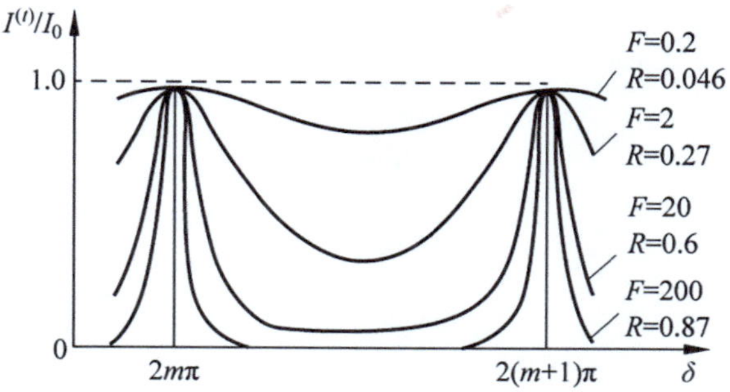

Fig. 1.7 Transmitted light intensity distribution of multi-beam interference

$$\Delta = 2nh\cos\theta_t = (2m+1)\frac{\lambda}{2} \tag{1.27}$$

there is a minimum value $I^{(t)}$:

$$I_{\min}^{(t)} = \frac{1}{1+F}I_0 = \frac{(1-R)^2}{(1-R)^2+4R}I_0 = \left(\frac{1-R}{1+R}\right)^2 I_0 \tag{1.28}$$

The above formula shows that when multiple beams interfere, the minimum value of transmitted light intensity is not zero, and its value is determined by the surface reflectivity *R*. The larger *R*, the closer the minimum value is to zero.

Characteristics of Interference Fringes

From the curves shown in Fig. 1.7, several characteristics of interference fringes can be seen:

1. The distribution of light intensity is determined by the reflectivity *R*. When *R* is very small, the interference light intensity does not change much, i.e., the interference fringes are extremely inconspicuous and have low visibility. When R increases, the fringe visibility improves. Therefore, controlling the value of *R* can change the distribution of light intensity.
2. The maximum intensity of interference light has nothing to do with *R* and is always equal to $\boldsymbol{I}_0$. The minimum value and the width of bright lines decreases with the increase of *R*, which is one of its great advantages. Therefore, when *R* is large, the interference fringes of the transmitted light are fine bright lines on a dark background. At the same time, the interference fringes of the reflected light are dark lines on a bright backgrounds, which are not easy to identify, so they are rarely used. It can be seen that the result of multi-beam interference is that when *R* is very large, most of the light is reflected, only when the maximum condition is met, the light can be fully transmitted, and the transmitted light intensity is independent of *R*.
3. Half-width of fringe

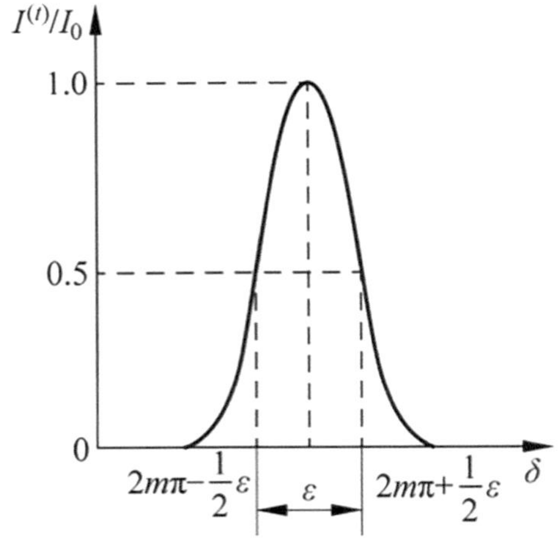

Fig. 1.8 FWHM of interference fringes ε

The sharpness of a fringe is measured by its width at half of its intensity peak, so called full-width-half-maximum (FWHM). In the case of transmitted light, FWHM refers to the distance between two points when the intensity on both sides of the maximum value drops to half the peak value. Because, when, the transmitted light intensity is the maximum value, i.e. In the present setting, the interference intensity drops to half of the maximum value. It is called the FWHM of the fringe, or simply the width of the fringe. Because when $\delta = 2m\pi$, the transmitted light intensity reaches the maximum, that is $I^{(t)}_{\max} = I_0$. Now assume that when $\delta = 2m\pi \pm \frac{\varepsilon}{2}$, the interference intensity drops to half of the maximum, that is $I^{(t)} = \frac{I_0}{2}$. As shown in Fig. 1.8. Obviously, the smaller ε, the sharper the fringes. At this time, according to formula (1.25):

$$\frac{I^{(t)}}{I_0} = \frac{1}{2} = \frac{1}{1 + F\sin^2\frac{1}{2}\left(2m\pi \pm \frac{\varepsilon}{2}\right)} = \frac{1}{1 + F\sin^2\left(\frac{\varepsilon}{4}\right)}$$

From this, we can obtain

$$F\sin^2\left(\frac{\varepsilon}{4}\right) = 1 \tag{1.29}$$

When F is large, i.e., R is large, ε must be very small, so

$$\sin\left(\frac{\varepsilon}{4}\right) \approx \frac{\varepsilon}{4}$$

and

$$F\left(\frac{\varepsilon}{4}\right)^2 = 1$$

Table 1.1 Relationship between R, F and N of multi-beam interference

R	0.5	0.8	0.9	0.94	0.96	0.98
F	8	80	360	1044	2400	9800
N	4.5	14.1	29.8	51	77	155

Finally,

$$\varepsilon = \frac{4}{\sqrt{F}} = \frac{2(1-R)}{\sqrt{R}} \tag{1.30}$$

ε is the expression for the width of the fringe, which is calculated in radians. The above formula shows that $\varepsilon \propto \frac{1}{\sqrt{F}}$, that is ε or the width of fringe decreases with R increases.

The sharpness of fringes is usually characterized by fineness N, which refers to the ratio of the interval between two adjacent fringes to the width of one fringe. Since the phase difference between two adjacent interference fringes is 2π, from the above definition, there is,

$$N = \frac{2\pi}{\varepsilon} = \frac{\pi}{2}\sqrt{F} = \frac{\pi\sqrt{R}}{1-R} \tag{1.31}$$

It can be seen that the thinner the bright fringe, the larger the N value, which is only determined by R. The relationship between R, F and N is shown in Table 1.1.

It should be noted that the above-mentioned fringe width ε is the width of a fringe of a monochromatic light produced by the multi-beam interference device, so it is also called the "instrument width", which is different from the spectral line width of a quasi-monochromatic light, which is determined by the characteristics of the light source itself, and the fringe width ε is determined by the interference device.

1.3 Low-Coherence Light Source Interferometry

"Low-coherence interferometry" (LCI) or "broad spectrum light source interferometry" refers to the interferometry using low coherence light sources. Examples of low-coherence light sources include light emitting diodes (LEDs), superluminescent diodes (SLD), Ti^{3+}: Sapphire lasers, and even white light lamps [3, 4]. Of course, the term "low-coherence" is quite vague, because even the most stabilized lasers are of finite coherence length. Generally, by low-coherence we understand a field that has a coherence length of the order of tens of microns (wavelengths) or less. Therefore, the name of white light interferometer is not accurate enough.

LCI aims at overcoming the fundamental flaw in a common interferometer that only managed to make a relative measurement other than the absolute one. Because the interferometer composed of monochromatic light source cannot calibrate the order of interference fringes, it is impossible to determine the absolute value of measurand. Only the variation of a physical quantity such as distance or length can be measured by the movement of interference fringe. This implies that an interferometer composed of a single wavelength can measure only a variation quantity rather than a static one. Therefore, people use the interference fringes obtained by light sources with a certain spectral width to determine the order of interference fringes (actually, the position of zero-order fringes is determined by equal optical path), so as to achieve the purpose of measuring quantities of state.

Figure 1.9 illustrates the schema of a LCI. The interferometer consists of two sets of Michelson interferometers: one for measurement and the other for reference. M_1/M_2/M_3/M_4 in Fig. 1.9 are total reflection mirrors, S_1 and S_2 are 1:1 beam splitter. Obviously, there are several optical paths in the interferometric structure will produce interference. Because the light source used is a wide spectrum light source, i.e., with poor coherence or a short interference length, at this time, only two light beams satisfy the equal-path requirement for interference by properly adjusting the mirror position. Conditions for equal optical path interference (i.e., producing zero-order fringes) are shown in Fig. 1.9. The two light beams are:

① (Red solid line) Light source $S \rightarrow$ Beam splitter $S_1 \rightarrow$ Mirror $M_2 \rightarrow$ Beam splitter $S_1 \rightarrow$ Beam splitter $S_2 \rightarrow$ Mirror $M_4 \rightarrow$ Beam splitter $S_2 \rightarrow$ Photodetector PD;

② (Green dot line) Light source $S \rightarrow$ Beam splitter $S_1 \rightarrow$ Mirror $M_1 \rightarrow$ Beam splitter $S_1 \rightarrow$ Beam splitter $S_2 \rightarrow$ Mirror $M_3 \rightarrow$ Beam splitter $S_2 \rightarrow$ Photodetector PD.

Obviously: When the two optical paths $(S_1 \rightarrow M_1) + (S_2 \rightarrow M_3)$ and $(S_1 \rightarrow M_2) + (S_2 \rightarrow M_4)$ are equal, zero-order interference fringes can be observed.

Therefore, when the mirror M_2 at position 1, the micro displacement platform is adjusted to move M_3 to (position 2) make the optical path ① and ② satisfy the

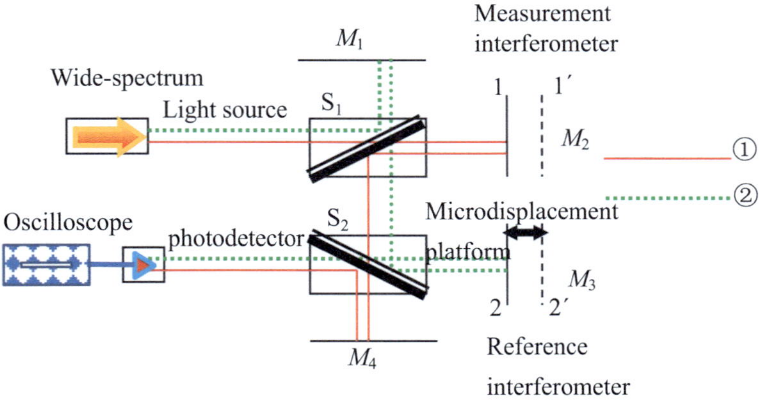

Fig. 1.9 Schematic diagram of broadband light source interferometer

equal optical path condition. At this time, the waveform shown in Fig. 1.10 will be observed on the oscilloscope, and the peak of the envelope corresponds to the zero-order interference fringe. When M_2 at position 1', M_3 is adjusted again to ensure that the equal optical path condition is still satisfied. At this time, the distance between position 1 and 2 can be obtained from the two readings of the microdisplacement platform. This is the operating principle of LCI.

The fundamental theory of the above-mentioned fiber interferometer for absolute quantity measurement is the interference principle of broadband light source. In the following section, we further analyze the double-beam interference theory of broadband light source and discuss the influence of light source characteristics on the performance of fiber interferometer system.

Let's discuss firstly the double-beam interference of monochromatic light. In Fig. 1.9, the monochromatic light emitted by the light source S is divided into two beams after passing through the semi-reflective mirror S_1, and then reflected by the mirrors M_1 and M_2 of the two arms of the Michelson interferometer, and the intensity of interference fringes is detected by the detector. The electric field intensity vector of the two beams received by the detector is:

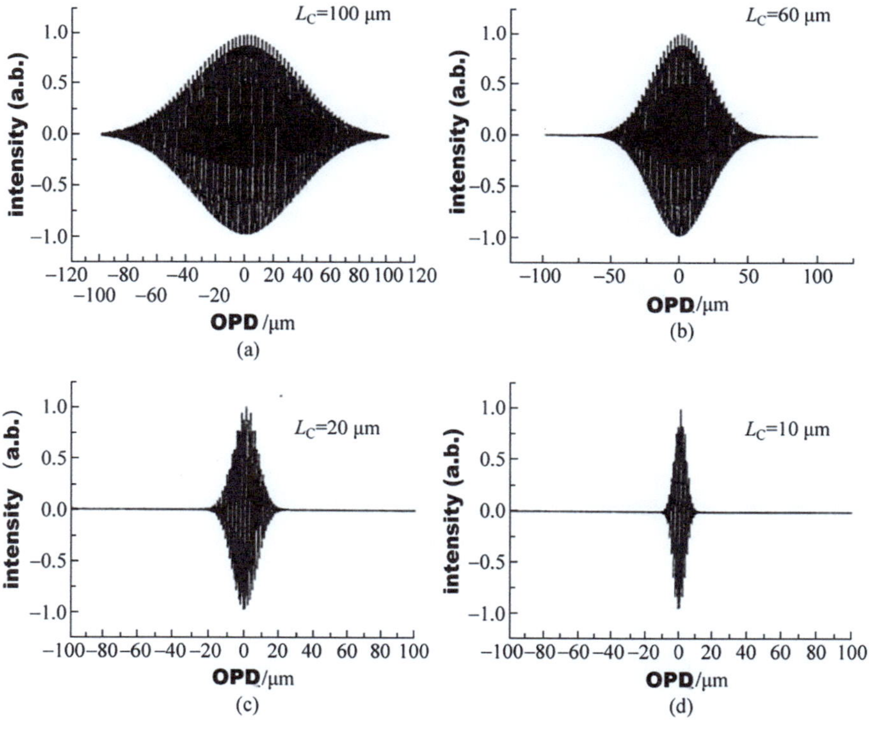

Fig. 1.10 Two-beam interference waveform of broadband light source **a** $L_C = 100$ μm **b** $L_C = 60$ μm **c** $L_C = 20$ μm **d** $L_C = 10$ μm

$$\boldsymbol{E}_1 = E_1 e^{i(\omega t+\phi_1)}\boldsymbol{e}_1$$
$$\boldsymbol{E}_2 = E_2 e^{i(\omega t+\phi_2)}\boldsymbol{e}_2 \tag{1.32}$$

where, $\boldsymbol{e}_1, \boldsymbol{e}_2$ are the unit vectors in the direction of the electric field intensity $\boldsymbol{E}_1$ and $\boldsymbol{E}_2$, respectively. Assuming that the polarization directions of the two beams are the same, the result of the vector synthesis is:

$$\boldsymbol{E} = \boldsymbol{E}_1 + \boldsymbol{E}_2 = \left[E_1 e^{i(\omega t+\phi_1)} + E_2 e^{i(\omega t+\phi_2)}\right]\boldsymbol{e}_1 \tag{1.33}$$

Then the interference light intensity I is:

$$\begin{aligned} I &= E \cdot E^* \\ &= \left[E_1 e^{i(\omega t+\phi_1)} + E_2 e^{i(\omega t+\phi_2)}\right]\left[E_1 e^{-i(\omega t+\phi_1)} + E_2 e^{-i(\omega t+\phi_2)}\right] \\ &= E_1^2 + E_2^2 + E_1E_2\left[e^{i(\phi_1-\phi_2)} + e^{-i(\phi_1-\phi_2)}\right] \\ &= I_1 + I_2 + 2\sqrt{I_1I_2}\cos(\Delta\phi) \end{aligned} \tag{1.34}$$

Among them, I_1 and I_2 are the intensity of the two light beams; $\Delta Ø$ is the phase difference between them, which is represented by the frequency v and optical path difference ΔL as:

$$\Delta\varphi = \frac{2\pi v}{c}\Delta L \tag{1.35}$$

For double-beam interference of a broadband light source, Eq. (1.34) is still applicable. The difference is that I_1 and I_2 in the formula are the light intensity components $i_1(v)$ and $i_2(v)$ of a certain frequency, and the superposition of light intensity between different frequency components is incoherent, that is

$$I = \int_0^\infty \left[i_1(v) + i_2(v) + 2\sqrt{i_1(v)i_2(v)}\cos(\Delta\phi)\right]dv \tag{1.36}$$

Assume the light intensity distribution of a broadband LED light source is Gaussian function, namely:

$$i_i(v) = I_i\frac{2}{\Delta v_D}\sqrt{\frac{ln2}{\pi}}exp\left[-\frac{4ln2(v-v_0)^2}{(\Delta v_D)^2}\right](i = 1, 2) \tag{1.37}$$

where Δv_D is the line width of the light source, and v_0 is the central frequency of the spectral line. Let $v - v_0 = \delta v$, then the phase difference $\Delta\varphi$ can be expressed as:

$$\Delta\phi = \frac{2\pi}{c}\Delta L(v_0 + \delta v)$$

Substitute Eqs. (1.37) and (1.38) into Eq. (1.36), and using the integral formula,

$$\int_{-\infty}^{\infty} e^{-\alpha^2 x^2} \cos(bx)dx = \frac{\sqrt{\pi}}{\alpha} e^{-\left(\frac{b}{2a}\right)^2} \quad (1.38)$$

$$\int_{-\infty}^{\infty} e^{-\alpha^2 x^2} dx = \frac{\sqrt{\pi}}{\alpha}$$

There is

$$I = I_1 + I_2 + \int_{-\infty}^{\infty} \frac{2\sqrt{I_1 I_2} 2}{\Delta \nu_D} \left(\frac{\ln 2}{\pi}\right)^{\frac{1}{2}} \exp\left[-\frac{4\ln 2(\delta\nu)^2}{\Delta\nu_D^2}\right] \cos\left[\frac{2\pi}{c}\Delta L(\nu_0 + \delta\nu)\right] d(\delta\nu) \quad (1.39)$$

make

$$L_C = \sqrt{\frac{16\, ln\, 2c^2}{\pi^2 \Delta\nu_D^2}} \quad (1.40)$$

And take the AC component,

$$I_{ac} = 2\sqrt{I_1 I_2} e^{-\left(\frac{2\Delta L}{L_C}\right)^2} \cos\left(\frac{2\pi \Delta L}{\lambda_0}\right) \quad (1.41)$$

If $\Delta L = 0$, the two beams have strictly equal light path, corresponding to zero-order interference fringe, whose intensity is:

$$I_{ac}(0) = 2\sqrt{I_1 I_2} \quad (1.42)$$

The intensity envelope of the interference light is:

$$I_{env} = 2\sqrt{I_1 I_2} e^{-\left(\frac{2\Delta L}{L_C}\right)^2} \quad (1.43)$$

When $|\Delta L| = L_C/2$, the interference light intensity envelope drops to 1/e of $I_{ac}(0)$. It can be considered that when $|\Delta L| > L_C/2$, the interference between the two beams $\frac{1}{e}$ is very weak, thus L_C is called the coherence length of the light source. From formula (1.40), it can be found that L_C is inversely proportional to the linewidth of the light source. The larger the linewidth of the light source, the worse its coherence $\Delta\nu_D$ and the smaller the coherence length L_C.

In addition, the interference fringe contrast V and the system minimum signal-to-noise ratio requirement $\mathrm{SNR}_{\mathrm{min}}$ are also two very important physical parameters of broadband light source interference theory. Their definitions are:

$$V = \frac{I_{\max} - I_{\min}}{I_{\max} + I_{\min}} = \frac{2\sqrt{I_1 I_2} e^{-\left(\frac{2\Delta L}{L_C}\right)^2}}{I_1 + I_2} \tag{1.44}$$

$$SNR_{min} = \frac{1}{1 - \frac{I^{(1)}}{I^{(0)}}} \tag{1.45}$$

In formula (1.45), $I^{(0)} = I_{ac}(0)$ is the intensity of zero-order interference peak; $I^{(1)}$ is the intensity of the first-order interference peak, satisfying

$$I^{(1)} = 2\sqrt{I_1 I_2} e^{-\left(\frac{2\lambda_0}{L_C}\right)^2} \tag{1.46}$$

Substitute formula (1.46) into formula (1.45), then we get

$$SNR_{min} = \frac{1}{1 - e^{-\left(\frac{2\lambda_0}{L_C}\right)^2}} \tag{1.47}$$

Expressed in dB as:

$$SNR_{min}^{(dB)} = 20\, lg(SNR_{min}) = -20\, lg\left\{1 - exp\left[-\left(\frac{2\lambda_0}{L_C}\right)^2\right]\right\} \tag{1.48}$$

From Eqs. (1.44) and (1.45), it can be seen that the contrast of interference fringes is maximum at $I_1 = I_2$, $\Delta L = 0$, and with the interference order increases, the contrast of the fringes decreases. In practical systems, due to the existence of various incoherent stray lights, the value of V is smaller than the theoretical value given by formula (1.44). $\mathrm{SNR}_{\mathrm{min}}$ reflects the anti-interference ability of the interference system. In order to accurately determine the position of the central fringe, the light intensity difference between the zero-order and the first-order interference peak $\Delta I = I^{(0)} - I^{(1)}$ should be greater than the light intensity fluctuation caused by system noise and external interference. The smaller $\mathrm{SNR}_{\mathrm{min}}$, the better the system performance. Figure 1.10a–d are the interference patterns obtained by numerical calculation at L_C = 100, 60, 20 and 10 μm, respectively. Figure 1.11 is the envelope of a typical LED light source interference waveform. Figure 1.12 is the relation curve of the $\mathrm{SNR}_{\mathrm{min}}$ of the interference signal to L_C.

The following conclusions can be drawn from the above analysis and calculation results:

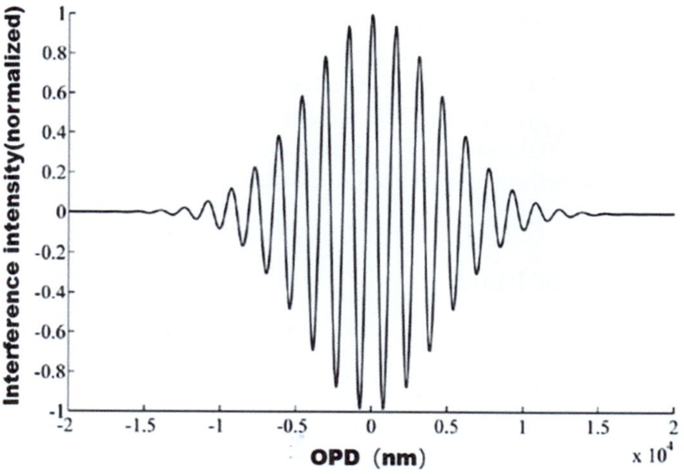

Fig. 1.11 Interference waveform of a LED light source

Fig. 1.12 Relationship of SNR_{min} to coherence length L of the light source

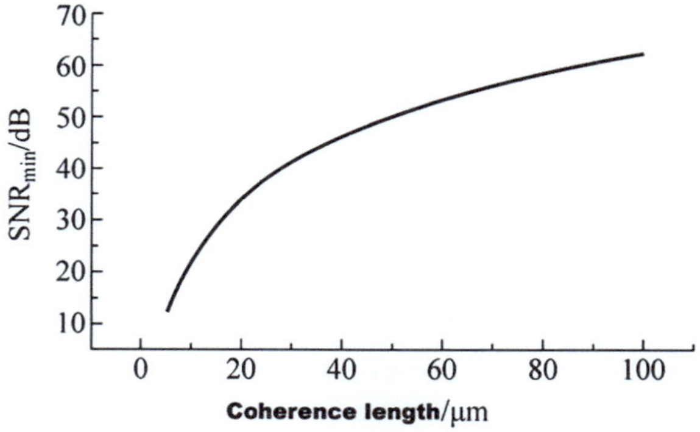

1. By changing the OPD ΔL between the two arms of the interferometer, the interference fringe of the two light paths can be obtained. The position of each interference peak is only related to the OPD, and the zero-order interference peak corresponds to the equal optical path of all wavelengths. The fluctuation of the light source intensity and external interference only affect the amplitude of interference fringe. Therefore, by scanning the position of the reference arm reflector of the interferometer, the position of the sensor arm reflector can be accurately located by the zero-order interference peak, thereby achieving absolute measurement.
2. The width of the interference intensity envelope depends on the coherence length of the light source. According to formula (1.43), the 1/e full width of the envelope $W_H = L_C$. The shorter the coherence length, the narrower the width of the interference envelope, and the fewer interference peaks it contains, which will be conducive to the accurate positioning of the central fringe. Therefore, the

light source with a shorter coherence length and wider spectral width should be selected to form a broadband optical fiber interferometer system.

3. When the central wavelength of the light source is constant, the system's minimum signal-to-noise ratio (SNR_{min}) requirement is only related to the coherence length of the light source. The smaller the coherence length and the SNR_{min}, the stronger the anti-interference ability of the system.

1.4 Coherence of Light

1.4.1 Overview

The aforementioned light interference is based on the coherence of the light field. Generally speaking, the light field changes with time and space, so the coherence of light has both temporal and spatial characteristics, which represents the degree of correlation of light disturbances between two points in time and space.

When investigating the interference of light waves emitted from a fixed point in the radiation field at two different time, what is being studied is the temporal coherence of the light field; and when investigating the interference of two different points in the light field at the same time, what is being studied is the spatial coherence of the light field.

Time coherence originates from the fact that the luminescence process of luminescent particles is not infinite, so the wave train emitted by the light source is of finite length, and there is no fixed phase relationship between the wave trains, which determines the maximum time delay (coherence time) and maximum OPD (coherence length) between two beams that meet the coherence condition. Temporal coherence is related to the spectral line width $\Delta\nu$ of the light source. Temporal coherence and monochromaticity are two different expressions of the same physical property of light. The spectral decomposition proves that the finite wave train length is not an ideal monochromatic wave of a single frequency, but a superposition of simple harmonics of various frequencies in a frequency band.

Spatial coherence originates from the fact that the luminous processes of different parts of the light source are independent of each other, that is, the phases of the wave trains emitted by different luminous points are independent. It shows that the light waves emitted by a light source with a certain line width can only interfere with two points within a certain distance on a certain wave front, which meets the coherence condition.

Coherence is an important property of light field, which directly affects the contrast of interference fringes. Two light waves are generally classified into three categories: **fully coherent**, **partially coherent** and **incoherent**. The vibrations of two fully coherent light waves are completely correlated and can completely interfere with each other; while partially coherent beams are only correlated within a certain time range or space range; and the phase fluctuations of two incoherent light beams are completely uncorrelated and will not produce interference fringe at all. In order to measure the degree of coherence between two light waves, that is, to provide an appropriate measure of the degree of correlation between the light vibrations of any two points and time in the light field, the mutual interaction function and complex coherence degree are introduced, and the basic concept of partial coherence theory is formed. The following is a brief introduction to this.

1.4.2 Coherence Time and Coherence Length

In a Michelson interferometer, as the reflector M_1 moves farther and farther away from M_2', the virtual image of M_2, (i.e., $L/2$ keeps increasing), the visibility of the interference fringes gradually deteriorates. When M_1 moving to a certain distance, i.e., when h increases to $L/2$, the interference fringes will completely disappear. At this time, there is no interference fringe in the field of view, leaving only a uniform illumination. This length of $L/2$ is related to the coherence of the light source used.

As mentioned above, the luminescence process of an actual light source does not last forever, but will be interrupted after a period of time (due to a large number of luminescent particles, the entire light source still emits continuous wave without interruption). Now assume that an independent luminous center in the light source emits the first wave train *A*, the second wave train *B*, and the third wave train *C*, etc., in chronological order, and the length of each wave train is *L*, the frequency of each wave train is exactly the same but without fixed phase relationship between each other, that is, the initial phases of each wave train are generally different. Now discuss the superposition of such wave trains in a Michelson interferometer.

When $h = 0$, the wave train is divided into two beams by S_1 (Fig. 1.9), and after being reflected by M_1 and M_2 respectively, they will arrive at the screen at the same time. At this time, when the wave train in path 1 (reflected by M_1) suddenly changes from *A* to *B*, the wave train in optical path 2 (reflected via M_2) also undergoes sudden changes at the same time, that is, the light waves from the two optical paths are always in the same phase when they meet, so the interference at the screen is always constructive. Therefore, in this case, the phase mutation between finite-length wave trains has no effect on the interference at the screen, as shown in Fig. 1.13a. When M_1 moving $\lambda/4$ ($h = \lambda/4$), there is an OPD of $\lambda/2$ between the two optical paths, and the wave trains reaching the screen is staggered by $\lambda/2$, as shown in Fig. 1.13b. As can be seen from the figure that during the period from C_A to D_A, the wave trains of the two optical paths belong to wave train *A*, and the phase difference between them is completely determined by their OPD. Therefore, at each time in this period, there is a fixed phase difference π between the two optical paths. This is also true within the time interval of C_BD_B, C_CD_C. However, during the time interval of C_BD_A (with diagonal lines in the figure), the situation is different. Wave train *A* comes from optical path 1, while wave train B comes from optical path 2. In this way, the two waves superimposed on the screen come from different wave train segments. As mentioned earlier, there is no fixed phase relationship between the two wave trains, so the time average of their interference terms during this period is zero. There will be no interference pattern appeared on the screen, just a uniform illumination. In the time intervals of D_BC_C, D_CC_D, …, the situation is the same. If the time interval with the diagonal lines is much smaller than that without diagonal lines, we can still see the interference fringes on the screen due to the stable interference effect on the screen in the latter time interval. However, due to the effect of uniform illumination, the visibility of interference fringes will drop a little. With the increase of h, the time interval without interference effect increases, on the other side, the time interval with interference effect shortens. As a result, the visibility of interference fringes gradually deteriorates, and the fringes become more and more blurred. When $h = L/2$ and the path difference between the two optical paths is L, the two light waves reaching the screen are completely staggered in time, that is, the two light beams superimposed on the screen come from two completely different light waves, as shown Fig. 1.13c. At this time, there is no interference effect on the screen at any moment, that is, the two light beams are completely unrelated.

a. The OPD is zero b. The OPD is less than the length of the wave train c. The OPD is greater than or equal to the length of the wave train.

Therefore, since the light source does not emit light infinitely, the moving distance of M_1 is also limited. This maximal OPD that can produce interference effect is called coherence length of the light source. The longer the coherence length, the better the coherence of the light source. Of course, the coherence length is only a rough value. It is followed that when two coherent waves are superimposed, their interference effect varies with the OPD. When the OPD is less than the coherence length, there is a constant phase difference between the two waves for a part of the time, so

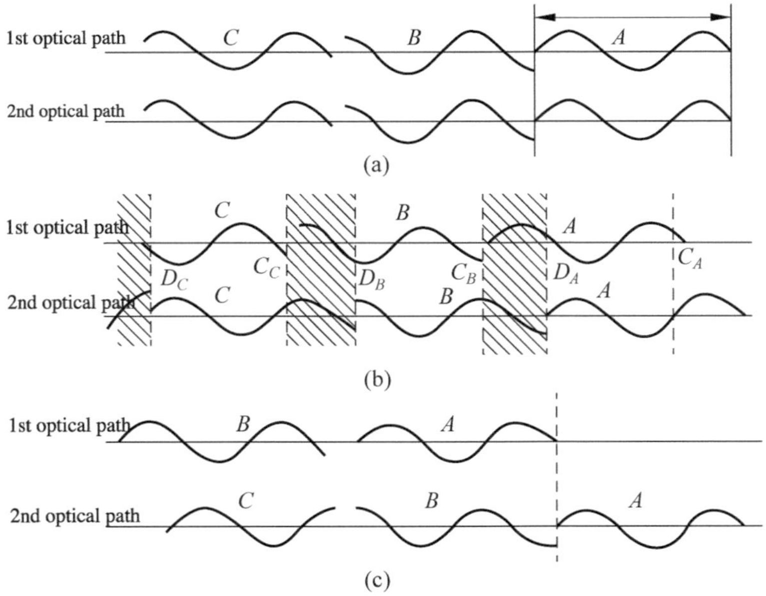

Fig. 1.13 Superposition of two finite length wave trains with different course differences

there is obvious interference effect. However, when the OPD is greater than the coherence length, there is no fixed phase relationship between the two waves in any time interval, so the interference effect cannot be seen. In addition, because of their different optical paths, these two coherent waves coming from the same wave train have a time difference when they meet at a certain point in space. So the interference of the two waves at this point is actually the interference of waves at two different points before and after time on the same wave train at this point. This interference effect of the wave emitted by the light source is called temporal coherence, and the quality of temporal coherence τ is measured by coherence time. Therefore, when two points with time difference Δt on a wave train meet at a certain point in space, there is an interference effect (i.e., there is a fixed phase relationship), otherwise there is no interference effect. Obviously, the longer the coherence time, the better the temporal coherence of the light source.

The relationship between coherence length L_C and coherence time τ is

$$L_C = \tau c$$

where c is the speed of light. Assume that the distance between two points a and b taken in the propagation direction of light is equal to (or less than) the coherence length L_C, then at each moment, the vibrations of the light wave field at two points have a fixed phase relationship, that is, it is coherent. However, since the vibration at point b propagates from point a (assuming that the wave propagation direction is a to b), the time required for propagation is $\tau = \frac{L}{c}$, so the phase relationship of the vibration at points a and b at the same time is the phase relationship of point b vibration at two different times t and $t + \tau$, there is a fixed phase relationship. Obviously, from the meaning of coherence time, this time τ is coherent time. Because there is no fixed phase relationship beyond this time τ.

1.4.3 Coherence Time and Spectral Linewidth

The wave train length of the light source is analyzed above, and its coherence can also be discussed from the monochromaticity of the light source. As mentioned earlier, the harmonics of a single wavelength are only ideal conditions, and any actual wave train has a certain line width. For example, incandescent lamps include all wavelengths in the visible light range (0.4–0.7 μm). Monochromatic light sources commonly used in laboratories: sodium lamps and mercury lamps, their spectral lines also include a small wavelength range, such as shown in Fig. 1.14. Intensity at the center of the spectral line maximum is I_0, and the corresponding wavelength is λ_0, the intensity on both sides of the center decreases symmetrically, and decreases to $I_0/2$ at $\lambda_0 \pm \frac{\Delta\lambda}{2}$. $\Delta\lambda$ is the width of the spectral line, which indicates the monochromaticity of the light source. The smaller $\Delta\lambda$, the higher the degree of monochromaticity. Therefore, the interference pattern obtained with this light source is actually the superposition of multiple sets of interference fringes with different wavelengths. But

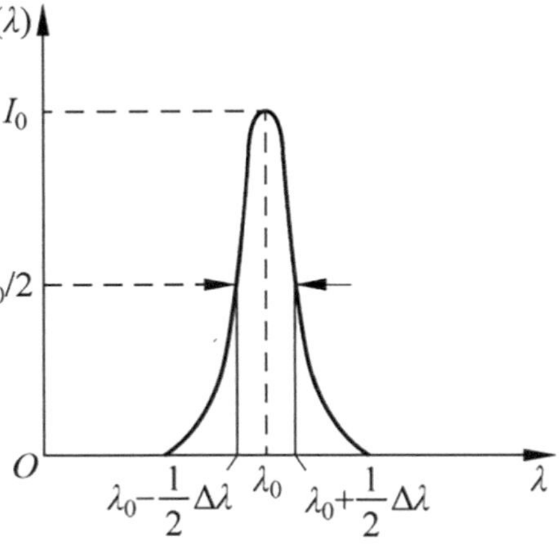

Fig. 1.14 Definition of spectral line width

as mentioned earlier, the spacing of interference fringes is related to wavelength: the longer wavelength, the larger fringe spacing (see the relevant formulas in previous sections). Therefore, in the interference pattern of quasi-monochromatic light, the fringes of different wavelengths cannot completely overlap, but slightly shift from each other. The only exception is zero-order interference fringes, because the OPD is zero, and all wavelengths of light produce constructive (or destructive) interference.

Imagine using two rulers with different scales to represent the interference fringes produced on the screen of two wavelengths. Assume that the spacing of the interference fringes produced by the light with wavelength λ_0 of 0.9 μm on the screen is 0.9 mm, and each grid of the scale represents a fringe; the spacing of interference fringes on the screen produced by the light with wavelength of $\lambda_0 \pm \Delta\lambda$ is 1.0 mm. If the zeros of the two rulers are aligned, thenit can be seen from Fig. 1.15 that with the increase of interference OPD and the interference level, the maximums of the same level of the two wavelengths λ_0 and $\lambda_0 \pm \Delta\lambda$ are increasingly staggered. For those wavelengths between λ_0 and $\lambda_0 \pm \Delta\lambda$, their levels of maximum are continuously distributed between the same levels of the maximum of λ_0 and $\lambda_0 \pm \Delta\lambda$, as shown in the shaded part of the figure. With the interference order increase, the bright area of interference caused by fringe staggering will expand, and the dark area will shrink. When the OPD reaches a certain value L, the $(m+1)$th maximum of λ_0 and the mth maximum of $\lambda_0 + \Delta\lambda$ will coincide with each other, as shown in Fig. 1.15, the 10th level of λ_0 and the 9th level of $\lambda_0 \pm \Delta\lambda$ coincide with each other. At this time, the maximum of each wavelength will join into one piece, and the interference fringes will completely disappear.

When the interference fringes disappear, there is $L = m(\lambda + \Delta\lambda) = (m+1)\lambda$ and $m = \frac{\lambda_0}{\Delta\lambda}$,

$$L = \frac{\lambda_0}{\Delta\lambda} m \tag{1.49}$$

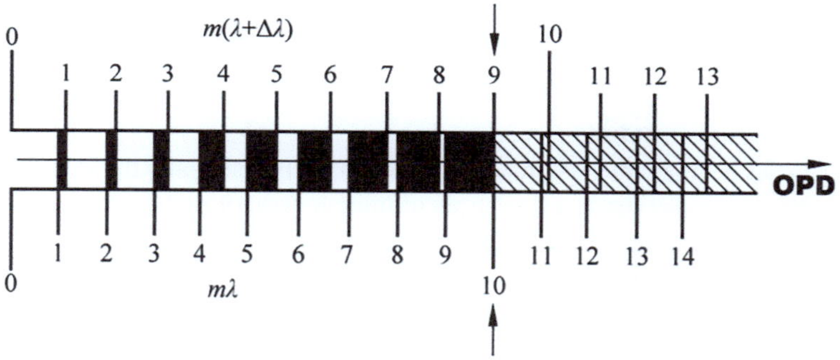

Fig. 1.15 The maximum relative positions in the range of λ_0 to $\lambda_0 + \Delta\lambda$

Formula (1.49) uses the relation $\Delta\lambda << \lambda$, and λ_0 is replaced by λ. This result shows that in the interference of quasi-monochromatic light ($\lambda_0 \pm \frac{\Delta\lambda}{2}$), the highest interference level is $\frac{\lambda_0}{\Delta\lambda}$ and the maximum OPD is $\frac{\lambda_0^2}{\Delta\lambda}$. As described below, this maximum OPD is the coherence length L. This is a consistent result obtained by analyzing the same thing from different aspects. Obviously, the better the monochromaticity of the light source, i.e., the smaller the $\Delta\lambda$, the larger the allowable OPD. For example, for the green light $\lambda_0 = 0.5$ μm, we get:

$\Delta\lambda \approx 0.01$ nm, $m = 5 \times 10^4$, $L = 2.5$ cm;
$\Delta\lambda = 0.001$ nm, $m = 5 \times 10^5$, $L = 25$ cm;
$\Delta\lambda = 0.0001$ nm, $m = 5 \times 10^6$, $L = 2.5$ m.

For white light, because $\Delta\lambda$ and λ_0 are of the same order of magnitude, only the maximum OPD of the order of wavelength λ_0 and low-order interference fringes can be obtained.

In the above discussion, it is supposed that the light emitted by each atom is a wave train of finite length, based on which the formula for calculating the coherence length is derived. On the other hand, it can also be considered that each wave train is the result of superposition of an infinite number of sinusoidal wave trains continuously distributed in a range, and this transformation can be performed using Fourier integration. This shows that the light emitted by each light source cannot be harmonics of a single wavelength (even lasers with excellent coherence), but quasi-monochromatic light with a certain wavelength range. From the calculation results of Fourier transform, it can be seen that the relationship between the spectral width Δv and time Δt of a finite length equal radiation wave train is $\Delta v \cdot \Delta t = 1$. Most of the intensity of the monochromatic light contained by the wave train is concentrated in the range of $(-\frac{1}{\Delta t} + v_0)$ to $(\frac{1}{\Delta t} + v_0)$. Δt is the time interval corresponding to the finite-length wave train. As can be seen from the above discussion, it is the coherence time τ. Because the relationship between the wave train length L and Δt is $L = c \cdot \Delta t$, if reused $\lambda v = c$, the relationship between L and $\Delta\lambda$ can be found as follows:

$$L = c \cdot \Delta t = c\frac{1}{\Delta \nu} = c\frac{\lambda_0^2}{c}\frac{1}{\Delta \lambda} = \frac{\lambda_0^2}{\Delta \lambda}$$

This is the formula that has been derived above.

Both views get the same result, which shows that it is essentially the same to regard the light emitted by atoms as a wave train of finite length, or as the superposition of a series of sine waves. Because a pure monochromatic light wave must be a harmonic wave that continues infinitely in time and space, and a wave train of finite length can only be the result of the superposition of a series of pure monochromatic waves. In fact, it is impossible to produce light waves with a single frequency by any method. The so-called monochromatic light just has a relatively small $\Delta \nu$.

If the wave train is an attenuated sine wave, its spectral relationship is $\Delta \nu \Delta t = 1/\pi$, i.e., the relationship $\Delta \nu \propto \frac{1}{\Delta t}$ still exists for the attenuated wave.

1.4.4 Mutual Coherence Function and Complex Correlation Coefficient

As mentioned earlier, since the light emitted by the actual physical light source cannot be strictly monochromatic, even the laser light source with the narrowest spectral linewidth has a certain spectral linewidth; Moreover, the actual light source is not an ideal point light source, but is composed of a large number of primitive radiators, so it has a certain size. In order to give a more accurate quantitative description of this temporal and spatial characteristic of actual light sources, it is necessary to introduce the concepts of mutual coherence function and complex correlation [5, 6]. The following takes Youngs double-hole interference as an example for discussion.

As shown in Fig. 1.16, S is an actual extended non-monochromatic light source. The sub-waves emitted by two points P_1, P_2 in the radiation field of this light source. The emitted wavelets reach the Q point on the observation screen as two coherent beams for coherent superposition. Therefore, the optical vibration at point Q can be expressed as the sum of the light vibrations of the emitted wavelets when they reach point Q, namely:

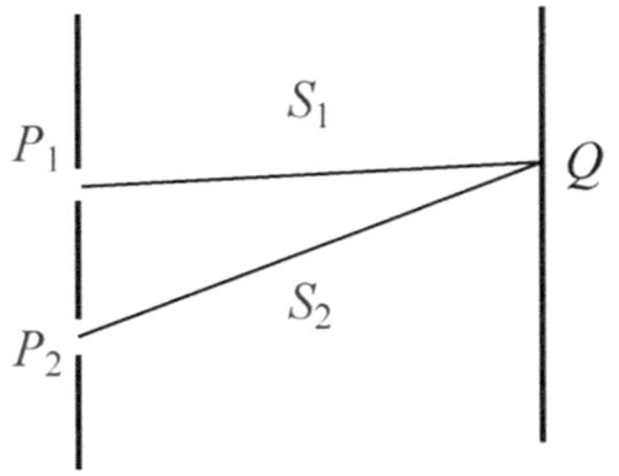

Fig. 1.16 Youngs interference schema of extended non-monochromatic light source

$$u_{Q(t)} = a_1 u_1\left(t - \frac{S_1}{c}\right) + a_2 u_2\left(t - \frac{S_2}{c}\right) \tag{1.50}$$

where, $u_1(t), u_2(t)$ denote the light vibration at two points P_1, P_2 respectively; S_1 and S_2 denotes the distance from two points P_1 and P_2 to point Q; c is the speed of light, a_1 and a_2 are the transmission factors, whose magnitudes are inversely proportional to the transmission distance S_1 and S_2. According to the diffraction theory of light, they are all imaginary numbers.

Use <> to represent the average value over time, and the average light intensity at point Q is:

$$\begin{aligned} I_Q &= \langle u_{Q(t)} u^*_{Q(t)} \rangle \\ &= I_1 + I_2 + 2R\langle a_1 a_2^* u_1\left(t - \frac{S_1}{c}\right) u_2^*\left(t - \frac{S_2}{c}\right)\rangle \end{aligned} \tag{1.51}$$

Let $t_1 = \frac{S_1}{c}$, $t_2 = \frac{S_2}{c}$, $\tau = t_2 - t_1$, then it can be expressed as $I_{(Q)}$:

$$I_Q = I_1 + I_2 + 2R_e\left[a_1 a_2^* \langle u_1(t+\tau) u_2^*(t)\rangle\right] \tag{1.52}$$

Let $\Gamma_{12}(\tau) = \langle u_1(t+\tau)u_2^*(t)\rangle$, where $\Gamma_{12}(\tau)$ is called the mutual interference function (note: the variables t in Eqs. (1.51) and (1.52) are different, and the two differ by a constant t_2).

According to the definition of the cross-correlation function of the stationary random process,

$$\Gamma_{12}(\tau) = \langle u_1(t+\tau)u_2^*(t)\rangle = \lim_{T\to\infty} \frac{1}{2T} \int_{-T}^{T} u_1(t+\tau)u_2^*(t)dt \tag{1.53}$$

Therefore, Eq. (1.52) can be expressed as:

$$I_{(Q)} = I_1 + I_2 + 2R_e\left[a_1 a_2^* \Gamma_{12}(\tau)\right] \tag{1.54}$$

As discussed above, the self-correlation function can be defined as:

$$\begin{cases} \Gamma_{11}(\tau) = \langle u_1(t+\tau)u_1^*(t)\rangle \\ \Gamma_{22}(\tau) = \langle u_2(t+\tau)u_2^*(t)\rangle \end{cases} \tag{1.55}$$

When the time is delayed $\tau = 0$, the self-correlation function is the light intensity:

$$\begin{cases} \Gamma_{11}(0) = \langle u_1(t)u_1^*(t)\rangle = \langle |u_1(t)|^2\rangle \\ \Gamma_{22}(0) = \langle u_2(t)u_2^*(t)\rangle = \langle |u_2(t)|^2\rangle \end{cases} \tag{1.56}$$

In Eq. (1.52), I_1 and I_2 are denoted by self-correlation functions as: $I_1 = |a_1|^2\Gamma_{11}(0)$, $I_2 = |a_2|^2\Gamma_{22}(0)$. In order to further clarify the physical meaning of the mutual coherence function $\Gamma_{12}(\tau)$, it can be normalized to:

$$\gamma_{12}(\tau) = \frac{\Gamma_{12}(\tau)}{\sqrt{\Gamma_{12}(0)\Gamma_{22}(\tau)}} \tag{1.57}$$

$\gamma_{12}(\tau)$ is called the degree of complex coherence, which is a complex number and can therefore be expressed as follows:

$$\gamma_{12}(\tau) = |\gamma_{12}(\tau)|e^{i\varphi_{12}(\tau)} \tag{1.58}$$

where $\varphi_{12}(\tau)$ is the amplitude angle of $\gamma_{12}(\tau)$, which generally consists of two parts, one part is the phase difference of the light vibration at two points P_1, P_2 and generally a function of the delay time τ; The other part is phase difference introduced by the two-point P_1, P_2 vibration propagating to the observation point Q. With the definition of complex coherence, Eq. (1.52), the interference light intensity at point Q is expressed as:

$$I_{(Q)} = I_1 + I_2 + 2\sqrt{I_1 I_2}|\gamma_{12}(\tau)|\cos\varphi_{12}(\tau) \tag{1.59}$$

Since the complex coherence is a normalized cross-correlation function, its magnitude $|\gamma_{12}(\tau)|$ varies between 0 and 1, so the coherence of the actual light source is nothing more than the following three cases:

(1) $|\gamma_{12}(\tau)| = 1$, completely coherent, interference fringes are distributed in the same way as those produced by strictly monochromatic plane waves.
(2) $|\gamma_{12}(\tau)| = 0$, completely irrelevant, the light intensity of the synthetic point is just a simple superposition of intensities, and the interference fringes disappear.
(3) $0 < |\gamma_{12}(\tau)| < 1$: partial coherence, if the total intensityof the light source I_{Total} is regarded as consisting of two parts: completely coherent I_{coh} and completely irrelevant I_{incoh}, then $|\gamma_{12}(\tau)|$represents the proportion of the coherent light fraction to the total intensity, i.e., $\frac{I_{coh}}{I_{\text{Total}}} = |\gamma_{12}(\tau)|$.

It can be seen from the above discussion that the mutual coherence function and complex correlation represent the coherence degree of two points in the light field. They determine the "clarity" of the interference effect when the beams are superimposed. It can be expected that they must be related to the contrast or clarity of the interference fringes. The contrast V of the interference fringes is defined as:

$$V = \frac{I_{\text{max}} - I_{\text{min}}}{I_{\text{max}} + I_{\text{min}}} \tag{1.60}$$

That is, the relative ratio of the two extreme values I_{min} and I_{max} of the interference fringe light intensity.

$$\begin{cases} I_{max} = I_1 + I_2 + 2\sqrt{I_1 I_2}|\gamma_{12}(\tau)| \\ I_{min} = I_1 + I_2 - 2\sqrt{I_1 I_2}|\gamma_{12}(\tau)| \end{cases} \tag{1.61}$$

At that time $I_1 = I_2$, it can be seen that the fringe contrast is the modulus of complex coherence, that is $V = |\gamma_{12}(\tau)|$, the specific method for determining the coherence function experimentally is given. Obviously, the contrast of interference fringes formed by completely coherent light is 1. For completely incoherent light, the contrast is zero, and the fringes disappear. The contrast of partially coherent light is between 0 and 1. In experiments, it is generally believed that as long as $|\gamma_{12}(\tau)| > 0.9$, the two beams are still completely coherent. In interference theory, it is taken $V_a = 1/\sqrt{2} = 0.707$ as the allowable value and $V = 0$ as the critical value for fringe disappearance.

1.4.5 Coherence Time and Coherence Length of a Light Source

(2) Temporal coherence of "quasi-monochromatic" light sources

In order to understand the relationship between partial coherence and the characteristics of the light source, consider first a hypothetical "quasi-monochromatic" light source, which has the following properties: the light vibration and its generated field change sinusoidally within a certain period of time τ_0, then change phase suddenly. As a result, it is difficult to accurately maintain repeatability. An illustration of such process is shown in Fig. 1.17. τ_0 is called coherence time. The phase change that occurs after each period of coherence time is considered to be a random distribution from 0 to 2π.

The time dependence of this quasi-monochromatic light field can be expressed as

$$u(t) = u_0(t)\exp[i(\omega t - \varphi(t))] \tag{1.62}$$

Where the phase angle $\emptyset(t)$ is a random step function as shown in Fig. 1.17. This type of field can be approximately regarded as the field of a luminous atom, and the sudden change of phase is the result of collision between atoms.

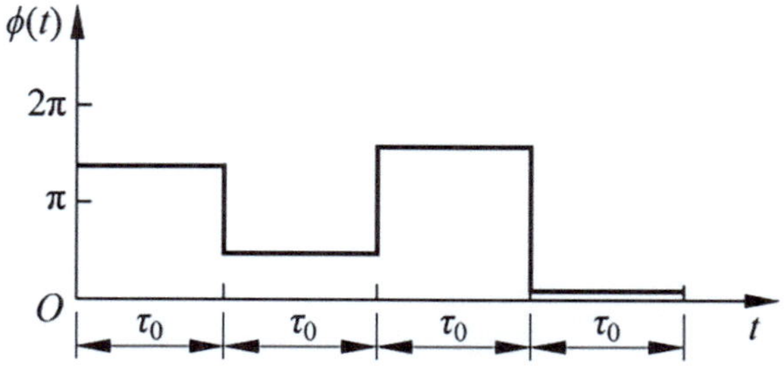

Fig. 1.17 Phase angle $\varphi(t)$ of a quasi-monochromatic light source

The light beam represented by Eq. (1.62) is divided into two beams, and the two beams of light interfere with each other, and assume

$$|u_1| = |u_2| = |u|$$

Since only self-coherence is involved here, the subscripts can be omitted and the normalized coherence function is written as,

$$\gamma(\tau) = \frac{\langle u(t)u^*(t+\tau)\rangle}{\langle |U|^2\rangle} \tag{1.63}$$

From formula (1.62),

$$\begin{aligned} \gamma(\tau) &= e^{i\omega\tau} e^{i[\varphi(t)-\varphi(t+\tau)]} \\ &= e^{i\omega\tau} \lim_{T\to\infty} \frac{1}{T}\int_0^T e^{i[\varphi(t)-\varphi(t+\tau)]dt} \end{aligned} \tag{1.64}$$

Now consider the quantities of $\varphi(t) - \varphi(t+\tau)$ in Fig. 1.18. For the first coherent time interval $0 < t < \tau_0$, suppose $\varphi(t) - \varphi(t+\tau) = 0$ at time $0 < t < \tau_0 - \tau$; and at that time $\tau_0 - \tau < t < \tau_0$, a random value between 0 and 2π is taken. The above assumptions apply to each subsequent coherent time interval.

The integration of Eq. (1.64) at the first coherence time interval has

$$\begin{aligned} \frac{1}{\tau_0}\int_0^{\tau_0} e^{i[\varphi(t)-\varphi(t+\tau)]}dt &= \frac{1}{\tau_0}\int_0^{\tau_0-\tau} dt + \frac{1}{\tau_0}\int_{\tau_0-\tau}^{\tau_0} e^{i\Delta}dt \\ &= \frac{\tau_0-\tau}{\tau_0} + \frac{\tau}{\tau_0}e^{i\Delta} \end{aligned} \tag{1.65}$$

where, Δ is the random phase difference.

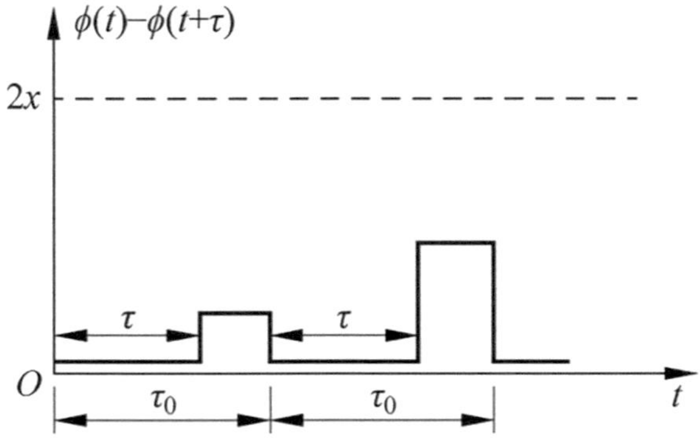

Fig. 1.18 Phase difference diagram of $\varphi(t) - \varphi(t+\tau)$

The same Δ results can be obtained for all subsequent time intervals, except that each interval is different. Because Δ is a random value, the time average of the containing term $e^{i\Delta}$ is zero. The remaining term $\frac{\tau_0-\tau}{\tau_0}$ is the same for all time intervals, so this term is equal to the mean of the integral. Obviously, if $\tau > \tau_0$, then the phase difference $\varphi(t) - \varphi(t+\tau)$ is always random, and as a result, the average value of the whole integral is zero.

From the above results, the normalized self-coherence function of a quasi-monochromatic light source can be obtained as

$$\gamma(\tau) = \left.\begin{matrix} \left(1 - \frac{\tau}{\tau_0}\right)e^{i\omega\tau} & \tau < \tau_0 \\ 0 & \tau \geq \tau_0 \end{matrix}\right\} \tag{1.66}$$

In the two-beam interference experiment, when the amplitudes of the light field are equal, the quantity $|\gamma|$ is equal to the visibility V of the fringe. Obviously, if τ is greater than the coherence time τ_0, the fringe visibility will drop to zero. Therefore, the OPD between the two light beams should be less than the coherence length L_C in order to obtain interference fringes.

In real luminous atoms, the time between two collisions is not constant, but changes randomly from one collision to the next. Therefore, the length of the wave train also changes in the same random manner. In a more realistic case, the coherence time can be defined as the average of the individual coherence times, and the same is true for the coherence length. Thus, the true mathematical form of coherence and fringe visibility will depend on a fine statistical distribution of the length of the individual wave train. In any case, when the OPD is less than the average coherence length, the fringe visibility is large. Conversely, when the OPD becomes greater than the average coherence length, the fringe visibility will become smaller and approach zero.

(3) Temporal coherence of typical light sources

As mentioned above, temporal coherence is described by the self-coherence function $\Gamma_{11}(\tau)$ and the complex coherence $\gamma_{11}(\tau)$. And τ is the time difference between the light passing through the two arms of the interferometer in Eq. (1.51), and the interference characteristics are completely determined by the complex correlation coefficient $\gamma_{11}(\tau)$ of the light source. The corresponding Eq. (1.57) can be expressed as:

$$\gamma_{11}(\tau) = \frac{\Gamma_{11}(\tau)}{\Gamma_{11}(0)} = \frac{\langle u_1^*(t)u_1(t+\tau)\rangle}{\langle u_1^*(t)u_1(t)\rangle} \tag{1.67}$$

If the light field $u(t)$ is expressed as a function in the frequency domain, that is, the Fourier transform is performed on $u(t)$ to obtain $V(\nu)$, let $I_{(\nu)} = V_{(\nu)}V^*_{(\nu)}$, express the spectral distribution of the intensity of the light source. From further mathematical derivation, it can be seen that the temporal coherence of the light field can be expressed as follows:

$$\gamma(\tau) = \gamma_{11}(\tau) = \frac{\int_{-\infty}^{\infty} I(\nu) e^{-i2\pi\nu\tau} d\nu}{\int_{-\infty}^{\infty} I(\nu) d\nu} \tag{1.68}$$

That is, the temporal coherence of the light field is the Fourier transform of the normalized power spectral density function.

The temporal coherence characteristics of several typical light sources are given below.

(1) Gaussian light source

For low-voltage gas discharge tube, its power spectrum distribution $I_{(\nu)}$ is a Gaussian curve, and broadening of its spectral line is mainly due to the Doppler effect caused by the motion of gas molecules. It can be expressed as follows:

$$I(\nu) = \frac{2\sqrt{ln2}}{\sqrt{\pi}\,\Delta\nu} exp\left[-4ln2\left(\frac{\nu - \nu_0}{\Delta\nu}\right)^2\right] \tag{1.69}$$

where $\Delta\nu$ is the half width of the emission spectral line of the light source. From Eq. (1.68):

$$\gamma(\tau) = exp\left[-\left(\frac{\pi\,\Delta\nu}{2\sqrt{ln2}}\right)^2\right] e^{-i2\pi\nu\tau} \tag{1.70}$$

(2) Lorentz light source

The spectral line broadening mechanism of high-voltage gas discharge tube is collision broadening caused by the collision of radiated atoms or molecules, and its spectral density function has a Lorentz curve shape:

$$I(\nu) = \frac{\Delta\nu}{2\pi\left[(\nu - \nu_0)^2 + (\frac{\Delta\nu}{2})^2\right]} \tag{1.71}$$

Its corresponding complex coherence is:

$$\gamma(\tau) = e^{-\pi\,\Delta\nu\tau} e^{-i2\pi\,\nu_0\tau} \tag{1.72}$$

(3) Rectangular spectrum light source

The monochromatic light from white light source through the monochromator output slit can be regarded as having a rectangular power spectrum curve. Assuming the center frequency and the spectral width is ν_0 and $\Delta\nu$, then the rectangular power spectrum curve can be expressed as (Figs. 1.19 and 1.20):

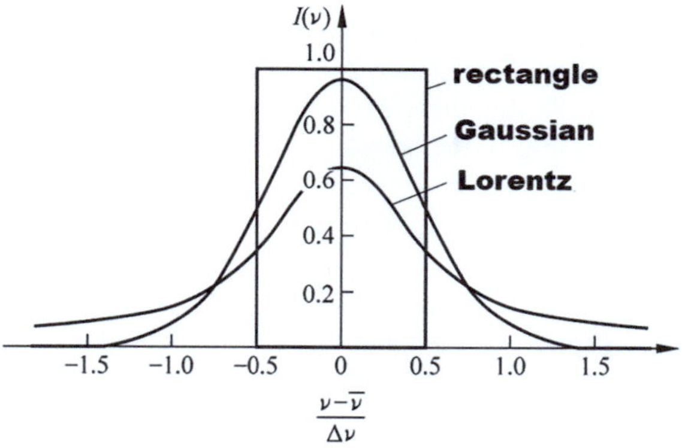

Fig. 1.19 $I(\nu)$ of three typical light sources

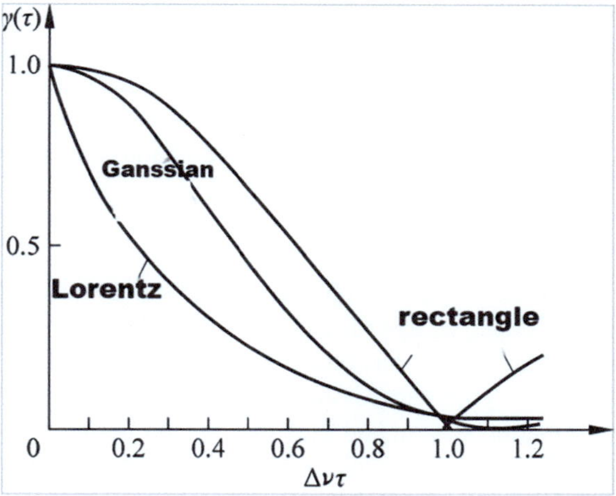

Fig. 1.20 $|\gamma(\tau)|$ of three typical light sources

$$I(\nu) = \begin{cases} I_0 \; \nu_0 - \frac{\Delta\nu}{2} \le \nu \le \nu_0 + \frac{\Delta\nu}{2} \\ 0 \; \nu < \nu_0 - \frac{\Delta\nu}{2}, or, \nu > \nu_0 + \frac{\Delta\nu}{2} \end{cases} \tag{1.73}$$

The corresponding temporal coherence is:

$$\gamma(\tau) = \text{sinc}(\Delta\nu\tau)e^{-i2\pi\nu_0\tau} \tag{1.74}$$

The coherence time can be precisely defined using complex coherence. Generally $|\gamma(\tau)| = \frac{1}{\sqrt{2}}$, the time difference between the two corresponding light beams can be defined as coherence time τ_c. The coherence times of these three typical light sources are: $\frac{0.312}{\Delta\nu}$, $\frac{0.110}{\Delta\nu}$, $\frac{0.444}{\Delta\nu}$, respectively. It can be seen that the coherence time of the light source τ_c is closely related to the spectral line width $\Delta\nu$ and has the same magnitude as $1/\Delta\nu$.

A typical situation encountered often is discussed below. That is, the spectral width of the light source $\Delta\nu$ and the average frequency of the light source $\bar{\nu}$ satisfy $\frac{\Delta\nu}{\bar{\nu}} << 1$. At this time, the intensity of the point Q in the interference pattern is (as shown in Fig. 1.16):

$$I_Q = I_1 + I_2 + 2\sqrt{I_1 I_2}|\gamma_{12}(\tau)|\cos[a_{12}(\tau) - \delta] \tag{1.75}$$

where $\tau = \frac{S_2 - S_1}{c}$, $\delta = 2\pi\bar{\nu}\tau$, $a_{12}(\tau) = \delta + \varphi_{12}(\tau)$, $a_{12}(\tau)$ represents the phase difference δ corresponding to a time delay of τ between the light vibration at two points P_1, P_2, which is the phase difference introduced when the light vibrations at the two points P_1, P_2 propagate to point Q respectively.

Under quasi-monochromatic conditions, $|\gamma_{12}(\tau)|$ and $a_{12}(\tau)$ are functions of τ, they all change slowly than $\cos 2\pi\bar{\nu}\tau$ and $\sin 2\pi\bar{\nu}\tau$, if the light intensity I_1 and I_2 can be considered uniform. Then there is a uniform background $I_1 + I_2$ near any point Q, on which a distribution of sinusoidal intensity is superimposed, and its amplitude is $2\sqrt{I_1 I_2}|\gamma_{12}(\tau)|$. In particular, when $I_1 = I_2$, it can be obtained that the visibility function V of the fringe is equal to the coherence of the light source $|\gamma_{12}(\tau)|$.

From Eq. (1.75), the maximum position near point Q is:

$$2\pi\bar{\nu}\tau - a_{12}(\tau) = 2m\pi \ (m = 0, \pm 1, \pm 2, \ldots)$$

This is completely consistent with the interference fringes formed when the phase of P_1 is delayed $a_{12}(\tau)$ relative to P_2 under strict monochromatic illumination with a frequency of $\bar{\nu}$. The phase delay of 2π corresponds to the fringes moved parallel to P_1P_2 by one stripe spacing of $\frac{D}{2l}\bar{\lambda}$. D is the distance from the pinhole to the observation screen, and $2l$ is the spacing between P_1 and P_2. Therefore, the fringe distribution of quasi-monochromatic interference has a shift amount relative to the interference fringes of the strictly monochromatic illumination pair with the light vibration of points P_1 and P_2 in the same phase:

$$x = \frac{a_{12}(\tau)}{2\pi} \cdot \frac{D\bar{\lambda}}{2l} \tag{1.76}$$

From this discussion, it can be seen that the amplitude and phase of the complex coherence of quasi-monochromatic light can be determined by measuring the visibility and position of interference fringes, respectively.

In fact, the time delay τ between two interfering beams is often very small. When it is much less than the coherence time τ_c (i.e. $\frac{1}{\Delta\nu}$), the differences between $|\Gamma_{12}(\tau)|$, $|\gamma_{12}(\tau)|$, $a_{12}(\tau)$ and $|\Gamma_{12}(0)|$, $|\gamma_{12}(0)|$, $a_{12}(0)$ respectively are very small. We set:

$$\begin{cases} J_{12} = \Gamma_{12}(0) = \langle u_1(t)u_2^*(t)\rangle \\ \mu_{12} = \gamma_{12}(0) = \frac{\Gamma_{12}(0)}{\sqrt{\Gamma_{11}(0)\Gamma_{22}(0)}} = \frac{J_{12}}{\sqrt{J_{11}J_{22}}} = \frac{J_{12}}{\sqrt{I_1 I_2}} \\ \beta_{12} = \alpha_{12}(0) \end{cases} \tag{1.77}$$

where, J_{12} represents the correlation between the light vibration of the points P_1, P_2 at the same time, which is only related to the positions of the two points and has nothing to do with the delay time τ, and is called mutual intensity. μ_{12} denotes the complex coherence of light vibration at two-point P_1, P_2 without time delay τ and is called complex coherence factor. β_{12} is the effective phase difference of μ_{12}. In this way, the law of interference can be expressed as:

$$I_Q = I_1 + I_2 + 2\sqrt{I_1 I_2}|\mu_{12}|\cos(\beta_{12} - \delta) \tag{1.78}$$

when the path difference between the two beams is much less than the coherence length, the mutual intensity J_{12} can be used instead of $\Gamma_{12}(\tau)$ to characterize the correlation of light vibrations between the two-point P_1 and P_2, which is independent of time difference τ.

1.4.6 Spatial Coherence

1. Spatial coherence and coherent area

The spatial coherence of light source aims to study the phase relationship of the light wave field at two points perpendicular to light propagation direction at each moment. If at any time, the light wave fields at these two points have a fixed phase relationship, the light waves "sampled" from these two points will interfere, or in other words, these two points in space are coherent, otherwise, they will be incoherent. Since this coherence of the light wave field belongs to two different spatial points at the same time, it is called spatial coherence. Double-slit interference is a typical experiment demonstrating this spatial coherence.

As mentioned earlier, in the double slit experiment, if an ordinary light source is used, a single slit must be added in front of the double slit (In Fig. 1.21) to limit the size of the light source S. Moreover, the width of this slit directly affects the clarity of interference fringes. When the slit is narrow, the interference fringe is clear, and when the slit gradually widens, the fringe becomes blurred. When the slit width reaches a certain degree, the fringes will disappear completely. A similar situation exists in some interference devices (such as wavefront-splitting interference devices). Taking double-slit interference as an example, we will analyze the influence of light source size S_0 (that is, the size of light source slit) on the visibility of interference fringes.

From Fig. 1.21, it can be seen that the distance of light emitted from different points on the light source slit S_0 (the slit that limits the size of the light source) to the double slits S_1 and S_2 are different. Now, compare the center point S_0 and edge point S_0' of the light source slit to see if there are differences in the interference fringes generated by the two point sources through the slits S_1 and S_2. When the light from the center S_0 illuminates the double slit S_1 and S_2, due to $S_0S_1 = S_0S_2$, there is no OPD between the two points. That is to say, if S_1 and S_2 are considered two coherent light sources, the initial phase difference between them is zero. However, if S_1 and

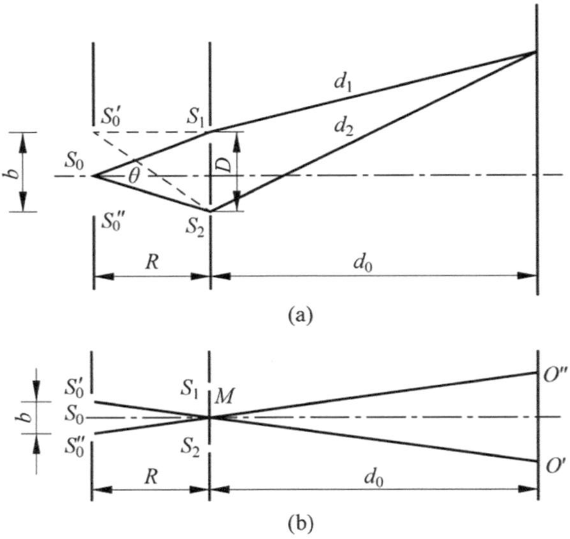

Fig. 1.21 Effect of light source size on interference fringes

S_2 are illuminated point light emitted from S_o', there is an additional initial phase difference δ_o' between the two light sources, which is determined by the additional OPD $\Delta_0 = S_0'S_2 - S_0'S_1$. From Fig. 1.21b,

$$\Delta_0 = S_0'S_2 - S_0'S_1 \approx \frac{bD}{2R} \tag{1.79}$$

where b is the width of the light source slit, that is, the effective size of the light source; D is the distance between the double slits; R is the distance between the double slit and the light source slit. Obviously, this OPD is an additional OPD, which is independent of the coordinates of the observation point P on the screen, and has the effect of shifting the interference fringes (relative to the fringes generated by the light coming from the point S_0), the offset Δ_0 is determined by the additional range difference. And when

$$\delta_0 = \frac{2\pi}{\lambda_0}\Delta_0 == \frac{2\pi bD}{\lambda_0 2R} = \pi \tag{1.80}$$

Instantly, when $b=\lambda_0 R/D$, the light emitted from different points on the light source slit passes through S_1 and S_2 and reaches P, the additional phase difference changes from $+\pi$ to $-\pi$, and the total light intensity of the point P is the superposition of these interference fringes caused by the interference light. From the expression of the interference light intensity of two beams, $I = I_1+I_2+2\sqrt{I_1I_2}\cos\delta$, that is, the value of δ is between $+\pi$ to $-\pi$, the average value of the interference term $J_{12} = 2\sqrt{I_1I_2}\cos\delta$ is zero, so the intensity of each point on the screen is equal. This result shows that when the light source slit S_0 is uniformly illuminated, each point on $S_0'S_0''$ (including the center point S_0) generates a set of interference fringes on the screen through the double slit S_1 and S_2. These fringes are staggered with each other, and their zero order is continuously distributed from point O' to point O'' on the screen, which are the intersection points of the extension line $S_0'M$ and $S_0''M$ on the screen respectively. From Fig. 1.21b, the wider $S_0'S_0''$ is, the more fringes participate in the superposition, the larger the staggered range, and the more uniform the intensity of each point after superposition, consequently the visibility of the fringes drops. When $O'O''$ is large enough to equal the fringe spacing Δx, O' and O'' just falls on the dark stripes of the grade of $m = -1/2$ and $m = +1/2$ on both sides of the point O, and the intensity on the screen becomes uniform without any fringes. At this time, according to the geometric relationship in Fig. 1.21b:

$$O'O'' = d_0 \frac{b}{R}$$

Also by Eq. (1.76):

$$O'O'' = \Delta x = \frac{\lambda_0 d_0}{D}$$

The maximum width of the light source can be obtained by combining the two equations:

$$b = \frac{\lambda_0 R}{D} \tag{1.81}$$

This is consistent with the result of Eq. (1.80). In order for double slit S_1 and S_2 to produce interference fringes, the width b of the slit S_0 should be less than the value given by Eq. (1.81).

On the other hand, this result shows that under certain circumstances of b, λ_0 and R, D is the maximum distance at which interference can be produced in the direction perpendicular to the propagation of light wave. When the distance between two points in space of the light wave is greater than D, the two points are irrelevant. For example, if a 500 nm green light incident to a double slit with 0.5 mm apart, and $R = 20cm$, the slit width that limits the size b of the light source should be $b \leq \lambda_0 \frac{R}{D} \approx 0.2mm$, that is, the slit width of the light source should not be greater than 0.2 mm. On the contrary, if the width of the light source is known to be 0.2 mm (wavelength $\lambda_0 = 500$nm), Eq. (1.81) can be used to prove that the light wave field is coherent at any time and any two points with distance D of (or less than) 0.5 mm in between on the cross section of $R = 20$mm from the light source. Therefore, the magnitude of the spacing D indicates the range of spatial coherence of the light source.

When discussing coherence issues, the concept of coherence area is sometimes used. To solve the problem, just generalize Eq. (1.81) from one dimension to two dimensions.

Because $b\theta \leq \lambda$, while $\theta = \frac{D}{R}$, there is

$$b^2\left(\frac{D}{R}\right)^2 \leq \lambda^2 \text{ or } D^2 \leq \frac{\lambda^2 R^2}{b^2} \tag{1.82}$$

Set $\Delta A_c = D^2$, $\Delta A_s = b^2$, then there is

$$\Delta A_C \leq \frac{\lambda^2 R^2}{\Delta A_s} \tag{1.83}$$

In the formula, ΔA_c is called the coherent area, and the light at all points on this cross section is coherent. ΔA_c is the area of the light source, which is the allowable size of the light source to produce interference effects. Obviously, the coherent area ΔA_c is inversely proportional to the area ΔA_s of the light source and directly proportional to the square of the distance R. Similar to the coherence length L, the coherence area ΔA_c is also a rough value.

So far, we have analyzed the coherence of light sources from two aspects, one is temporal coherence and the other is spatial coherence. If two points are selected along the propagation direction of the light wave, the coherence length can be used to describe the temporal coherence of the light source; if the same point is "sampled" at different times at the same point, the coherence time can be used to describe the coherence of the light source. If the two selected points in space are on a plane perpendicular to the propagation direction, the coherent area, whose size is D^2, is used to describe the spatial coherence on the normal cross section of the beam. As mentioned earlier, double-slit interference is a typical experiment to demonstrate this spatial coherence, while Michelson experiment is a typical experiment to demonstrate the temporal coherence of light.

2. Influence of light source size on interference effect

The coherence of light source is one of the important factors to be considered in interferometry. As mentioned above, its spatial coherence limits the actual usable area of the light source, while its temporal coherence limits the maximum OPD that can be obtained in an interferometer.

Due to the requirement of spatial coherence, the requirement for the size of the light source is: the additional path difference caused by the size of the light source should be less than $\frac{\lambda}{2}$, that is, $\Delta<\frac{\lambda}{2}$. This is an extreme situation. In fact, in order to keep the interference fringes clear, the requirements on the light source are more stringent, and the additional range difference caused by the light source should be less than $\frac{\lambda}{4}$, i.e., $\Delta \leq \frac{\lambda}{4}$.

However, from the perspective of use, a larger size of the light source is always desirable to increase the intensity of incident light for the convenience of observation. Therefore, for each specific interference device, the allowable light source size should be estimated in advance according to the measured object. For example, the requirement of the light source in double-slit interference is: $b \cdot 2\theta \leq \lambda$ or $D^2 \leq \lambda^2 \frac{R^2}{b^2}$, which limits the effective area D^2 of the light source.

For equal thickness interference, the limitation of spatial coherence on the size of light source can be estimated from the condition $\Delta \leq \frac{\lambda}{4}$. Figure 1.22 is a schematic

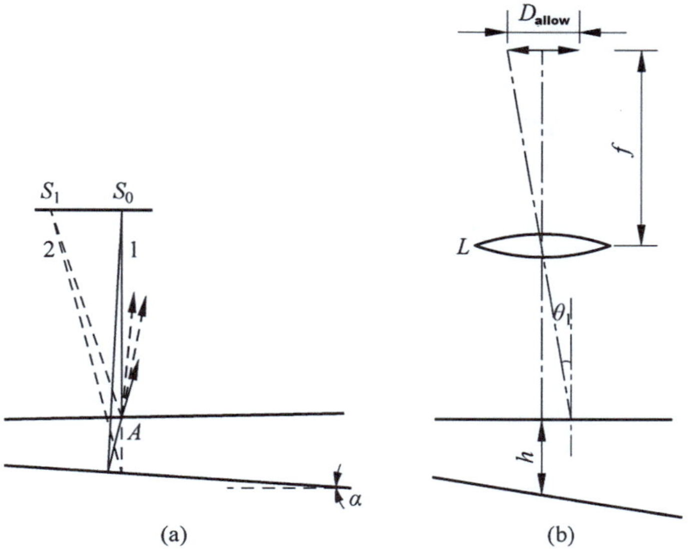

Fig. 1.22 Size effect of light source on interference fringes

diagram of producing equal-thickness interference fringes. Assuming that light beam 1 incidents vertically on the surface of the film, the two reflected beams will intersect at point A near the upper surface of the film, but the OPD is

$$\Delta_1 = 2nh$$

where n is the refractive index of the thin plate material and h is the thickness of the film at point A. Since the light source has a certain size, the light beam 2 emitted from another point S_1 on the surface of the light source is incident on the surface of the film at an angle θ_1, and the two reflected lights also meet near point A on the surface of the film, but the OPD is

$$\Delta_2 = 2h\sqrt{n^2 - n_0^2 \sin^2\theta_1} \approx 2h\sqrt{n^2 - n_0^2\theta_1^2}$$

Because θ_1 is actually very small, $\theta_1 \approx \sin\theta_1$ will applied. The OPD at point A of interference between light from different points S_0 and S_1 on the light source is different, which caused the movement of fringes. And the difference in the OPD of two pairs of coherent light satisfying,

$$\Delta = \Delta_1 - \Delta_2 \leq \frac{\lambda}{4}$$

Then the offset of the fringe is allowed.

And $\Delta_1 - \Delta_2 = \frac{\Delta_1^2 - \Delta_2^2}{\Delta_1 + \Delta_2} \approx \frac{\Delta_1^2 - \Delta_2^2}{2\Delta_1} = \frac{4h^2 n_0^2 \theta_1^2}{4nh} = \frac{h}{n} n_0^2 \theta_1^2$. So there is, $\Delta = \Delta_1 - \Delta_2 \approx \frac{h}{n} n_0^2 \theta_1^2 \leq \frac{\lambda}{4}$, or

$$\theta_1 \leq \frac{1}{2n_0}\sqrt{\frac{n\lambda}{h}} \tag{1.84}$$

This is the limitation of equal thickness interference on the size of the light source: the opening angle of the light source to the observation point θ_1 should not exceed the calculated value of Eq. (1.84). Where n is the refractive index of the film material, h is the thickness of the film at the observation point, λ is the light wavelength, and n_0 is the refractive index of the medium where the film is located. When $n_0 = 1.0$, $n = 1.5$, $\lambda = 0.55$ μm, θ_1 to various values of h are shown in Table 1.2.

Generally, in the observation device of equal thickness interference, parallel incident light is used. At this time, the light source locates on the focal plane of the converging lens L in Fig. 1.22b. Therefore, from Eq. (1.84), the allowable diameter D_{allow} of the light source can be found as

$$D_{\text{allow}} = 2f\theta_1 = \frac{f}{n_0}\sqrt{\frac{n\lambda}{h}} \tag{1.85}$$

From this, the allowable size of the light source can be estimated. It can be seen that the allowable dimension D_{allow} of the light source is related to the thickness of the measured film, so the thinner the film to be measured, the larger the used light sources. Therefore, in some interference devices, a variable aperture is placed in front of the light source to measure films with various thicknesses. Finally, it should be noted that the calculation of D_{allow} is only an estimation and needs to be corrected according to the specific instrument. In addition, if a single-mode laser is used as the light source, there is no limitation of the size of light source because all the points on beam cross section are coherent.

3. Extended light source

This section uses the aforementioned mutual intensity J_{12} and complex correlation coefficients to illustrate the spatial coherence of the extended light source. As shown in Fig. 1.23, the extended quasi-monochromatic primary light source illuminates the two points P_1 and P_2 on the screen, assuming that the medium between the light source and the screen is uniform, the linear dimension of the light source σ is much smaller than the distance R between the light source and the screen, and the distance between the two points P_1 and P_2 is also much smaller than R. Now discuss the correlation between the light vibration of two points P_1 and P_2.

Table 1.2 θ_1 to various values of h

h(mm)	100	10	1	0.1	0.01	0.001
θ_1	5′	16′	50′	2°40′	8°20′	26′

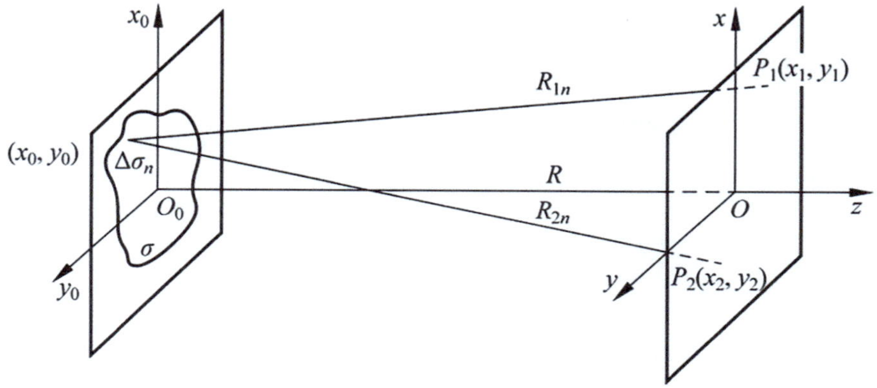

Fig. 1.23 Mutual intensity of extended quasi-monochromatic light sources

Assuming that the light source is divided into many surface elements $\Delta\sigma_1, \Delta\sigma_2 \cdots$, the centers of which are respectively at $S_1, S_2 \cdots$, and the linear dimensions are much smaller than the average wavelength$\bar{\lambda}$. Assume the complex disturbance caused by the surface elements $\Delta\sigma_n$ at P_1 and P_2 is represented by $u_{1n}(t)$ and$u_{2n}(t)$, the entire light field at two points P_1 and P_2 can be written as: $u_1(t) = \sum_n u_{1n}(t)$, $u_2(t) = \sum_n u_{2n}(t)$.

Where u_{1n} and u_{2n} can be written as:

$$\begin{cases} u_{1n}(t) = A_n\left(t - \frac{R_{1n}}{V}\right)\frac{e^{-i2\pi\bar{\nu}\left(t - \frac{R_{1n}}{\nu}\right)}}{R_{1n}} \\ u_{2n}(t) = A_n\left(t - \frac{R_{2n}}{V}\right)\frac{e^{-i2\pi\bar{\nu}\left(t - \frac{R_{2n}}{\nu}\right)}}{R_{2n}} \end{cases} \tag{1.86}$$

where A_n is the radiation intensity of the nth surface element; v is the speed of light in the medium between the light source and the screen. Assuming that the light vibrations caused by different surface elements of the light source are statistically independent, the mutual intensities $J_{(P_1,P_2)}$ of the two points P_1 and P_2 can be obtained from the above two equations as follows

$$J_{(P_1,P_2)} = \langle u_{1(t)} u_2^*(t)\rangle = \sum_n \langle A_n(t)A_n^*(t)\rangle \frac{\exp\left[\frac{i2\pi\bar{\nu}(R_{1n}-R_{2n})}{\nu}\right]}{R_{1n}R_{2n}} \tag{1.87}$$

$\langle A_n(t)A_n^*(t)\rangle$ represents the radiation intensity of the surface element$\Delta\sigma_n$. Since the light source itself is continuous, if $I_{(s)}$ represent the intensity of the light source per unit area, R_1 and R_2 represent the distance between the representative point on the light source to two points P_1 and P_2, then Eq. (1.87) is expressed as an integral as follows:

$$J_{(P_1,P_2)} = \int\limits_{\sigma} I_{(s)} \frac{\exp\left(\frac{i2\pi\bar{\nu}(R_{1n}-R_{2n})}{\nu}\right)}{R_{1n}R_{2n}} ds \tag{1.88}$$

By the Eqs. (1.77) and (1.88), the complex correlation coefficient can be obtained as follows:

$$\mu(R_1, R_2) = \frac{1}{\sqrt{I_{(P_1)}I_{(P_2)}}} \int\limits_{\sigma} I_{(s)} \frac{exp[i2\pi\bar{\nu}(R_{1n} - R_{2n})/\nu]}{R_{1n}R_{2n}} ds \tag{1.89}$$

wherein, $I_{(P_1)} = J_{(P_1,P_2)} = \int\limits_{\sigma} \frac{I(s)}{R_1^2} ds; I_{(P_2)} = J_{(P_1,P_2)} = \int\limits_{\sigma} \frac{I(s)}{R_2^2} ds.$

Equations (1.88) and (1.89) give the correlation of the vibrations of the two points P_1 and P_2 on the plane illuminated by an extended quasi-monochromatic light source, which describes the relationship between the spatial complex correlation and the intensity distribution of the light source, and is called Van Cittert-Zernike theorem, from which it can be seen that even a completely incoherent light source may have a high degree of coherence between two points P_1 and P_2 in its radiation field after propagation. This theorem will be used to discuss the spatial coherence of several typical extended light sources in detail.

(1) Spatial coherence of a circular light source

Suppose a circular light source with a radius of r, whose light intensity is evenly distributed, and each point on the light source radiates incoherent quasi-monochromatic light. According to the Van Cittert-Zernike theorem, the complex coherence of the two points P_1 and P_2 under the radiation field illumination can be obtained. The degree is:

$$\mu_{12} = \frac{2J_1(\nu)}{\nu} e^{i\phi} \tag{1.90}$$

where, $\nu = \frac{2\pi}{\bar{\lambda}} \cdot \frac{r}{R}\sqrt{(x_1 - x_2)^2 + (y_1 - y_2)^2}; \phi = \frac{2\pi}{\bar{\lambda}} \cdot \frac{r}{R}\left[\frac{(x_2^2+y_1^2)+(x_2^2+y_2^2)}{2R}\right]$; $J_1(\nu)$ represents a first-order Bessel function.

When$\nu = 0$, $\left|\frac{2J_1(\nu)}{\nu}\right| = 1$, the degree of coherence is the best and decreases with the distance between the two points increases, that is, as the distance between the two points increases. At that time of$P_1P_2 = \sqrt{(x_1 - x_2)^2 + (y_1 - y_2)^2} = \frac{0.61R\bar{\lambda}}{r}$, these two points becomes completely irrelevant, i.e., $\left|\frac{2J_1(0)}{\nu}\right| = 0$ and$\nu = 3.83$. When $\nu = 1$ the degree of coherence deviates from the maximum 1 by 12%. At this time, $P_1P_2 = \frac{0.16R\bar{\lambda}}{r}$. So the light field within the circular area with a radius of P_1P_2 can be considered approximately completely coherent, which is called coherent area.

(2) Spatial coherence of a slit light source or a line light source

Many discharge tubes are slit light sources, and line light sources can also be obtained by imaging ordinary light source onto rectangular slit through lens. Assuming that the line light source with width of a emits light uniformly, its complex correlation coefficient is:

$$\mu_{12} = \operatorname{sinc}\left(\frac{da}{\bar{\lambda}l}\right) \tag{1.91}$$

In the equation, l is the distance between the light source and the screen where the two points P_1 and P_2 are located; d is the distance between points P_1 and P_2.

It can be seen from the Eq. (1.91) that the degree of coherence between the two points will gradually decrease as the distance between them increases.

The above two examples prove that for a thermal light source, which is completely incoherent in space, a certain spatial coherence can be obtained in its radiation field after the radiation travels over a long distance. The reason is that the light vibration at the same point on the light source contributes to the vibration of the observation point P_1 and P_2 on the screen respectively, although the points on the surface of the light source are not related to each other, so the points P_1 and P_2 will have a certain degree of coherence.

1.4.7 Interference Characteristics of Partially Coherent Light

The concept and basic functions describing the coherence of light sources have been discussed in previous sections. It can be seen that the coherence degree can be used to classify light sources into three categories: completely coherent, partially coherent and completely incoherent. The actual light source not only has a certain limit, but also has a certain spectral width, so it is always partially coherent. The characteristic quantities to deal with the interference problem of partially coherent light are mutual interference function and complex coherence.

Equation (1.75) shows the intensity distribution of partially coherent light [6] interference. It can be found that the contrast of the experimentally observable light intensity distribution is related to the theoretical complex coherence. Complex coherence is a measure of the coherence of a light source, which represents the proportion of the coherent light part to the total intensity.

Based on the analysis of the temporal coherence and spatial coherence of the light source, it can be drawn that the fringe contrast is related to the size of the light source and the spectral width when observing the interference fringes of partially coherent light. The coherence conditions can only be satisfied under certain conditions, and the interference fringes observed differs from completely coherent light whose fringe contrast can reach to 1 and the interference light field is the superposition of complex amplitudes. It also differs from completely incoherent light whose fringe contrast is equal to 0. The intensity of the superposition light field is a simple addition of light intensity. The fringe contrast of partially coherent light is between 0–1. The superposition light field has both completely coherent parts and completely incoherent parts.

1.4.8 Laser Coherence

Similar to ordinary light sources, laser coherence also includes two aspects: temporal coherence and spatial coherence. In order to understand the coherence characteristics of laser, we should first understand the basic principle of laser generation.

Laser is the abbreviation of "light amplification by stimulated emission of radiation". Its luminescence process is the population inversion of laser working material under external stimulation, and stimulated emission takes place for the particles at high energy level with the effect of radiation field. Since the active materials holds in an optical resonant cavity, stimulated amplification and enhancement can be obtained from the feedback of the stimulated radiation by continuously passes through the laser materials. And finally produces a stable laser output. The characteristic of stimulated radiation is that the luminous particles are in a state of synchronous luminescence, and the emitted photons have the same state. In addition, the resonant cavity is also selective on frequency and direction of light radiation, so laser has excellent coherence and directionality.

1. Temporal coherence of laser

The ideal laser is a pure monochromatic oscillation, i.e., only a single frequency of light will satisfy the resonance condition, and the generated light field can be expressed as: $\mu(t) = A\cos(2\pi \nu_0 t - \varphi)$. Due to various linewidth broadening effects, the general single longitudinal mode laser still has a certain spectral width. For example, the spectral width $\Delta\nu$ of a single longitudinal mode He–Ne laser is 10^6 Hz about equivalent to its coherence time of 1 μs and its coherence length of 300 m.

The outputs of many lasers have multiple longitudinal modes, in which multiple light outputs with different wavelengths satisfy the resonant frequency. At this time, the light field can be expressed as: $\mu(t) = \sum_{i=1}^{N} A_i\cos[2\pi \nu_i t - \varphi_i(t)]$. The light of each frequency (mode) vibrates independently in common cases. Even the spectral width of a laser with multiple longitudinal modes is still very narrow. For example, the spectral width of multi-longitudinal mode He–Ne laser is about 1.5×10^9Hz, the corresponding coherence length and time are about 20 cm and 1 ns.

Compared with ordinary light sources, laser has a narrow spectral width $\Delta\nu$ and very long coherence time Δt that leading to good time coherence. Even if the two light beams from a laser light source have a large path difference, the interferometer can still obtain clear interference fringes.

2. Spatial coherence of laser

The spatial coherence of laser mainly depends on the structure of the transverse mode, i.e., the distribution of the output beam in space. The resonant cavity limits the distribution of the beam in space to some extent, and laser can be excited only by beams travelling back and forth in the cavity which can gain enough amplification. The intensity patterns of several transverse modes of the resonant cavity with circular mirror are shown in Fig. 1.24.

A laser with a multi-transverse mode structure is equivalent to a combination of multiple light sources, whose frequencies, polarizations and positions are different, so

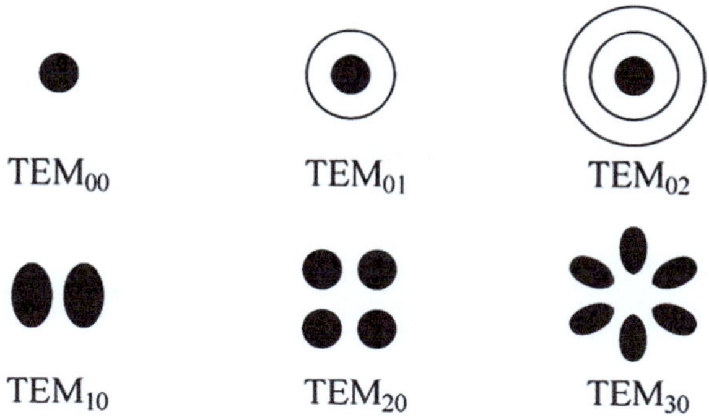

Fig. 1.24 Intensity pattern of circular mirror cavity mode

they are incoherent with each other. However, each transverse mode is an excellent coherent light source and spatially coherent throughout its cross-section, i.e., the vibrations of each point on the cross-section are related.

In order to obtain nice beam quality, various methods are proposed to select modes. For example, output of single transverse mode can be obtained using parallel planar cavity or adding small holes in the cavity. Single transverse mode laser has excellent spatial coherence, and clear interference fringes can be observed on the screen directly when the laser beam hits the double holes.

1.5 Typical Multi-beam Interferometer: Fabry–Perot Interferometer

1.5.1 Structure of a Fabry–Perot Interferometer

Such as shown in Fig. 1.25, a Fabry–Perot interferometer (FPI) is mainly composed of two parallel glass plates (or quartz plates) P_1, P_2, which are coated with multiple layers of dielectric films (or metal films) to produce high reflectivity. The processing technology of the opposite surfaces of the plate requires a very high level of precision, and the requirement of flatness is generally better than $\lambda/20$. In order to obtain equal-tilted interference fringes, the instrument is equipped with a precise adjustment device to ensure that the two planes are strictly parallel. In addition, there is a small included angle between the two surfaces of each plate in order to eliminate the interference caused by light reflected from the outer surfaces (non-working surfaces) of the two plates P1, P2. The light emitted from the expansion light source S incident in all directions passes through lens L_1, and becomes a parallel light beam. The transmitted light passes through lens L_2 and formed equal-inclination interference fringes on the focal plane of lens L_2.

If the optical path between the two plates can be adjusted, this interference device is called Fabry–Perot interferometer, or FP interferometer for short. Generally, the

Fig. 1.25 Optical path diagram of a FPI

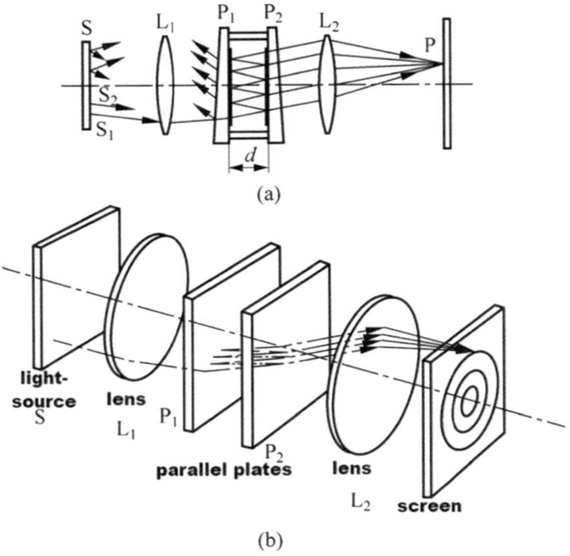

method of changing the optical path is to fix one of the two plates on a precision guide rail and move it to adjust the distance h between the two plates. Change of the refractive index of the medium between the two plates P_1, P_2 (such as evacuating or adding high-pressure gas) can also change the optical path. When the distance between the two plates is fixed by a spacer, this interference device is called Fabry–Perot etalon, or FP etalon for short. Spacer is generally made of invar steel or fused silica cylinder, and both ends are grounded and polished to be exactly parallel to each other. Some spacers are made of piezoelectric ceramic, that the spacing can be slightly changed when various voltages applied. It is also possible to make a single etalon by grinding both sides of a single glass plate parallel to each other, then coating with a highly reflective film. The thickness h of the plate is pre-designed according to the specific application.

1.5.2 *Performance Parameters of Interferometer*

Equations (1.24) and (1.25) describe the intensity distribution of the multi-beam equal-inclination interference fringes produced by a FP interferometer. Because of the high reflectivity of the two flat surfaces, the width of the bright stripes is extremely narrow. Therefore, FP is a high-resolution optical spectroscopic instrument, which is most used to study the fine structure of spectrum. With the development of opto-electronic technology, FP interferometer expands more applications, such as laser resonator.

1. Transmission wavelength and frequency under normal incidence

By Eq. (1.26), the maximum transmission condition of the FPI at normal incidence is ($\theta_1 = 0$)

$$2nh = m\lambda \tag{1.92}$$

If the spectrum of incident light is continuous, i.e., including many wavelengths, such as white light, only the lights with wavelengths satisfying the above Eq. (1.92) (as described below, these transmitted wavelengths still have a certain wavelength range) can pass through the interferometer, while the other wavelengths will be reflected. The FPI functions as a filter in this case. However, Eq. (1.92) can be satisfied by more than one light wavelength. It is necessary to find the suitable wavelength range for an actual FPI, i.e., the difference between the two adjacent wavelengths passing through the instrument, which can be obtained from Eq. (1.92). Assume two wavelengths of λ and $\lambda' = \lambda + \Delta\lambda$ satisfy Eq. (1.92), and the order difference of the interference is 1, so

$$m\lambda = (m-1)(\lambda + \Delta\lambda)$$

At that time, $m\rangle\rangle 1$

$$\Delta\lambda = \frac{\lambda}{m} = \frac{\lambda^2}{2nh} \tag{1.93}$$

This is the wavelength difference $\Delta\lambda$ between two adjacent wavelengths that pass through the interferometer. This wavelength difference is generally called etalon constant or free spectral range, which is an important constant of the instrument and an actual wavelength range that interferences of different orders do not overlap.

If the light incidents at various angles, the light of each wavelength will produce independently a set of equal-inclination interference rings in the wavelength range of λ and $\lambda + \Delta\lambda$ without order overlap, while the interference rings produced by the light outside this wavelength range will have order overlap.

Because the transmitted wavelength satisfies $\lambda \propto \frac{1}{m}$, the wavelengths passing through the interferometer are not equally spaced (regardless of nh value) at normal incidence. But the transmission frequencies are equally spaced. The frequency of the transmitted light can be obtained from Eq. (1.92) as

$$\nu = \frac{c}{\lambda} = \frac{m}{2nh}c \tag{1.94}$$

It is proportional to m, therefore

$$\Delta\nu = \nu_{m+1} - \nu_m = \frac{c}{2nh} \tag{1.95}$$

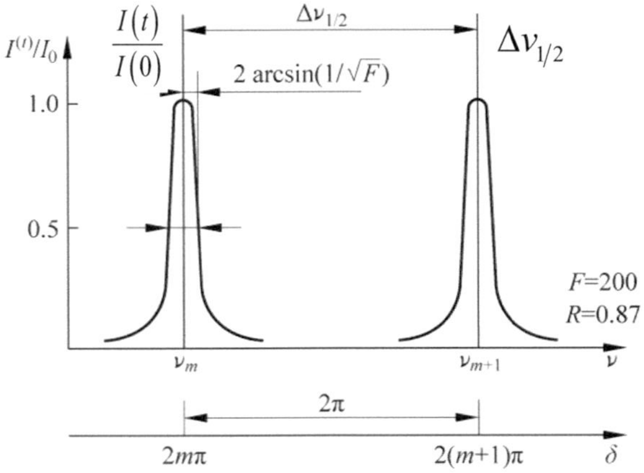

Fig. 1.26 Transmission spectrum of a FPI

It shows obviously that $\Delta\nu$ has nothing to do with the order m. since $\lambda \propto \frac{1}{nh}$, the larger nh, the narrower the transmission frequency interval $\Delta\nu$, and correspondingly the smaller the wavelength interval $\Delta\lambda$.

In the case of normal incidence of the beam, draw the variation curve of the relative transmitted light intensity $\frac{I^{(t)}}{I_0}$ of the Fabry–Perot interferometer with frequency ν (Fig. 1.26). By comparison with the variation curve of $\frac{I^{(t)}}{I_0}$ with δ (Fig. 1.7), it can be seen that the two curves have the same shapes with different abscissa, which results from the phase difference that is proportional to the frequency:

$$\delta = \frac{4\pi}{\lambda_0}nh = \frac{4\pi}{c}nh\nu$$

Each transmission spectrum itself still has a certain half width $\Delta\nu_{1/2}$ that can also be drawn from the curve in Fig. 1.26, which define as the frequency interval between the two points where the relative intensity $\left(\frac{I^{(i)}}{I_0}\right)$ on both sides of the center of the spectral line drops to 1/2. Because the corresponding phase difference of this half width is ε (fringe half width, Eq. (1.30)), so

$$\Delta\nu_{\frac{1}{2}} = \frac{c}{2\pi nh}\frac{1-R}{\sqrt{R}} \tag{1.96}$$

This is the frequency width of the transmitted spectral lines. The corresponding wavelength width can be obtained using $\lambda = \frac{c}{\nu}$

$$\left|\Delta\lambda_{\frac{1}{2}}\right| = \frac{\lambda}{c}\Delta\nu_{\frac{1}{2}} = \frac{\lambda^2}{2\pi nh}\frac{1-R}{\sqrt{R}} \tag{1.97}$$

Fig. 1.27 Two sets of interference rings produced by a FPI

2. Resolution of instrument

When the light incidents on instrument has several wavelengths, the solid angles of the same order of interference between any two wavelengths (such as the maximum of *m* th order) are different, which follows:

$$m = \frac{2nh\cos\theta_t}{\lambda} = \frac{2nh\cos\theta'_t}{\lambda'}$$

The smaller the wavelength is, the larger the θ_t will be, i.e., the radius of the interference ring increases, and the larger the wavelength $\left(\lambda' = \lambda + \delta\lambda\right)$ is, the smaller the solid angle θ'_t, i.e., the radius of the interference ring decreases. Therefore, the interference rings of these two wavelengths will be slightly separated, and the extend of separation depends on the wavelength difference $\delta\lambda$, as shown in Fig. 1.27. Even though the fringes produced by multi-beam interference in a FPI are narrow, there still is a certain width. So the two sets of the interference rings overlaps each other and cannot be distinguished when the wavelength difference $\delta\lambda$ of the incident beams is too small.

The current problem to be discussed is: what is the minimum wavelength difference that can be separated by a FPI? To this end, it is necessary to first establish a standard of separation before calculating the corresponding wavelength difference. A simple criterion for judgement used to adopt is: when the distance between the interference maximums of the two wavelength is not less than the half width ε of the fringe, the two wavelengths is considered to just be separated. Because the two peaks can be still distinguishable after superimpose, such as shown in Fig. 1.27. According to this criterion, the minimal wavelength difference $\delta\lambda$ that a FPI is able to distinguish can be obtained.

The phase difference of the corresponding beam of the *m*th interference ring with wavelength of λ_0 is:

$$\delta = \frac{4\pi}{\lambda_0} nh\cos\theta_t = 2\pi m \tag{1.98}$$

For light with a wavelength of $\lambda_0 + \delta\lambda$, the phase difference between coherent lights along the same direction θ_t is:

$$\delta' = \frac{4\pi}{\lambda_0 + \delta\lambda_0} nh \cdot \cos\theta_t \tag{1.99}$$

According to the above judgment criteria, when

$$\Delta\delta = \delta - \delta' = \varepsilon \tag{1.100}$$

The interference rings formed by the two wavelengths can be resolved (Fig. 1.28). Substitute Eqs. (1.98), (1.99) into Eq. (1.100):

$$\begin{aligned}\Delta\delta &= 4\pi nh\cos\theta_t\left(\frac{1}{\lambda_0} - \frac{1}{\lambda_0 + \delta\lambda_0}\right) \\ &= 4\pi nh\cos\theta_t\frac{\delta\lambda_0}{\lambda_0^2} = 2\pi m\frac{\delta\lambda_0}{\lambda_0}\end{aligned}$$

wherein, utilizing $\delta\lambda_0 \langle\langle \lambda_0$, Eq. (1.98), and Eq. (1.100) has:

$$\frac{\lambda_0}{\delta\lambda_0} = \pi m\frac{\sqrt{R}}{1-R} = \frac{2\pi nh \cdot \cos\theta_t}{\lambda_0}\frac{\sqrt{R}}{1-R} \tag{1.101}$$

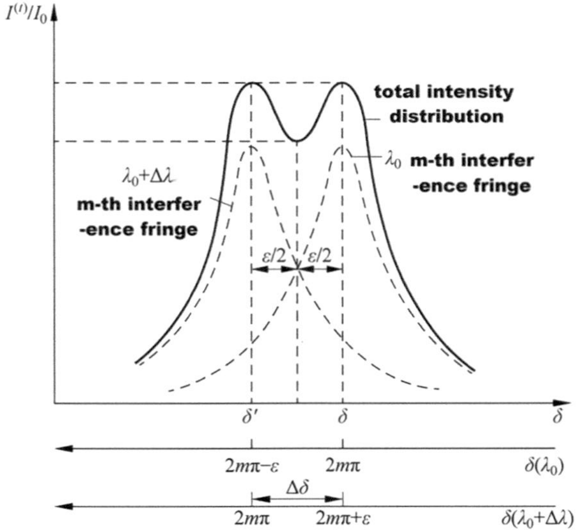

Fig. 1.28 Resolution criterion of two interference loops

where, $\delta\lambda_0$ is the minimum wavelength difference corresponding to the two interference rings that can be distinguished by the interferometer. The smaller the $\delta\lambda_0$, the greater the resolution of instrument, so $A = \frac{\lambda_0}{\delta\lambda_0}$ is used to define the resolution of optical instrument. Equation (1.101) is the expression of the resolution of the FPI. It shows that A increases with the interference order m and R.

According to Eqs. (1.31), (1.101) is also rewrite as

$$A = \frac{\lambda}{\delta\lambda} = mN \tag{1.102}$$

where, N is the fineness of the interferometer. In the vicinity of normal incidence, $\theta_t \approx 0°$, $\cos\theta_t \approx 1$, the corresponding interference order $m = \frac{2nh}{\lambda}$, and the resolution is

$$A = \frac{\lambda}{\delta\lambda} = \frac{2\pi nh}{\lambda}\frac{\sqrt{R}}{1-R} \tag{1.103}$$

The minimum resolvable wavelength difference is

$$\delta\lambda = \frac{\lambda^2}{2\pi nh}\frac{1-R}{\sqrt{R}}$$

If expressed in terms of frequency, δv is determined using $\nu = \frac{c}{\lambda}$,

$$\delta\nu = \frac{c}{2\pi nh}\frac{1-R}{\sqrt{R}}$$

This is the mimimum frequency difference that the Fabry–Perot interferometer can resolve in the vicinity of normal incidence. It shows that the resolving power of the F-P interferometer raises with h. Apparently, $\Delta\lambda_{\frac{1}{2}}$ and $\Delta\nu_{\frac{1}{2}}$ equal to $\delta\lambda$ and $\delta\nu$, respectively.

For example, when $R = 0.9$ and $nh = 10$ mm, for red light of $\lambda = 0.6934$ μm, $A = 8.6 \times 10^{5}$;$\delta\lambda = 0.0008$ nm; $\delta v = 5 \times 10^8$ Hz (500 MHz).

3. The angular dispersion of instrument

The main purpose of spectroscopic instruments is to decompose polychromatic light into monochromatic light. Therefore, angular dispersion, that describes the extend of separation of different wavelengths is one of the important performance indicators. The angular dispersion of instrument refers to the angle separated by the unit wavelength interval after passing through the spectroscopic instrument, which is defined as $d\theta/d\lambda$. The larger the angular dispersion $d\theta/d\lambda$ of instrument, the separation angle of different wavelengths will be greater after passing through the instrument. For F-P interferometer,

$$2nh\cos\theta = m\lambda$$

Take the differential of both sides (regardless of the dispersion of the plate material)

$$\frac{d\theta}{d\lambda} = -\frac{\cos\theta}{\lambda}$$

The above formula shows that the smaller the angle θ, the greater the angular dispersion of instrument $\frac{d\theta}{d\lambda}$. Therefore, the central spectrum of the interference ring is the purest, and this feature is often used in spectral analysis with the F-P interferometer, and has nothing to do with parameters h and R of the instrument.

References

1. Born M, Wolf E (1964) Principles of optics. Pergamon Press, London
2. Tolansky S (1948) Multiple-beam interferometry. Clarendon Press, Oxford
3. Shajenko P (1982) Remote sensing with white light. Appl Opt 21:2095–2095
4. Shajenko P (1982) Multimode optical fibers as sensing devices. Appl Opt 21:4185–4186
5. Parrent, George B Jr (1960) Contributions to the theory of partial coherence [D]. United Kingdom: University of Manchester
6. Gene R (1970) A theoretical study of the mutual coherence function associated with the fields generated within a laser cavity schell [D]. University of Houston, Houston

Chapter 2
Transmission and Control of Polarized Light in Optical Fiber

2.1 Introduction

An optical fiber is a dielectric cylindrical waveguide that operates in optical band. It constrains the energy of electromagnetic waves in the form of light within its interface and proceed along the direction of the fiber axis based on the principle of total reflection. The transmission characteristics of optical fibers are determined by their structure and materials.

The basic structure of optical fiber is two layers of cylindrical medium, the inner layer is the core and the outer layer is the cladding; The refractive index of the core n_1 is slightly larger than that of the cladding n_2. When the incidence condition of total reflection is met, the light wave can propagate forward along the core. Figure 2.1 is a structure diagram of an optical fiber. Outside of the cladding of an actual fiber also has one or more protective layers to keep optical fiber from environmental pollution, moist and mechanical damage, as well as more complex optical fiber structures that meet different requirements in use.

When light wave transmits in optical fiber, the solution of electromagnetic field is discrete due to the restriction of core-cladding boundary. Such discrete field solutions are called modes. There are a variety classification of optical fibers. According to the number of transmission modes, it can be divided into single-mode fiber (SMF) and multi-mode fiber (MMF). An optical fiber that can only transmit one mode is a single-mode optical fiber, and an optical fiber that can transmit multiple modes at the same time is called a multi-mode optical fiber. The main differences between single-mode fibers and multi-mode fibers are core size and core-cladding refractive index difference. The core diameter and the core-cladding refractive index difference of a multi-mode fiber ($2a = 50 \sim 500\mu m$ and $\Delta = (n_1 - n_2)/n_1 = 0.01 \sim 0.02$) are larger than those of a SMF ($2a = 2 \sim 12\mu m$ and $\Delta = 0.005 - 0.01$).

According to the refractive index distribution of the core, it can be divided into step index fiber and gradient index fiber. The core refractive index of the former is uniform, and the refractive index changes abruptly (i.e. step) at the interface between

Y. Liao et al., *Fiber Optic Interferometer*, Springer-TUP Physics Series,
https://doi.org/10.1007/978-981-95-2558-4_2

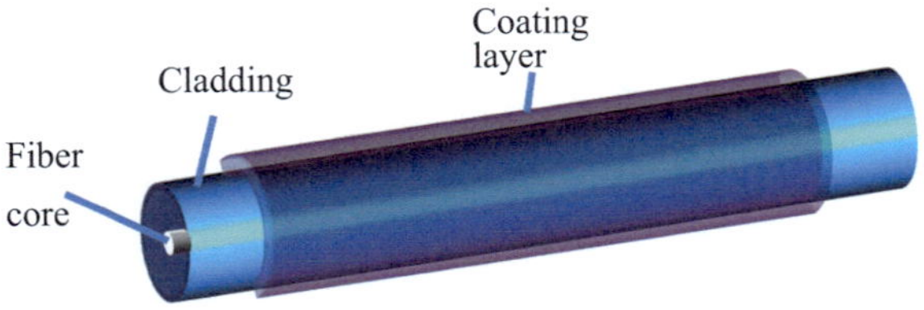

Fig. 2.1 Optical fiber structure diagram

the core and the cladding; The core refractive index of the latter is a function of radial distance. Schematic diagrams and typical dimensions of these two types of optical fibers are given in Fig. 2.2. Figure 2.2a is a single-mode step-index fiber, Fig. 2.2b and c are multimode step index fibers and gradient index fibers, respectively.

According to the transmission polarization state, SMF can be further classified into non-polarization-maintaining optical fiber (referred to as non-PMF) and polarization-maintaining optical fiber (referred to as PMF). The difference is that the former cannot maintain transmission of polarized light, while the latter can. Polarization-maintaining fibers are further divided into four types: single-polarization fiber, high-birefringence fiber, low-birefringence fiber and polarization-circular-maintaining fiber. An optical fiber that can transmit only one polarization mode is called a single polarization optical fiber; The optical fiber that can transmit two orthogonal polarization modes and has a large difference in propagation speeds is a high birefringence

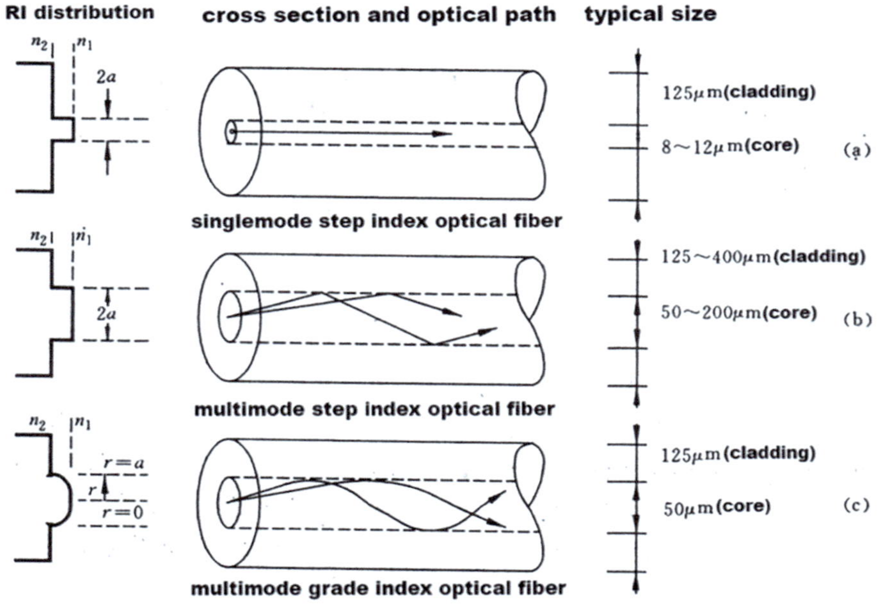

Fig. 2.2 Schematic diagram of single-mode and multi-mode fiber structure

optical fiber, and if the propagation speeds are nearly equal, it is a low birefringence optical fiber; Those that can transmit circularly polarized light are called circular birefringence fiber.

Optical fibers can also be classified differently according to manufacturing materials, structures, and applications. There are three categories of methods to analyze the transmission of light waves in optical fibers: geometric optics, wave optics, and numerical analysis. Based on wave theory, this chapter will discuss the transmission and characteristics of polarized light in SMF.

2.2 Wave Theory of Uniform Refractive Index Fiber

2.2.1 Maxwell's Equations and Wave Equations

1. Maxwell's Equations

Light is electromagnetic wave, so the basic properties of light wave transmission in an optical fiber can be deduced from the basic equations of the electromagnetic field, that are Maxwell's equations.

The electromagnetic field in vacuum is described by two vectors: electric field intensity $\boldsymbol{E}$ and magnetic induction intensity $\boldsymbol{B}$. In order to describe the effect of field on matter, such as the propagation of light wave in transparent medium, it is necessary to introduce three vectors: induction intensity $\boldsymbol{D}$, magnetic field intensity $\boldsymbol{H}$ and current density $\boldsymbol{j}$. At each point in the field, the relationship of these five vectors with time and space is given by the following Maxwell equations:

$$\nabla \times \boldsymbol{H} = j + \frac{\partial \boldsymbol{D}}{\partial t} \tag{2.1}$$

$$\nabla \times \boldsymbol{E} = -\frac{\partial \boldsymbol{B}}{\partial t} \tag{2.2}$$

$$\nabla \cdot \boldsymbol{B} = 0 \tag{2.3}$$

$$\nabla \cdot \boldsymbol{D} = \rho \tag{2.4}$$

where ρ is the charge density in the field; For Hamiltonian operator ∇: $\nabla = l_x \frac{\partial}{\partial x} + l_y \frac{\partial}{\partial y} + l_z \frac{\partial}{\partial z}$; $\boldsymbol{l_x}$, $\boldsymbol{l_y}$, $\boldsymbol{l_z}$ is the unit vector along x, y, z directions.

Due to Eqs. (2.3) and (2.4) are derived from Eqs. (2.1) and (2.2), and the law of conservation of charge ($\nabla \cdot j = -\frac{\partial \rho}{\partial t}$), the Eqs. (2.1) and (2.2) are the most basic equations. However, in order to solve $\boldsymbol{E}$, $\boldsymbol{D}$, $\boldsymbol{B}$, and $\boldsymbol{H}$, it is necessary to connect their matter equations, which vary with the medium in the electromagnetic field, to these two basic equations. If the medium is anisotropic and inhomogeneous, there is

$$\boldsymbol{D}(r) = \boldsymbol{\varepsilon}(r) \cdot \boldsymbol{E}(r) \tag{2.5}$$

$$\boldsymbol{B}(r) = \boldsymbol{\mu}(r) \cdot \boldsymbol{H}(r) \tag{2.6}$$

where $\boldsymbol{\varepsilon}$(r) and $\boldsymbol{\mu}$(r) are the permittivity tensor and permeability tensor of the medium, respectively, representing the anisotropy of the material. For example, for ferrite

$$\mu = \begin{bmatrix} \mu_0\mu_r & -ik\mu_0 & 0 \\ ik\mu_0 & \mu_0\mu_r & 0 \\ 0 & 0 & 1 \end{bmatrix}$$

For biaxial crystals

$$\varepsilon = \begin{bmatrix} \varepsilon_{11} & 0 & 0 \\ 0 & \varepsilon_{22} & 0 \\ 0 & 0 & \varepsilon_{33} \end{bmatrix}$$

where, μ_0 is the magnetic permeability in vacuum.

$$\mu_r = 1 + \frac{\omega_0\omega_M}{\omega_0^2 - \omega^2},\ k = \frac{\omega_0 - \omega_M}{\omega_0^2 - \omega^2}$$

ω is the operating frequency; ω_0 is the ferrite precession frequency; ω_M is the ferrite eigen-frequency. For anisotropic species, the directions of $\boldsymbol{E}$ and $\boldsymbol{D}$ are different, as well as the directions of $\boldsymbol{H}$ and $\boldsymbol{B}$. For isotropic media,

$$\boldsymbol{D} = \varepsilon \boldsymbol{E} \tag{2.7}$$

$$\boldsymbol{B} = \mu \boldsymbol{H} \tag{2.8}$$

where $\varepsilon = \varepsilon(r)$, $\mu = \mu(r)$. At this time, $\boldsymbol{D}$ and $\boldsymbol{E}$ are in the same direction, and $\boldsymbol{H}$ and $\boldsymbol{B}$ are in the same direction.

2. Wave Equation

Maxwell's equation only gives the relationship between the field and the field source, that is, the interrelationship between $\boldsymbol{E}, \boldsymbol{D}, \boldsymbol{B}$ and $\boldsymbol{H}$. In order to find the propagation law of light waves in optical fiber, we should further solve the variation law of each quantity with time and space, i.e., solve the variation relationship between $\boldsymbol{E}$ and $\boldsymbol{H}$ in time and space from Maxwell's equations.

The following deductions are made for isotropic media. To this end, using Eqs. (2.1) and (2.2)

$$\nabla \times (\nabla \times \boldsymbol{E}) = -\nabla \times \frac{\partial \boldsymbol{B}}{\partial t} = -\frac{\partial(\nabla \times \mu \boldsymbol{H})}{\partial t} \tag{2.9}$$

$$\nabla \times (\nabla \times \boldsymbol{H}) = \frac{\partial(\nabla \times \varepsilon \boldsymbol{E})}{\partial t} + \nabla \times \boldsymbol{j} \tag{2.10}$$

While $\nabla \times \nabla \times \boldsymbol{E} = \nabla(\nabla \cdot \boldsymbol{E}) - \nabla^2 \boldsymbol{E}$, the Eq. (2.9) is transformed into

$$\nabla(\nabla \cdot \boldsymbol{E}) - \nabla^2 \boldsymbol{E} = -\frac{\partial(\mu \nabla \times \boldsymbol{H} + \nabla \mu \times \boldsymbol{H})}{\partial t} \tag{2.11}$$

By the eqs. (2.4) and (2.7) and vector identities, we can obtain

$$\nabla \cdot \boldsymbol{E} = \frac{\rho}{\varepsilon} - \frac{\boldsymbol{E} \cdot \nabla \varepsilon}{\varepsilon} \tag{2.12}$$

Then from the Eqs. (2.1) and (2.2), we can get

$$\frac{\partial(\mu \nabla \times \boldsymbol{H})}{\partial t} = \mu \frac{\partial(\nabla \times \boldsymbol{H})}{\partial t} = \mu \varepsilon \frac{\partial^2 \boldsymbol{E}}{\partial t^2} + \mu \frac{\partial \boldsymbol{j}}{\partial t} \tag{2.13}$$

$$\frac{\partial(\nabla \mu \times \boldsymbol{H})}{\partial t} = \nabla \mu \times \frac{\partial \boldsymbol{H}}{\partial t} = -\nabla \mu \times \frac{\nabla \times \boldsymbol{E}}{\mu} \tag{2.14}$$

where μ is assumed to be independent of time. Substitute Eqs. (2.12)~(2.14) into Eq. (2.11), we get

$$\nabla^2 \boldsymbol{E} + \nabla\left(\boldsymbol{E} \cdot \frac{\nabla \varepsilon}{\varepsilon}\right) - \nabla\left(\frac{\rho}{\varepsilon}\right) + \frac{\nabla \mu}{\mu} \times \nabla \times \boldsymbol{E} = \mu \varepsilon \frac{\partial^2 \boldsymbol{E}}{\partial t^2} + \mu \frac{\partial \boldsymbol{E}}{\partial t} \tag{2.15}$$

Similarly for vector $\boldsymbol{H}$, there is

$$\nabla^2 \boldsymbol{H} + \nabla\left(\frac{\nabla \varepsilon}{\varepsilon} \cdot \boldsymbol{H}\right) + \frac{\nabla \varepsilon}{\varepsilon} \times \nabla \times \boldsymbol{H} = \mu \varepsilon \frac{\partial^2 \boldsymbol{H}}{\partial t^2} + \frac{\nabla \varepsilon}{\varepsilon} \times \boldsymbol{j} - \nabla \times \boldsymbol{H} \tag{2.16}$$

Equation (2.15) and (2.16) are the wave equations in isotropic and inhomogeneous media, which are rather complicated equations.

Since we are currently concerned with the transmission of light waves in transparent media (optical fiber), where $\mu = \mu_0$, $\rho = 0$, $\nabla \cdot \boldsymbol{E} = 0$ and $j = 0$, the above wave equations simplify to

$$\nabla^2 \boldsymbol{E} + \nabla\left(\boldsymbol{E} \cdot \frac{\nabla \varepsilon}{\varepsilon}\right) = \mu_0 \varepsilon \frac{\partial^2 \boldsymbol{E}}{\partial t^2} \tag{2.17}$$

$$\nabla^2 \boldsymbol{H} + \frac{\nabla \varepsilon}{\varepsilon} \times \nabla \times \boldsymbol{H} = \mu_0 \varepsilon \frac{\partial^2 \boldsymbol{H}}{\partial t^2} \tag{2.18}$$

Obviously, the above two Eqs. (2.17) and (2.18) are still very complicated and difficult to solve. Considering two cases of uniform fibers (ε is a constant, and $\nabla \varepsilon = 0$)

or fiber with slow variation of ε (ε is not a constant, but $\nabla\varepsilon \approx 0$), the above Eqs. (2.17) and (2.18) can be further simplified as

$$\nabla^2 \boldsymbol{E} = \mu_0 \varepsilon \frac{\partial^2 \boldsymbol{E}}{\partial t^2} \tag{2.19}$$

$$\nabla^2 \boldsymbol{H} = \mu_0 \varepsilon \frac{\partial^2 \boldsymbol{H}}{\partial t^2} \tag{2.20}$$

This is the simplest wave equation.

For monochromatic light waves:

$$\boldsymbol{E}(r) = E_0(r) \exp[-ik_0\phi(r)] \exp[i\omega t]$$

$$\boldsymbol{H}(r) = H_0(r) \exp[-ik_0\phi(r)] \exp[i\omega t]$$

i.e., ${}^{\partial}/_{\partial t} = i\omega$, ${}^{\partial^2}/_{\partial t^2} = -\omega^2$. Thus, Eqs. (2.17) and (2.18) become

$$\nabla^2 \boldsymbol{E} + \nabla\left(\boldsymbol{E} \cdot \frac{\nabla\varepsilon}{\varepsilon}\right) = -k^2 \boldsymbol{E} \tag{2.21}$$

$$\nabla^2 \boldsymbol{H} + \frac{\nabla\varepsilon}{\varepsilon} \times \nabla \times \boldsymbol{H} = -k^2 \boldsymbol{H} \tag{2.22}$$

Accordingly, Eqs. (2.19) and (2.20) become

$$\nabla^2 \boldsymbol{E} + k^2 \boldsymbol{E} = 0 \tag{2.23}$$

$$\nabla^2 \boldsymbol{H} + k^2 \boldsymbol{H} = 0 \tag{2.24}$$

In above equations, taking advantage of $\omega^2 \mu_0 \varepsilon = \omega^2 \mu_0 \varepsilon_0 \varepsilon_r = \varepsilon_r k_0^2 = n^2 k_0^2 = k^2$. where ε_r and n are the relative permittivity and refractive index of optical fiber; $k_0 = 2\pi/\lambda$ is the wave number in vacuum; λ is the wavelength of light waves in vacuum. The above two equations are Helmholtz equations for vectors. In the rectangular coordinate system, all of x, y, and z components of $\boldsymbol{E}$ and $\boldsymbol{H}$ satisfy the scalar Helmholtz equation:

$$\nabla^2 \psi + k^2 \psi = 0 \tag{2.25}$$

where ψ represents the components of $\boldsymbol{E}$ or $\boldsymbol{H}$.

2.2.2 Vector Mode

From the derivation of the wave equations above, it can be seen that the spatial distribution of refractive index mainly affects the transmission characteristics of optical waveguide. In above discussion, we have assumed that the refractive index distribution is linear (please refer to *Nonlinear Fiber Optics* for nonlinearity in optical fibers), time-invariant, and isotropic, i.e., $n = n(x, y, z)$. To this end, optical waveguides can be classified according to the spatial distribution of the refractive index as follows:

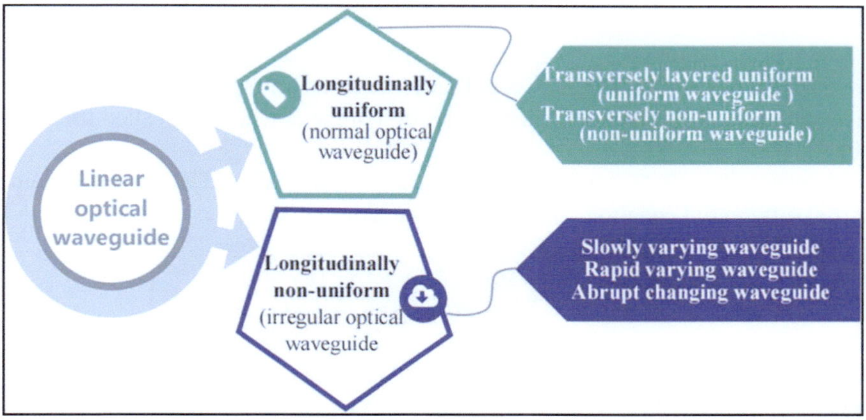

This classification method is convenient for theoretical analysis, different types of optical waveguides solve the corresponding types of differential equations. Every optical fiber can be classified into one of these categories according to the application.

Mode

In order to solve the wave equation, we should also pay attention to the characteristics of optical fiber structure, i.e., the directional difference between the longitudinal (axial direction of optical fiber, that is the direction of light transmission) and the transverse. The fundamental characteristic of optical fiber determines the difference in field solutions between the longitudinal and transverse modes in an optical fiber. A normal optical waveguide exhibits obvious light-guiding properties. The concept of mode derived from a normal optical waveguide is the most basic concept in optical waveguide theory.

As mentioned above, a normal optical waveguide refers to a waveguide whose refractive index profile is constant along the longitudinal direction (z-direction), and its mathematical description is $n(x, y, z) = n(x, y)$. It can be shown that the optical field in a normal optical waveguide can be expressed in the form of separated transverse and longitudinal components as follows,

$$\begin{bmatrix} E \\ H \end{bmatrix}(x, y, z, t) = \begin{bmatrix} e \\ h \end{bmatrix}(x, y)e^{i(\omega t - \beta z)} \tag{2.26}$$

If the nonlinear problem of optical fiber is not involved, ω remains unchanged when the light wave transmit in the optical fiber. In this case, the term e^{iwt} can be ignored, and the above formula can be simplified to

$$\begin{bmatrix} E \\ H \end{bmatrix}(x, y, z) = \begin{bmatrix} e \\ h \end{bmatrix}(x, y)e^{-i\beta z} \tag{2.27}$$

where β is the phase shift constant or propagation constant; both $\boldsymbol{e}(x, y)$ and $\boldsymbol{h}(x, y)$ are complex vectors, which have amplitude, phase and direction, and represent the distribution of $\boldsymbol{E}$ and $\boldsymbol{H}$ along the cross-section of the optical fiber, called **the mode field**.

Substitute Eq. (2.27) into the Helmholtz Eqs. (2.21) and (2.22), we can obtain a partial differential equation with only two variables (x, y):

$$\left.\begin{aligned} \left[\nabla_t^2 + \left(k^2 - \beta^2\right)\right]\boldsymbol{e} + \nabla_t\left(\boldsymbol{e} \cdot \tfrac{\nabla_t \varepsilon}{\varepsilon}\right) + i\beta e_z\left(\boldsymbol{e} \cdot \tfrac{\nabla_t \varepsilon}{\varepsilon}\right) = 0 \\ \left[\nabla_t^2 + \left(k^2 - \beta^2\right)\right]\boldsymbol{h} + \tfrac{\nabla_t \varepsilon}{\varepsilon} \times (\nabla_t \times \boldsymbol{h}) + i\beta h_z\left(\boldsymbol{h} \cdot \tfrac{\nabla_t \varepsilon}{\varepsilon}\right) = 0 \end{aligned}\right\} \tag{2.28}$$

where the subscript t represents the transverse direction perpendicular to the z direction. According to the theory of partial differential equations, the above equations have infinite discrete eigen-solution and can be sorted for given boundary conditions. Each eigen-solution can be written as $\begin{bmatrix} e_i \\ h_i \end{bmatrix}(x, y)e^{-i\beta_i z}$, so a eigen-solution of the above Eq. (2.28) is called a mode, and the total light field distribution in a fiber is a linear combination of these modes:

$$\begin{bmatrix} E \\ H \end{bmatrix} = \sum_i \begin{bmatrix} a_i\ e_i \\ b_i\ h_i \end{bmatrix}(x, y)e^{-i\beta z}$$

where a_i, b_i are the decomposition coefficients and represent the relative magnitude of the mode. A series of modes can be regarded as the spatial spectrum of the field distribution of an optical waveguide.

Mode is a basic concept in optical waveguide, which has the following characteristics:

- Stability. When a mode propagates in the longitudinal direction, its field distribution remains unchanged, i.e., there is a stable distribution along the z direction.
- Order. A mode is a series of eigen solutions of the wave equations, which are discrete and can be sorted. There are two sorting methods. One is to sort by the propagation constant β, the larger β, the smaller the sequence number; The other is to sort by the two independent arguments (x, y), so there are two columns of sequence numbers.

- Superposition. The overall field distribution in the optical waveguide is a linear superposition of these modes.
- Orthogonality. Different modes of a normal optical waveguide satisfy the orthogonal relationship. For an optical waveguide, let (*E*, *H*) be one mode and (*E*', *H*') be another mode, which respectively satisfy

$$\begin{cases} \nabla \times \boldsymbol{E} = -i\omega\mu_0 \boldsymbol{H} \\ \nabla \times \boldsymbol{H} = i\omega\varepsilon \boldsymbol{E} \end{cases}, \begin{cases} \nabla \times E' = -i\omega\mu_0 H' \\ \nabla \times H' = i\omega\varepsilon E' \end{cases}$$

Let $(\boldsymbol{E}, \boldsymbol{H})$ be the i-th mode and $(\boldsymbol{E}', \boldsymbol{H}')$ be the k-th mode, thus,

$$\begin{bmatrix} \boldsymbol{E} \\ \boldsymbol{H} \end{bmatrix} = \begin{bmatrix} \boldsymbol{e}_i \\ \boldsymbol{h}_i \end{bmatrix} (x, y) e^{-i\beta_i z}, \begin{bmatrix} \boldsymbol{E}' \\ \boldsymbol{H}' \end{bmatrix} = \begin{bmatrix} \boldsymbol{e}_k \\ \boldsymbol{h}_k \end{bmatrix} (x, y) e^{-i\beta_k z}$$

Then it can be proved that the following formula holds

$$\int\limits_{A \to \infty} \left(\boldsymbol{e}_i \times \boldsymbol{h}_k^*\right) \cdot dA = \int\limits_{A \to \infty} \left(\boldsymbol{e}_k^* \times \boldsymbol{h}_i\right) \cdot dA = 0 \quad (i \neq k)$$

where A is the integral range; The superscript * indicates conjugation. This is the mathematical expression of model orthogonality.

Longitudinal and Transverse Components of Modal Fields

Because the longitudinal (z-direction) and transverse structure of the optical fiber are quite different, the light field in an optical fiber can be decomposed into the sum of the longitudinal component and the transverse component:

$$\begin{cases} \boldsymbol{E} = \boldsymbol{E}_t + \boldsymbol{E}_z \\ \boldsymbol{H} = \boldsymbol{H}_t + \boldsymbol{H}_z \end{cases}$$

The subscripts z and t correspond to the longitudinal and transverse directions, respectively. The differential operator ∇ can also be expressed as a superposition of the longitudinal and transverse directions, i.e., $\nabla = \nabla_t + \boldsymbol{l}_z \frac{\partial}{\partial z}$, Where $\boldsymbol{l}_z$ is the unit vector in the z direction. Substitute the isotropic Maxwell equations and make the longitudinal and transverse components of both sides of the equation equal, then we get

$$\nabla_t \times \boldsymbol{E}_t = -i\omega\mu_0 \boldsymbol{H}_z \tag{2.29}$$

$$\nabla_t \times \boldsymbol{H}_t = i\omega\varepsilon \boldsymbol{E}_z \tag{2.30}$$

$$\nabla_t \times \boldsymbol{E}_z + \boldsymbol{l}_z \times \frac{\partial \boldsymbol{E}_t}{\partial z} = -i\omega\mu_0 \boldsymbol{H}_t \tag{2.31}$$

$$\nabla_t \times \boldsymbol{H}_z + \boldsymbol{l}_z \times \frac{\partial \boldsymbol{H}_t}{\partial z} = i\omega\varepsilon \boldsymbol{E}_t \tag{2.32}$$

Equations (2.29) and (2.30) show the distribution of the transverse component in the cross section is always rotational and depends on the corresponding longitudinal component; And Eqs. (2.31) and (2.32) indicate that the longitudinal component is determined by the transverse distribution of the cross-section, so its rotation not only depends on the corresponding (electric/magnetic) transverse components, but also the respective (magnetic/electric) transverse component.

Obviously, for three-dimensional mode fields, there are also

$$\begin{aligned} \boldsymbol{e} &= \boldsymbol{e}_t + \boldsymbol{e}_z \\ \boldsymbol{h} &= \boldsymbol{h}_t + \boldsymbol{h}_z \end{aligned} \tag{2.33}$$

So,

$$\begin{bmatrix} \boldsymbol{E}_t \\ \boldsymbol{E}_z \\ \boldsymbol{H}_t \\ \boldsymbol{H}_z \end{bmatrix} = \begin{bmatrix} \boldsymbol{e}_t \\ \boldsymbol{e}_z \\ \boldsymbol{h}_t \\ \boldsymbol{h}_z \end{bmatrix} e^{-i\beta z} \tag{2.34}$$

Substitute Eq. (2.34) into the relationship (2.29)~(2.32) between the longitudinal component and the transverse component of the light field of any optical waveguide,

$$\nabla_t \times \boldsymbol{e}_t = -i\omega\mu_0 h_z \tag{2.35}$$

$$\nabla_t \times \boldsymbol{h}_t = i\omega\varepsilon \boldsymbol{e}_z \tag{2.36}$$

$$\nabla_t \times \boldsymbol{e}_z + i\beta \boldsymbol{e}_z \times \boldsymbol{e}_t = -i\omega\mu_0 \boldsymbol{h}_t \tag{2.37}$$

$$\nabla_t \times \boldsymbol{h}_z + i\beta \boldsymbol{h}_z \times \boldsymbol{h}_t = i\omega\varepsilon \boldsymbol{e}_t \tag{2.38}$$

By Eqs. (2.37) and (2.38), using $\nabla_t \times \boldsymbol{e}_z = -\boldsymbol{l}_z \times \nabla_t \boldsymbol{e}_z$, and $\nabla_t \times \boldsymbol{h}_z = -\boldsymbol{l}_z \times \nabla_t \boldsymbol{h}_z$, we get

$$\begin{aligned} i\beta \boldsymbol{l}_z \times \boldsymbol{e}_t + i\omega\mu_0 \boldsymbol{h}_t &= \boldsymbol{l}_z \times \nabla_t \boldsymbol{e}_z \\ i\beta \boldsymbol{l}_z \times \boldsymbol{h}_t - i\omega\varepsilon \boldsymbol{e}_t &= \boldsymbol{l}_z \times \nabla_t \boldsymbol{h}_z \end{aligned}$$

Reuse $\boldsymbol{l}_z \times [\boldsymbol{l}_z \times \boldsymbol{l}_t] = -\boldsymbol{l}_t$, and get

$$\left.\begin{aligned} e_t &= \tfrac{i}{\omega^2\mu_0\varepsilon-\beta^2}[\omega\mu_0 \boldsymbol{l}_z \times \nabla_t \boldsymbol{h}_z + \beta\nabla_t \boldsymbol{e}_z] \\ h_t &= \tfrac{i}{\omega^2\mu_0\varepsilon-\beta^2}[-\omega\varepsilon \boldsymbol{l}_z \times \nabla_t \boldsymbol{e}_z + \beta\nabla_t \boldsymbol{h}_z] \end{aligned}\right\} \tag{2.39}$$

From Eq. (2.39), It can be seen that the transverse component of the mode field can be uniquely determined by the distribution of the longitudinal component along the cross-section. It can be proved that $\boldsymbol{e}_z$ and $\boldsymbol{h}_z$ are temporal in phase. It can also be concluded from Eq. (2.39) that if $\boldsymbol{e}_z$ and $\boldsymbol{h}_z$ are real numbers, then $\boldsymbol{e}_t$ and $\boldsymbol{h}_t$ must be purely imaginary numbers, i.e., there is a phase difference of 90° between the longitudinal and transverse components. This phase relationship shows that the normal optical waveguide has obvious light guiding properties.

Because in $p = \boldsymbol{E} \times \boldsymbol{H}^* = (\boldsymbol{e}_t + \boldsymbol{e}_z) \times \left(\boldsymbol{h}_t^* + \boldsymbol{h}_z^*\right) = \boldsymbol{e}_t \times \boldsymbol{h}_t^* + \boldsymbol{e}_z \times \boldsymbol{h}_t^* + \boldsymbol{e}_t \times \boldsymbol{h}_z^* + 0$, the first item is a real number and represents the transmission power along the z direction; The latter two terms are pure imaginary numbers, and the direction is transverse, indicating that there is power oscillation in the transverse direction, but it is not transmitted.

According to the directional characteristics of the mode field in space or the longitudinal components it contains, modes are usually divided into three categories:

(1). **TEM mode**: having only transverse component, no longitudinal component, i.e., $\boldsymbol{e}_z = \boldsymbol{h}_z = 0$
(2). **TE mode or TM mode**: there is only one longitudinal component, i.e., TE mode: $e_z = 0$, $h_z \neq 0$; TM mode: $h_z = 0$, $e_z \neq 0$.

For TE mode, since $e_z = 0$, it can be obtained from Eq. (2.39)

$$\boldsymbol{e}_t = \frac{\omega\mu_0}{\beta}\boldsymbol{l}_z \times \boldsymbol{h}_t \tag{2.40}$$

The above Eq. (2.41) illustrates that the transverse component of the electric and magnetic fields $\boldsymbol{e}_t$ and $\boldsymbol{h}_t$ are perpendicular to each other and opposite in phase (Under the right-hand screw rule of the three items $\boldsymbol{e}_t$, $\boldsymbol{h}_t$ and $\boldsymbol{l}_z$), their amplitudes are proportional to each other with a coefficient of $\omega\mu_0/\beta$ which has the dimension of impedance, which is defined as the wave impedance of the TE mode.

For TM mode, since $\boldsymbol{h}_z = 0$, it can be obtained from Eq. (2.39)

$$\boldsymbol{e}_t = \frac{\beta}{\omega\varepsilon}\boldsymbol{l}_z \times \boldsymbol{h}_t \tag{2.41}$$

Equation (2.41) illustrates that transverse components of electric field $\boldsymbol{e}_t$ is perpendicular to that of the magnetic fields $\boldsymbol{h}_t$, in the same phase, and proportional in amplitude. The proportional coefficient $\beta/\omega\varepsilon$ is the wave impedance. However, the wave impedance of the TM mode is different at each point in the waveguide because $\varepsilon = \varepsilon(x, y)$, which is different from that of the TE mode.

(3). **HE mode or EH mode**: Both longitudinal components are non-zero, i.e., $\boldsymbol{h}_z \neq 0$, $\boldsymbol{e}_z \neq 0$. From Eq. (2.39), we get

$$\boldsymbol{e}_z \cdot \boldsymbol{h}_t = \frac{1}{\omega^2 \mu_0 \varepsilon - \beta^2} (\nabla_t \boldsymbol{e}_z) \cdot (\nabla_t \boldsymbol{h}_z)$$

Since neither $\boldsymbol{e}_z$, $\boldsymbol{h}_z$ is zero nor constant, $\boldsymbol{e}_t \cdot \boldsymbol{h}_t \neq 0$, i.e., $\boldsymbol{e}_{\rm t}$ is not perpendicular to $\boldsymbol{h}_{\rm t}$. And there is no concept of wave impedance.

It can be seen that there is a clear difference between the light field distribution in the optical fiber and that in the free space: first, there is no transverse light waves in the optical fiber; Second, the field solution in the optical fiber is discrete. For the former, TEM mode is proved nonexistent in an optical waveguide. Nevertheless, these modes are sometimes treated as TEM modes for the convenience of analysis in the case of $|e_z| << |e_t|$ and $|h_z| << |h_t|$, which is satisfied often. Several typical fiber structures are discussed below: the specific form of field solution with uniform/non-uniform refractive index distribution and single-layer/double-layer structure.

There are two general methods to solve the field mode in uniform refractive index fiber: vector method and scalar method. Vector method is an accurate method to solve the three components of the characteristic parameters $\boldsymbol{E}$ and $\boldsymbol{H}$. The scalar law holds that the transverse component of the mode field in the optical fiber is non-oriented, that is equal opportunities in all directions. The solving process of vector method and scalar method is different, so the results obtained and expression of mode field are also different. More attention should be paid in consulting references.

The vector method finds first the longitudinal component (z component) before solving the transverse component (x, y component) from the obtained longitudinal component. The scalar rule finds the transverse component first, then find the longitudinal component from the transverse component. They will be described separately below.

Vector Pattern

A uniform refractive index fiber is a cylindrical uniform waveguide that has the general characteristics of a uniform waveguide described above.

1. Existing Transmission Mode

There must be a transmission mode in a uniform refractive index fiber, that is, the field distribution in the fiber can be separated into a two-dimensional mode field distributed with the cross section and a longitudinal fluctuation term $e^{-i\beta z}$, and its mathematical expression is

$$\begin{bmatrix} E \\ H \end{bmatrix} = \begin{bmatrix} e \\ h \end{bmatrix} (x, y) e^{-i\beta z}, \text{ Wherein, } \begin{cases} e = \boldsymbol{e}_t + \boldsymbol{e}_z \\ h = \boldsymbol{h}_t + \boldsymbol{h}_z \end{cases}$$

Two coordinate systems can be selected for ($\boldsymbol{e}_\mathrm{t}$, $\boldsymbol{h}_\mathrm{t}$) in optical fibers. Different coordinate systems correspond to different equations, and obtain different mode sequences.

Assume that a cylindrical coordinate system is selected,

$$\left.\begin{aligned} \boldsymbol{e}_t &= \boldsymbol{e}_r + \boldsymbol{e}_\phi \\ \boldsymbol{h}_t &= \boldsymbol{h}_r + \boldsymbol{h}_\phi \end{aligned}\right\} \tag{2.42}$$

The modes obtained in a cylindrical coordinate system are consistent with the boundary shape (circle) of fiber, generally called vector mode.

If the rectangular coordinate system is selected, there are

$$\left.\begin{aligned} \boldsymbol{e}_t &= \boldsymbol{e}_x + \boldsymbol{e}_y \\ \boldsymbol{h}_t &= \boldsymbol{h}_x + \boldsymbol{h}_y \end{aligned}\right\} \tag{2.43}$$

Each component of the mode obtained in the rectangular coordinate system has a fixed polarization direction, which is called linear polarization mode or LP mode (linear polarization mode) for short. Obviously, due to the spatial symmetry of the optical fiber, either vector mode ($\boldsymbol{e}_r, \boldsymbol{e}_\phi, \boldsymbol{e}_z,\ \boldsymbol{h}_r, \boldsymbol{h}_\phi, \boldsymbol{h}_z$) or scalar mode ($e_x, e_y, e_z, h_x, h_y, h_z$) should take a circular symmetric distribution form,

$$\begin{bmatrix} \boldsymbol{e} \\ \boldsymbol{h} \end{bmatrix}(r,\phi) = \begin{bmatrix} \boldsymbol{e} \\ \boldsymbol{h} \end{bmatrix}(r)e^{-im\phi} \quad (\mathrm{m} = 0, \pm 1, \pm 2, \ldots)$$

2. The Mode Field Satisfies the Homogeneous Wave Equation

The mode field in a uniform refractive index fiber satisfies the following wave equation

$$\left[\nabla_t^2 + \left(k_0^2 n^2 - \beta^2\right)\right]\begin{bmatrix} \boldsymbol{e} \\ \boldsymbol{h} \end{bmatrix} = 0$$

The above equation can be decomposed into

$$\left[\nabla_t^2 + \left(k_0^2 n^2 - \beta^2\right)\right] e_z = 0 \tag{2.44}$$

$$\left[\nabla_t^2 + \left(k_0^2 n^2 - \beta^2\right)\right] h_z = 0 \tag{2.45}$$

$$\left[\nabla_t^2 + \left(k_0^2 n^2 - \beta^2\right)\right] e_t = 0 \tag{2.46}$$

$$\left[\nabla_t^2 + \left(k_0^2 n^2 - \beta^2\right)\right] h_t = 0 \tag{2.47}$$

Equations (2.44) and (2.45) are scalar equations, while (2.46) and (2.47) are vector equations. Since the two vector equations cannot be decomposed into a scalar equation of a single component ($\boldsymbol{e}_r, \boldsymbol{e}_\phi, \boldsymbol{h}_r, \boldsymbol{h}_\phi$) in the cylindrical coordinate system, the solution process of these four components can only be: first obtain (longitudinal component e_z, h_z) from the first two scalar equations, and then obtain these four components (transverse component) from the relationship between the longitudinal and transverse components below, so it is called "vector method".

3. Relationship Between the Longitudinal and Transverse Components of the Mode Field

The vector mode in the cylindrical coordinate system,

$$\nabla_t \psi = \frac{\partial \psi}{\partial r} r + \frac{1}{r}\frac{\partial \psi}{\partial \phi}\phi \tag{2.48}$$

So,

$$\begin{aligned}
e_r &= \frac{-i}{\omega^2\mu_0\varepsilon - \beta^2}\left[\beta\frac{\partial e_z}{\partial r} + \frac{\omega\mu_0}{r}\frac{\partial h_z}{\partial \phi}\right] \\
e_\phi &= \frac{-i}{\omega^2\mu_0\varepsilon - \beta^2}\left[\frac{\beta}{r}\frac{\partial e_z}{\partial r} - \omega\mu_0\frac{\partial h_z}{\partial \phi}\right] \\
h_r &= \frac{-i}{\omega^2\mu_0\varepsilon - \beta^2}\left[\beta\frac{\partial h_z}{\partial \varphi} - \frac{\omega\varepsilon}{r}\frac{\partial e_z}{\partial r}\right] \\
h_\phi &= \frac{-i}{\omega^2\mu_0\varepsilon - \beta^2}\left[\frac{\beta}{r}\frac{\partial h_z}{\partial \phi} + \omega\varepsilon\frac{\partial e_z}{\partial r}\right]
\end{aligned} \tag{2.49}$$

Considering the circular symmetry of the mode field:

$$\left.\begin{aligned}
e_r(r,\varphi) &= e_r(r)e^{-im\varphi} \\
e_\varphi(r,\varphi) &= e_\varphi(r)e^{-im\varphi} \\
h_r(r,\varphi) &= h_r(r)e^{-im\varphi} \\
h_\varphi(r,\varphi) &= h_\varphi(r)e^{-im\varphi}
\end{aligned}\right\} (\mathrm{m} = 0,\ \pm 1,\ \pm 2,\ \ldots) \tag{2.50}$$

Available,

$$\left.\begin{aligned}
e_r(r) &= \tfrac{-i}{\omega^2\mu_0\varepsilon-\beta^2}\left[\beta\tfrac{de_z(r)}{dr} - \tfrac{im\omega\mu_0}{r}h_z(r)\right] \\
e_\phi(r) &= \tfrac{i}{\omega^2\mu_0\varepsilon-\beta^2}\left[\tfrac{im\beta}{r}e_z(r) + \omega\mu_0\tfrac{dh_z(r)}{dr}\right] \\
h_r(r) &= \tfrac{-i}{\omega^2\mu_0\varepsilon-\beta^2}\left[\beta\tfrac{dh_z(r)}{dr} + \tfrac{im\omega\varepsilon}{r}e_z(r)\right] \\
h_\phi(r) &= \tfrac{i}{\omega^2\mu_0\varepsilon-\beta^2}\left[\tfrac{im\beta}{r}h_z(r) - \omega\varepsilon\tfrac{de_z(r)}{dr}\right]
\end{aligned}\right\} \tag{2.51}$$

This is the general relationship between the longitudinal component and the transverse component of the vector mode, and is also the fundamental equations to solve the field distribution. Possible vector modes are discussed in detail below.

4. Possible Vector Modes

(1) Transverse mode

When $m = 0$, the corresponding TE modes and TM modes in the fiber can be obtained from Eq. (2.51). For TE Mode ($e_z = 0$):

$$\left.\begin{aligned} e_\phi &= \frac{i}{\omega^2\mu_0\varepsilon-\beta^2}\omega\mu_0 h_z'(r) \\ h_r &= \frac{i}{\omega^2\mu_0\varepsilon-\beta^2}\beta h_z'(r) \end{aligned}\right\} \tag{2.52}$$

While $e_r = h_\phi = 0$, then $e_t = e_\phi$, $h_t = h_r$, the two are perpendicular to each other, and $e_\phi = \left(\frac{\omega\mu_0}{\beta}\right)h_r$.

For TM Mode ($h_z = 0$):

$$\left.\begin{aligned} h_\phi &= -\frac{i\omega\varepsilon}{\omega^2\mu_0\varepsilon-\beta^2}e_z'(r) \\ e_r &= -\frac{\beta}{\omega\varepsilon}h_\phi \end{aligned}\right\} \tag{2.53}$$

While $e_\phi = h_r = 0$, then $e_t = e_r$, $h_t = h_\phi$, the two are perpendicular to each other.

(2) Mixed mode

When both e_z, h_z are not zero, the values can be obtained by the following equations:

$$\left.\begin{aligned} e_z'' + \frac{1}{r}e_z' + \left[k_0^2 n^2 - \beta^2 - \frac{m^2}{r^2}\right]e_z = 0 \\ h_z'' + \frac{1}{r}h_z' + \left[k_0^2 n^2 - \beta^2 - \frac{m^2}{r^2}\right]h_z = 0 \end{aligned}\right\},\ (m = 0,\ \pm 1,\ \pm 2\cdots) \tag{2.54}$$

The solutions of the above equations are different combinations of four Bessell functions:

$$\begin{bmatrix} e_z \\ h_z \end{bmatrix}(r) = \begin{bmatrix} A & B \\ C & D \end{bmatrix}\begin{bmatrix} J_m\left(\sqrt{k_0^2 n^2 - \beta^2}\, r\right) \\ N_m\left(\sqrt{k_0^2 n^2 - \beta^2}\, r\right) \end{bmatrix} \quad (nk > \beta) \tag{2.55}$$

Or

$$\begin{bmatrix} e_z \\ h_z \end{bmatrix}(r) = \begin{bmatrix} A & B \\ C & D \end{bmatrix}\begin{bmatrix} I_m\left(\sqrt{\beta^2 - k_0^2 n^2}\, r\right) \\ K_m\left(\sqrt{\beta^2 - k_0^2 n^2}\, r\right) \end{bmatrix} \quad (nk < \beta) \tag{2.56}$$

2.2.3 Linear Polarization Mode and Scalar Method

From the above discussion, it can be seen that the mode field distribution in the optical fiber is extremely complex. In order to simplify the calculation, Gloge et al. proposed a scalar approximation method, the basis of which is the linear polarization mode. As described in Sect. 2.2.1, when the mode field is decomposed in a rectangular coordinate system, each component has a fixed linear polarization direction. These modes are linear polarization modes, which can be divided into two groups: $[0, e_y, e_z, h_x, h_y, h_z]$ and $[e_x, 0, e_z, h_x, h_y, h_z]$.

In the rectangular coordinate system, there are

$$\left.\begin{array}{c} \frac{\partial e_y}{\partial x} - \frac{\partial e_x}{\partial y} = -i\omega\mu_0 h_z \\ \frac{\partial h_y}{\partial x} - \frac{\partial h_x}{\partial y} = i\omega\varepsilon e_z \\ \frac{\partial h_z}{\partial y} + i\beta h_y = i\omega\varepsilon e_x \\ i\beta h_x + \frac{\partial h_z}{\partial y} = -i\omega\varepsilon e_y \\ \frac{\partial e_x}{\partial y} + i\beta e_y = -i\omega\mu_0 h_x \\ i\beta e_x + \frac{\partial e_z}{\partial x} = i\omega\mu_0 h_y \end{array}\right\} \quad (2.57)$$

At this time, if we take the first group of modes, i.e., $e_x = 0$, and assuming e_y is known, the remaining 4 variables can be obtained from 4 of the above 6 equations. For example, solve the first 4 equations to get

$$\left\{\begin{array}{l} h_z = \frac{1}{i\omega\mu_0}\frac{\partial e_y}{\partial x} \\ h_y = -\frac{1}{\omega\beta\mu_0}\frac{\partial^2 e_y}{\partial x \partial y} \\ h_x = -\frac{1}{\omega\mu_0\beta}\frac{\partial^2 e_y}{\partial x^2} - \frac{\omega\varepsilon}{\beta} e_y \\ e_z = \frac{i}{\beta}\frac{\partial e_y}{\partial y} \end{array}\right. \quad (2.58)$$

The scalar approximation method considers that the refractive index of each layer of the optical fiber changes little, so it is assumed that in the expression of mode field, the change rate of refractive index above the second order can be ignored. This kind of fiber with a very small change of ε is called a weakly guided fiber. So the scalar approximation can also be called weakly guided approximation. And the above equation is simplified to

$$\left\{\begin{array}{l} h_z = \frac{1}{i\omega\mu_0}\frac{\partial e_y}{\partial x} \\ h_y \approx 0 \\ h_x \approx \frac{-\omega\varepsilon}{\beta} e_y \\ e_z = \frac{i}{\beta}\frac{\partial e_y}{\partial y} \end{array}\right. \quad (2.59)$$

Therefore, the components of the two groups of linear polarization modes under the scalar approximation are: $[0, e_y, e_z; h_x, 0, h_z]$ and $[e_x 0, e_z; 0, h_y, h_z]$. This linear polarization mode has the following characteristics: the transverse components are perpendicular to each other, the amplitudes are proportional, and the proportional coefficient is the wave impedance, so it is very similar to the TE mode and TM mode in the vector method, but e_z, h_z are not zero at this time.

The linear polarization mode under the scalar approximation still has circular symmetry, i.e., $e_y(r,\phi) = e_y(r)e^{-im\phi}(m = 0, \pm 1 \cdots)$. But this m has different meanings from the m in the vector method, and $m = 0$ here does not refer to TE and TM mode anymore.

In brief, the mode fields ($\boldsymbol{e}, \boldsymbol{h}$) in the optical fiber have different decomposition in different coordinate systems, corresponding to different classifications of vector modes and linear polarization modes, respectively. That is

$$\begin{bmatrix} e \\ \boldsymbol{h} \end{bmatrix}(r) = \begin{bmatrix} e_t + e_z \\ h_t + h_z \end{bmatrix}$$

$$= \begin{cases} \left\{ \begin{bmatrix} e_y + e_z \\ h_x + h_z \end{bmatrix} \atop \begin{bmatrix} e_x + e_z \\ h_y + h_z \end{bmatrix} \right\} \rightarrow \text{scalar linear polarization modes} \\ \begin{bmatrix} e_r + e_\phi + e_z \\ \boldsymbol{h}_r + \boldsymbol{h}_\phi + \boldsymbol{h}_z \end{bmatrix} = \left\{ \begin{array}{l} \begin{bmatrix} \boldsymbol{e}_\phi \\ \boldsymbol{h}_r + \boldsymbol{h}_z \end{bmatrix} \rightarrow TE \text{ mode} \\ \begin{bmatrix} \boldsymbol{e}_r + \boldsymbol{e}_z \\ \boldsymbol{h}_\phi \end{bmatrix} \rightarrow TM \text{ mode} \\ \begin{bmatrix} e_r + e_\phi + e_z \\ \boldsymbol{h}_r + \boldsymbol{h}_\phi + \boldsymbol{h}_z \end{bmatrix} \rightarrow HE, EH \text{ mode} \end{array} \right\} \text{vector modes} \end{cases}$$

The vector method solves two equations:

$$\left[\nabla_1^2 + \left(k_0^2 n^2 - \beta^2\right)\right]\begin{bmatrix} \boldsymbol{e}_z \\ \boldsymbol{h}_z \end{bmatrix} = 0$$

Other components are obtained by solving Eq. (2.51), which describes the relationship between the perpendicular and the horizontal components. While the scalar method only solves one equation

$$\left[\nabla_1^2 + \left(k_0^2 n^2 - \beta^2\right)\right]e_y = 0$$

Other components are obtained by solving Eq. (2.58), which describes the relationship between the perpendicular and the horizontal components. Either the vector method or the scalar method comes ultimately down to solving the Bessel equation.

$$\frac{\mathrm{d}^2\psi}{\mathrm{d}r^2} + \frac{1}{r}\frac{\mathrm{d}\psi}{\mathrm{d}r} + \left[k_0^2 n^2 - \beta^2 - \frac{m^2}{r^2}\right]\psi = 0 \tag{2.60}$$

2.2.4 Two-Layer Uniform Optical Fiber

Two-layer uniform optical fiber (step index optical fiber) has only a core and a cladding. It is the simplest structure and the most basic structure of optical fiber. Its refractive index distribution is $n(\mathrm{r}) = n_1(r < a)$ or $n_2(r > a)$, and $n_1 > n_2$, where a is the radius of fiber core.

(1) Vector method

Using the discussion in Sect. 2.2.2, the following results can be directly written from Eqs. (2.55) and (2.56):

$$e_z(r) = \begin{cases} A\,J_m\left(\frac{U}{a}r\right) \\ B\,K_m\left(\frac{W}{a}r\right) \end{cases} \quad r < a \tag{2.61}$$

$$h_z(r) = \begin{cases} C\,J_m\left(\frac{U}{a}r\right) \\ D\,K_m\left(\frac{W}{a}r\right) \end{cases} \quad r > a \tag{2.62}$$

where, $U^2 = \left(k_0^2 n_1^2 - \beta^2\right)a^2$, $W^2 = \left(\beta^2 - k_0^2 n_2^2\right)a^2$, and $V^2 = U^2 + W^2$.

Equations (2.61) and (2.62) are special solutions selected from the general solutions of the wave Eq. (2.60) in response to the requirements of application of optical fiber. Because the purpose of the optical fiber design is to confine the energy of light in the fiber core and transmit forward along the fiber axis (i.e., z-axis). There is no light wave transmission in the cladding (which is of course an ideal situation). Therefore, the principle of selecting the field solutions according to this requirement is: the oscillating solution $\boldsymbol{J}_\mathrm{m}$ is in the core, and the decay solution $\boldsymbol{K}_\mathrm{m}$ is in the cladding.

Using the vertical and horizontal relationship Eq. (2.51), the remaining components can be obtained. In the core ($r < a$),

$$\begin{bmatrix} \boldsymbol{e}_r \\ \boldsymbol{e}_\phi \\ \boldsymbol{h}_r \\ \boldsymbol{h}_\phi \end{bmatrix}\left(\frac{r}{a}\right) = \frac{ia}{U^2}\begin{bmatrix} -\beta U J_m'\left(U\frac{r}{a}\right) & i\frac{m\omega\mu_0}{r}aJ_m\left(U\frac{r}{a}\right) \\ i\frac{m\beta a}{r}J_m\left(U\frac{r}{a}\right) & \omega\mu_0 U J_m'\left(U\frac{r}{a}\right) \\ -\frac{im\omega\varepsilon_1 a}{r}J_m\left(U\frac{r}{a}\right) & -\beta U J_m'\left(U\frac{r}{a}\right) \\ -\omega\varepsilon_1 U J_m'\left(U\frac{r}{a}\right) & i\frac{m\beta a}{r}J_m\left(U\frac{r}{a}\right) \end{bmatrix}\begin{bmatrix} A \\ C \end{bmatrix} \tag{2.63}$$

In the cladding ($r > a$) there are:

$$\begin{bmatrix} \boldsymbol{e}_r \\ \boldsymbol{e}_\phi \\ \boldsymbol{h}_r \\ \boldsymbol{h}_\phi \end{bmatrix}\left(\frac{r}{a}\right) = \frac{ia}{W^2}\begin{bmatrix} -\beta W K_m'\left(W\frac{r}{a}\right) & i\frac{m\omega\mu_0 a}{r}K_m\left(W\frac{r}{a}\right) \\ i\frac{m\beta a}{r}K_m\left(W\frac{r}{a}\right) & \omega\mu_0 W K_m'\left(W\frac{r}{a}\right) \\ -\frac{im\omega\varepsilon_2 a}{r}K_m\left(W\frac{r}{a}\right) & -\beta W K_m'\left(W\frac{r}{a}\right) \\ -\omega\varepsilon_2 W K_m'\left(W\frac{r}{a}\right) & \frac{im\beta a}{r}K_m\left(W\frac{r}{a}\right) \end{bmatrix}\begin{bmatrix} B \\ D \end{bmatrix} \tag{2.64}$$

This is the analytical expression of the mode field of a two-layer uniform fiber. The field diagram of the mode field can be plotted according to this analytical expression. The so-called field diagram is the distribution diagram of electromagnetic fields on the cross-section of the optical fiber drawn with electric and magnetic lines.

(1) Eigen equation

The eigen equation takes advantage of the consecutive condition of the field on the boundary to solve β. Since there are only 4 unknowns (A, B, C, D) in the analytical formula of the mode field, only 4 consecutive conditions need to be taken, namely

$$\begin{bmatrix} \boldsymbol{e}_\phi \\ \boldsymbol{h}_\phi \\ \boldsymbol{e}_r \\ \boldsymbol{h}_r \end{bmatrix}_{\substack{\text{core} \\ r=a}} = \begin{bmatrix} \boldsymbol{e}_\phi \\ \boldsymbol{h}_\phi \\ \boldsymbol{e}_r \\ \boldsymbol{h}_r \end{bmatrix}_{\substack{\text{cladding} \\ r=a}} \tag{2.65}$$

In the core layer:

$$\begin{bmatrix} \boldsymbol{e}_\phi \\ \boldsymbol{h}_\phi \\ \boldsymbol{e}_z \\ \boldsymbol{h}_z \end{bmatrix}_{r-a} = \begin{bmatrix} -\frac{m\beta a}{U^2} J_m & i\frac{\omega\mu_0 a}{U} J'_m \\ -i\frac{\omega\varepsilon_1 a}{U} J'_m & -\frac{m\beta a}{U^2} J_m \\ J_m & 0 \\ 0 & J_m \end{bmatrix} \begin{bmatrix} A \\ C \end{bmatrix} \equiv G(U) \begin{bmatrix} A \\ C \end{bmatrix}$$

In the cladding:

$$\begin{bmatrix} \boldsymbol{e}_\phi \\ \boldsymbol{h}_\phi \\ \boldsymbol{e}_z \\ \boldsymbol{h}_z \end{bmatrix}_{r=a} = \begin{bmatrix} -\frac{m\beta a}{W^2} K_m & i\frac{\omega\mu_0 a}{W} K'_m \\ -i\frac{\omega\varepsilon_2 a}{W} K'_m & \frac{m\beta a}{W^2} K_m \\ K_m & 0 \\ 0 & K_m \end{bmatrix} \begin{bmatrix} B \\ D \end{bmatrix} \equiv H(W) \begin{bmatrix} B \\ D \end{bmatrix}$$

And the above two are equal:

$$G(U) \begin{bmatrix} A \\ C \end{bmatrix} - H(W) \begin{bmatrix} B \\ D \end{bmatrix} = 0$$

For this homogeneous equation to have a non-zero solution, its determinant must be zero:

$$\begin{vmatrix} -\frac{m\beta a}{U^2} J_m(U) & i\frac{\omega\mu_0 a}{U} J'_m(U) & -\frac{am\beta}{W^2} K_m(W) & i\frac{\omega\mu_0 a}{W} K'_m(W) \\ -i\frac{\omega\varepsilon_1 a}{U} J'_m(U) & -\frac{m\beta a}{U^2} J_m(U) & -i\frac{\omega\varepsilon_2 a}{W} K'_m(W) & \frac{am\beta}{W^2} K_m(W) \\ J_m(U) & 0 & K_m(W) & 0 \\ 0 & J_m(U) & 0 & K_m(W) \end{vmatrix} = 0$$

After simplification, we get

$$m^2 \left[\frac{1}{U^2} + \frac{1}{W^2} \right] \left[\frac{n_1^2}{U^2} + \frac{n_2^2}{W^2} \right] = \left[\frac{1}{U} \frac{J'_m(U)}{J_m(U)} + \frac{1}{W} \frac{K'_m(W)}{K_m(W)} \right]$$

$$\left[\frac{n_1^2 J_m'(U)}{UJ_m(U)} + \frac{n_2^2 K_m'(W)}{WK_m(W)}\right] \tag{2.66}$$

Or

$$\beta^2 m^2 \left[\frac{1}{U^2} + \frac{1}{W^2}\right]^2 = \left[\frac{1}{U}\frac{J_m'(U)}{J_m(U)} + \frac{1}{W}\frac{K_m'(W)}{K_m(W)}\right]\left[\frac{k_0^2 n_1^2 J_m'(U)}{UJ_m(U)} + \frac{k_0^2 n_2^2 K_m'(W)}{WK_m(W)}\right] \tag{2.67}$$

This is the eigen equation derived with the vector method. Let

$$\mathfrak{J} = \frac{J_m'(U)}{UJ_m(U)}, \mathfrak{R} = \frac{K_m'(W)}{WK_m(W)}$$

Substitution into Eq. (2.66), and solve for

$$\begin{aligned}
&\mathfrak{J} \\
&= -\frac{1}{2}\left[1 + \frac{n_2^2}{n_1^2}\right]\mathfrak{R} \\
&\pm \frac{1}{2}\sqrt{\left[1 + \frac{n_2^2}{n_1^2}\right]^2 \mathfrak{R}^2 - 4\left[\frac{n_2^2}{n_1^2}\mathfrak{R}^2 - m^2\left(\frac{1}{U^2} + \frac{n_2^2}{n_1^2}\frac{1}{W^2}\right)\left(\frac{1}{U^2} + \frac{1}{W^2}\right)\right]}
\end{aligned} \tag{2.68}$$

When the second term on the right side of Eq. (2.68) takes a positive sign, it is defined as EH_{mn} Mode. When taking a negative sign, it is defined as HE_{mn} Mode. For weakly guided fibers, there are $n_1 \approx n_2$, so the above Eq. (2.68) is simplified to

$$\mathfrak{J} = -\mathfrak{R} \pm m\left[\frac{1}{U^2} + \frac{1}{W^2}\right] \tag{2.69}$$

Equations (2.67) and (2.68) are the general field solutions in optical fiber. But it is still difficult to find the specific solution of optical fiber (i.e., to find the values of U and β) from it. For this reason, the specific methods of the field solutions are discussed below for two important special cases: cutoff and away from cutoff. The so-called cut-off means the light wave transmitted in optical fiber just comply with the critical condition of total reflection at the interface between the core and the cladding. Light waves that do not meet the total reflection condition cannot propagate along the fiber axis in the optical fiber, but leak into the cladding. So the cutoff condition is $W \to 0$, $U \to V$. Far from the cut-off means that the light wave in the core propagates in a direction close to the fiber axis, which can always meet the total reflection condition. Therefore the conditions far away from cutoff is: $V \to \infty$, $W \to \infty$, $U \to$ finite values. These two cases are discussed separately below. This is the basic method to deal with field solutions in optical fibers.

(2) Cut-off Conditions

The cutoff condition is a special form of the eigen equation when $W \to 0$, $U \to V$.

(1) $m = 0$, for TE mode, Eq. (2.69) is simplified to

$$\frac{1}{U}\frac{J_0'}{J_0} + \frac{1}{W}\frac{K_0'}{K_0} = 0$$

Taking the limit on both sides of the above Eq. (2.69), since $W \to 0$,

$$K_0(W) \to -\ln\left(\frac{W}{2}\right),\ K_1(W) \to \frac{1}{W}$$

So,

$$\frac{WK_0(W)}{K_1(W)} \approx -W^2\ln\left(\frac{W}{2}\right) \to 0$$

Therefore, we have $\frac{UJ_0(U)}{J_1(U)} \to 0$. This is the cutoff condition of the TE mode. Since $U = 0$, $\frac{UJ_0(U)}{J_1(U)} \to 1$, thus $U = 0$ is not its root, and the cutoff condition should be

$$J_0(U) = 0 \tag{2.70}$$

Similarly available this is also the cut-off condition of TM mode. The corresponding roots are, in turn,

$U_1 = 2.4048$	$U_2 = 5.5201$	$U_3 = 8.0537$	...
↓	↓	↓	
TE_{01}, TM_{01}	TE_{02}, TM_{02}	TE_{03}, TM_{03}	

The U_{0n} values equal to the corresponding V values at the cutoff of TE_{0n} and TM_{0n} modes, i.e., at the cutoff, the two wave modes are degenerate. But away from cutoff, the eigen equations of the two modes are different, so their U_{0n} and β_{0n} are also different, and will be separated from each other.

(2) $m > 1$, HE_{mn} mode

HE_{mn} Mode is defined as the negative sign is taken before the square root of Eq. (2.68). The weakly guided approximation is adopted for simplicity. Starting from the Eq. (2.69), using the respective functional relationships of abnormal Bessel function and Bessel function, we can obtain

$$K_m'(W) = -K_{m-1}(W) - \frac{m}{W}K_m(W) = -K_{m+1}(W) + \frac{m}{W}K_m(W)$$

$$J_m'(U) = -\frac{m}{U}J_m(U) + J_{m-1}(U) = -\frac{m}{U}J_m(U) - J_{m+1}(U)$$

So there is,

$$\frac{-K_m'(W)}{WK_m(W)}=\frac{K_{m-1}(W)+\frac{m}{W}K_m(W)}{WK_m(W)}=\frac{K_{m-1}(W)}{WK_m(W)}+\frac{m}{W^2}$$

$$\frac{J_m'(U)}{UJ_m(U)}=-\frac{m}{U^2}+\frac{J_{m-1}(U)}{UJ_m(U)}$$

Substitute the above two expressions into Eq. (2.69), there is

$$\frac{J_{m-1}(U)}{UJ_m(U)}=\frac{K_{m-1}(W)}{WK_m(W)}$$

taking the approximate formula of $K_{m\text{-}1}(W)$ and $K_m(W)$ at $W\to 0$:

$$K_m(x)\sim\frac{1}{2}\Gamma(m)\left(\frac{2}{x}\right)^m,\ m>0$$

Simplification, finally get the cutoff condition:

$$\frac{J_{m-1}(U)}{J_m(U)}=\frac{U}{2(m-1)},\ m>0 \tag{2.71}$$

If starting deriving from Eq. (2.68), then get

$$\frac{J_{m-1}(U)}{J_m(U)}=\frac{U}{m-1}\frac{n_2^2}{n_1^2+n_2^2},\ m>0 \tag{2.72}$$

The U value can be calculated by using Eqs. (2.71) and (2.72) at the cutoff of HE_{mn}, i.e. U_{mn}. But it is only applicable to $m>1$, that is, only suitable for the wave modes of HE_{2n}, HE_{3n},….

(3) $m=1$, fundamental mode HE_{11} and HE_{1n}

Still Start from Eq. (2.69) of weakly guided condition, where $m=1$, then

$$J_1'(U)=-\frac{1}{U}J_1(U)+J_0(U)$$

$$K_1'(W)=-K_0(W)-\frac{1}{W}K_1(W)$$

When there is $W\to 0$,

$$K_0(W)\approx -I_0(W)In\left(\frac{W}{2}\right);\ K_1(W)\approx\frac{1}{2}\Gamma(1)\left(\frac{2}{W}\right)$$

While $I_0(W)=1$. Therefore, Eq. (2.69) becomes:

$$\frac{J_0(U)}{UJ_1(U)} = \frac{K_0(W)}{WK_1(W)} \approx \underset{W\to 0}{In}\left(\frac{W}{2}\right) = \infty$$

Therefore, $J_1(U) = 0$ is the equation of HE_{1n} at the cutoff including the fundamental mode HE_{11}. The corresponding roots are, in order:

$U_{11} = 0$	$U_{12} = 3.8317$	$U_{13} = 7.0155$	…
↓	↓	↓	
HE_{11}	HE_{12}	HE_{13}	…

Note that roots zero are included here. Because $U_{11} = 0$, $V = 0$, $\lambda \to \infty$, that is, the cutoff wavelength is infinite, which means there is no cutoff frequency. The cutoff value of TE_{01} and TM_{01} modes are $U_{01} = 2.4048$, $V = U$, so they are the second wave modes hard to reach cutoff. As long as $V < 2.4048$, a single-mode of HE_{11} exists and transmits in the fiber, and the HE_{11} mode corresponding to the "0" root is called the fundamental mode.

(4) $m > 0$, EH_{mn} wave mode

For simplicity, we still use the weakly guided approximation (2.69), because

$$J_m'(U) = \frac{m}{U}J_m(U) - J_{m+1}(U)$$

$$K_m'(W) = -K_{m+1}(W) + \frac{m}{W}K_m(W)$$

So Eq. (2.69) becomes

$$-\frac{J_{m+1}(U)}{UJ_m(U)} = \frac{K_{m+1}(W)}{WK_m(W)}$$

For $m \geq 1$, and W→0, the above equation becomes

$$\frac{-J_{m+1}(U)}{UJ_m(U)} = \frac{2m}{W^2} \to \infty$$

So, $J_m(U) = 0$ is the equation of EH_{mn} mode at cutoff. Each solution corresponds to an EH_{mn} mode, but no zero.

(3) Far away from cutoff

When the transmission direction of the light wave closes to the fiber axis, $V \to \infty$, $W \to \infty$, $U \to$ finite value, which is the case of far away from cutoff. As mentioned earlier, under the weakly guided approximation, the approximate eigen equations of HE_{mn} and EH_{mn} modes are:

$$\mathrm{HE}_{mn}: \quad \frac{J_{m-1}(U)}{UJ_m(U)} = \frac{K_{m-1}(W)}{WK_m(W)}$$

$$\mathrm{EH}_{\mathrm{mn}}: \quad \frac{-J_{m+1}(U)}{UJ_m(U)} = \frac{K_{m+1}(W)}{WK_m(W)}$$

When $W \to \infty$ is reused,

$$K_m(x) \approx \sqrt{\frac{\pi}{2x}} e^{-x}$$

So,

$$\frac{K_{m-1}(W)}{WK_m(W)} = 0$$

While there are $\boldsymbol{J}_{m-1}(U) = 0$ and $\boldsymbol{J}_{m+1}(U) = 0$ for HE_{mn} and EH_{mn} respectively. These are the equations when the two modes are far away from cutoff, by which the range of U can be obtained. For example, HE_{11} mode has $J_0(U) = 0$, its first solution is 2.4048, so the U value of HE_{11} varies between 0 and 2.4048. TE_{01} and TM_{01} modes have $J_1(U) = 0$, their first solution is known to be 3.8317, so their U values range from 2.4083 to 3.8317.

(2) Scalar method

e_y(r), satisfying the Bessel equation, can be solved by using the results in Sect. 2.4.2 and Eq. (2.60),

$$\frac{d^2 e_y}{dr^2} + \frac{1}{r}\frac{de_y}{dr} + \left[k_0^2 n_i^2 - \beta^2 - \frac{m^2}{r^2}\right] e_y = 0$$

Immediately available

$$e_y(r, \varphi) = \begin{cases} AJ_m\left(\frac{U}{a}r\right)e^{-im\varphi}, r < a \\ BK_m\left(\frac{W}{a}r\right)e^{-im\varphi}, r > a \end{cases}$$

Only one of the two integral constants is independent, which is then determined by

$$e_z = \frac{i}{\beta}\frac{\partial e_y}{\partial y} = \frac{i}{\beta}\left[\frac{\partial e_y}{\partial r}\frac{\partial r}{\partial y} + \frac{\partial e_y}{\partial \phi}\frac{\partial \phi}{\partial y}\right] = \frac{i}{\beta}\left[\sin\phi\frac{\partial e_y}{\partial r} + \frac{\cos\phi}{r}\frac{\partial e_y}{\partial \phi}\right]$$

Available:

$$e_z(r,\phi)=\begin{cases}A\frac{i}{\beta}e^{im\phi}\left[\frac{U\sin\phi}{a}J'_m\left(\frac{U}{a}r\right)+i\frac{m\cos\phi}{r}J_m\left(\frac{U}{a}r\right)\right] & r<a\\ B\frac{i}{\beta}e^{im\phi}\left[\frac{W\sin\phi}{a}K'_m\left(\frac{W}{a}r\right)+i\frac{m\cos\phi}{r}K_m\left(\frac{W}{a}r\right)\right] & r>a\end{cases}\tag{2.73}$$

Similarly, according to

$$h_z=\frac{1}{i\omega\mu_0}\frac{\partial e_y}{\partial x}=-\frac{i}{\omega\mu_0}\left[\cos\phi\frac{\partial e_y}{\partial r}+\frac{\sin\phi}{r}\frac{\partial e_y}{\partial\phi}\right]$$

So,

$$h_z(r,\phi)=\begin{cases}A\frac{-i}{\omega\mu_0}e^{im\phi}\left[\frac{U\cos\phi}{a}J'_m\left(\frac{U}{a}r\right)-i\frac{m\sin\phi}{r}J_m\left(\frac{U}{r}a\right)\right] & r<a\\ B\frac{-i}{\omega\mu_0}e^{im\phi}\left[\frac{W\cos\phi}{a}K'_m\left(\frac{W}{a}r\right)-i\frac{m\sin\phi}{r}K_m\left(\frac{W}{r}a\right)\right] & r>a\end{cases}\tag{2.74}$$

(1) Eigen equation

Utilizing the boundary conditions of continuity of e_y and e_z, let $r=a$, the following can be obtained from the above Eqs. (2.73) and (2.74):

$$\begin{cases}AJ_m(U)-BK_m(W)=0\\ A\left[U\sin\varphi_m J'_m(U)+im\cos\varphi J_m(U)\right]\\ -B\left[W\sin\varphi K'_m(W)+im\cos\varphi K_m(W)\right]=0\end{cases}$$

Since its coefficient determinant equal to zero, the eigen equation can be obtained after simplification:

$$\frac{UJ'_m(U)}{J_m(U)}=\frac{WK'_m(W)}{K_m(W)}$$

Take advantage of, $\begin{cases}K'_m(W)=\frac{m}{W}K_m(W)-K_{m+1}(W)\\ J'_m(U)=\frac{m}{U}K_m(U)-J_{m+1}(U)\end{cases}$, $\begin{cases}K'_m(W)=-\frac{m}{W}K_m(W)-K_{m-1}(W)\\ J'_m(U)=-\frac{m}{U}J_m(U)-J_{m-1}(U)\end{cases}$
Available,

$$\frac{UJ_{m+1}(U)}{J_m(U)}=\frac{WK_{m+1}(W)}{K_m(W)};\ \frac{UJ_{m-1}(U)}{J_m(U)}=-\frac{WK_{m-1}(W)}{K_m(W)}\tag{2.75}$$

This is the common eigen equations of LP modes. Obviously, it is much simpler than the eigen equations of the vector method (2.67).

(2) Cut-off Conditions

Using the cut-off conditions $W\to0$, $U\to V$ and Eq. (2.75), we can obtain:

(1) When $m \neq 0$, the cutoff condition is (excluding $U = 0$)

$$J_{m-1}(U) = 0 \tag{2.76}$$

(2) When $m = 0$, the cut-off condition is (including $U = 0$)

$$J_1(U) = 0 \tag{2.77}$$

The corresponding solutions are in order:

$m = 0$	$J_1(\mathrm{U}) = 0$	0	3.83	7.01	…
		LP_{01}	LP_{02}	LP_{03}	
$m = 1$	$J_0(\mathrm{U}) = 0$	2.40	5.32		…
		LP_{11}	LP_{12}		
$m = 2$	$J_1(\mathrm{U}) = 0$	3.83	7.01		…
		LP_{21}	LP_{22}		

Thereby the order of $\mathrm{LP}_{\mathrm{mn}}$ is: LP_{01}, LP_{11}, LP_{02}, LP_{21}, LP_{12},…

(3) Away from cutoff

From the conditions far away from cutoff, $W \to \infty$, $V \to \infty (U \to$ finite value), the equation for determining U value far away from cutoff is

$$J_m(U) = 0 \tag{2.78}$$

3. Relation of Linear Polarization Modes and Vector Modes

Because the scalar approximation is weakly guided approximation, the relationship between the eigen equations of the scalar approximation and that of the vector mode when $n_1 \sim n_2$ can be obtained by comparison of the two. The eigen equation of the vector modes when $n_1 \sim n_2$ is

$$\frac{1}{U}\frac{J_m'(U)}{J_m(U)} + \frac{1}{W}\frac{K_m'(W)}{K_m(W)} = \pm m\left(\frac{1}{U^2} + \frac{1}{W^2}\right)$$

When $m = 0$, the above formula becomes

$$\frac{1}{U}\frac{J_1(U)}{J_0(U)} + \frac{1}{W}\frac{K_1(W)}{K_0(W)} = 0$$

This is consistent with the equation of the scalar case ($m = 1$), so the vector modes $\mathrm{TE}_{0\mathrm{n}}$ and $\mathrm{TM}_{0\mathrm{n}}$ have approximately the same β as the linear scalar mode LP_{11}.

When $m \neq 0$, for the vector $\mathrm{HE}_{\mathrm{mn}}$ mode, the equation takes the "−" sign, there is

$$\frac{1}{U}\frac{J_{m-1}(U)}{J_m(U)} - \frac{1}{W}\frac{K_{m-1}(W)}{K_m(W)} = 0$$

It is the same eigen equations with that of the $LP_{m+1,\,n}$ modes.

For vector mode EH_{mn}, the equation takes the " + " sign, there is

$$\frac{1}{U}\frac{J_{m+1}}{J_m} + \frac{1}{W}\frac{K_{m+1}}{K_m} = 0$$

It is associated with the eigen equation of $LP_{m-1,\,n}$. The LP mode is degenerated by a set of vector modes with very close propagation constants β. Table 2.1 gives the names of the lower order LP modes and the corresponding vector modes, degeneracy degree, cutoff and U values far from the cutoff, U_0 and U_∞. Figure 2.3 gives the relation curves of β/k_0 and V value of an uniform fiber. It explains that after the V value is determined (the V value is determined by the fiber structure), for each specific mode (field solution), β/k_0 value can be found from the curve in the figure. In addition, for the same V value (i.e., the same optical fiber), different modes correspond to different β values, that is, different modes have different transmission characteristics. For the same V value, U values are different for different modes. From the curve, we can see that when $V < 2.4048$, there is only one fundamental mode LP_{01} (HE_{11} mode) transmit in the optical fiber, and all other modes are cut off. When $2.405 < V < 3.832$, the LP_{11} mode (TE_{01}, TM_{01} and HE_{21} Mode) can be excited. Figure 2.4 shows the relationship curve of the variation range of U values of low-order modes and Bessel function.

Table. 2.1 Names, degeneracy degrees, and U values of lower-order LP modes and corresponding vector modes

LP mode	Name × number of vector modes	Degree of degeneracy	U_0	U_∞
LP_{01}	$HE_{11} \times 2$	2	0	2.4048
LP_{11}	$HE_{21} \times 2$, TE_{01}, TM_{01}	4	2.4048	3.8317
LP_{21}	$EH_{11} \times 2$, $HE_{31} \times 2$	4	3.8317	5.1356
LP_{02}	$HE_{12} \times 2$	2	3.8317	5.5200
LP_{31}	$EH_{21} \times 2$, $HE_{41} \times 2$	4	5.1356	6.3801
LP_{12}	$HE_{22} \times 2$, TE_{02}, TM_{02}	4	5.5200	7.0155
LP_{41}	$EH_{31} \times 2$, $HE_{51} \times 2$	4	6.3801	7.5883
LP_{22}	$EH_{12} \times 2$, $HE_{32} \times 2$	4	7.0155	8.4172
LP03	HE13 × 2	2	7.0155	8.6537
LP51	EH41 × 2, HE61 × 2	4	7.5883	8.7714

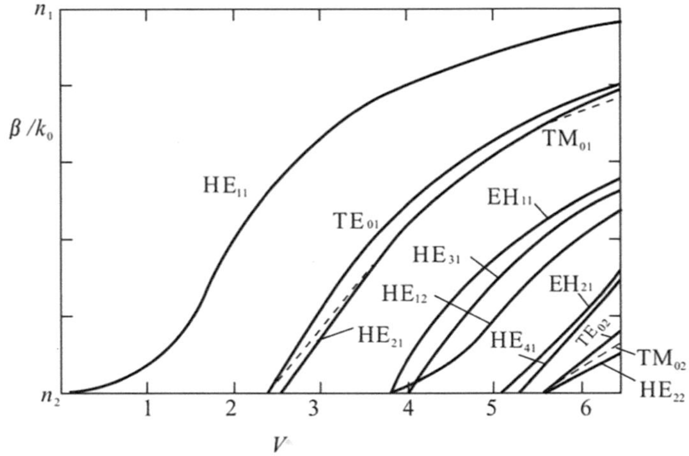

Fig. 2.3 Relationship curve of $V \sim \beta/k_0$ of uniform step index fiber

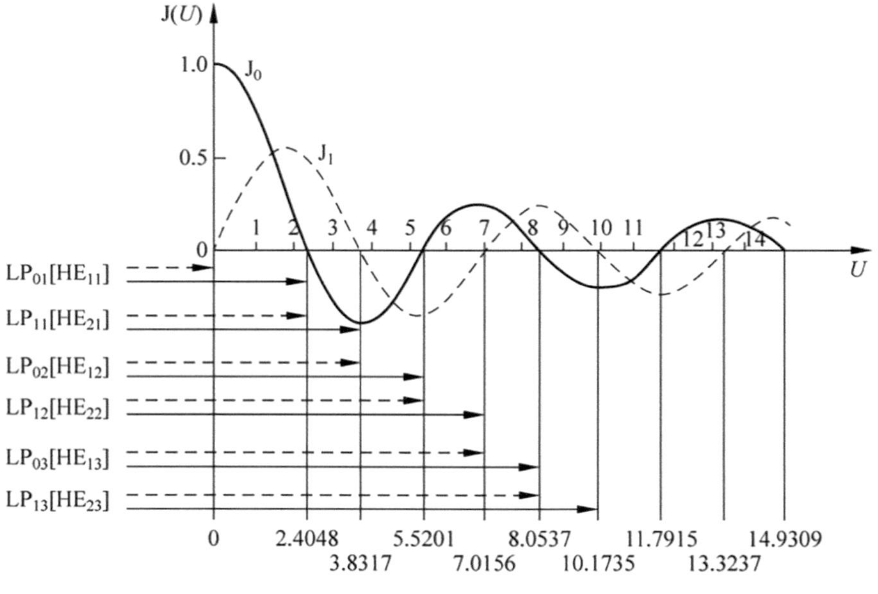

Fig. 2.4 The relation curve of the variation range of U value of low-order modes and Bessel function

2.3 Transmission and Control of Polarized Light in Optical Fiber

2.3.1 *Polarization Characteristics of Single-Mode Optical Fibers*

Description of Polarization Characteristics of Single-Mode Fibers

Generally, the following physical quantities are used to describe the polarization state of light vector or the birefringence of a single-mode fiber.

1. Mode Birefringence or Polarization Birefringence $\delta\beta$

Defined as the difference of the propagation coefficient between two mutually orthogonal polarization fundamental modes HE_{11x} and HE_{11y} in a single-mode fiber, when transmitting along the direction of the fiber axial

$$\delta\beta = \beta_x - \beta_y = \frac{2\pi\left(n_x - n_y\right)}{\lambda} \tag{2.79}$$

where λ is the wavelength of light in free space, and n_x and n_y are the effective refractive index of two orthogonal polarization fundamental modes HE_{11x} and HE_{11y}.

2. Normalized Birefringence B

It is defined as the difference of effective refractive indexf two orthogonal polarization modes, which directly describes the magnitude of the birefringence of a single-mode fiber.

Where $n_1 = \left(n_x + n_y\right)/2$, is the average value.

$$B = \frac{n_x - n_y}{n_1} \tag{2.80}$$

3. Beat Length L_B

After the degeneracy of the transmission mode in single-mode fiber, the phase difference between these two orthogonal polarization modes varies along the fiber length. When incident light is linearly polarized, the output light composed from two orthogonal polarization modes is polarized light on the output end face of the fiber, whose polarization direction changes with the length of the optical fiber, and the polarization state periodically evolves between linear polarization, elliptical polarization and circular polarization. The polarization evolution period is the cross-sectional spacing L_B with a phase difference of 2π, i.e., beat length L_B.

Beat length L_B is also known as coupling length. In a longitudinally uniform single-mode fiber, the polarization of light varies periodically. Therefore, The definition of beat length L_B is: in a longitudinally uniform single-mode fiber, when the phase difference between two mutually orthogonal polarization modes HE_{im} and HE" is 2π, the travelling distance by light in optical fiber is the length of one beat. That is

$$L_B = \frac{2\pi}{\delta\beta} \tag{2.81}$$

Among the above three physical quantities, mode birefringence or polarization birefringence reflects the cause of birefringence of a single-mode fiber. The normalized birefringence *B* reflects the relationship between birefringence and refractive index distribution. Beat length L_B reflects the magnitude of birefringence of single-mode fiber. To understand the magnitude of birefringence or the polarization state of light vector, the first two physics usually measured through the beat length L_B of a single-mode fiber since direct measurement is difficult.

The relation between beat length L_B and the normalized birefringence *B* is:

$$L_B = \frac{\lambda}{n_1 B} \tag{2.82}$$

(4) Mode coupling coefficient η

Mode coupling coefficient η is used to characterize the polarization- maintaining ability of a fiber and determined by the extinction ratio η of the single-mode fiber.

$$\eta = 10\lg\left(\frac{P_x}{P_y}\right) = 10\lg[\tanh(\eta L)] \tag{2.83}$$

where P_x, P_y are the mode power of corresponding polarization, respectively; L is the fiber length.

Classification of Polarizing Optical Fibers

An optical fiber with $B < 10^{-6}$ refers as a low birefringence fiber (LBF); The optical fiber with $B > 10^{-5}$ is a high birefringence optical fiber (HBF). The *B* value of a single-mode optical fiber for communication is between 10^{-6} and 10^{-5}. LBF and HBF are collectively referred to as polarization-maintaining optical fibers.

For HBF, it can be divided into single polarization fiber (SPF) (transmitting only one of the two orthogonal polarization modes) and dual polarization fiber (TPF). According to the way the inherent birefringence in the optical fiber is produced, polarization optical fiber can be divided into: geometry effect (GE) and stress-induced optical fiber (SE). The main structures of polarization-maintaining optical fibers are shown in the following classification structure:

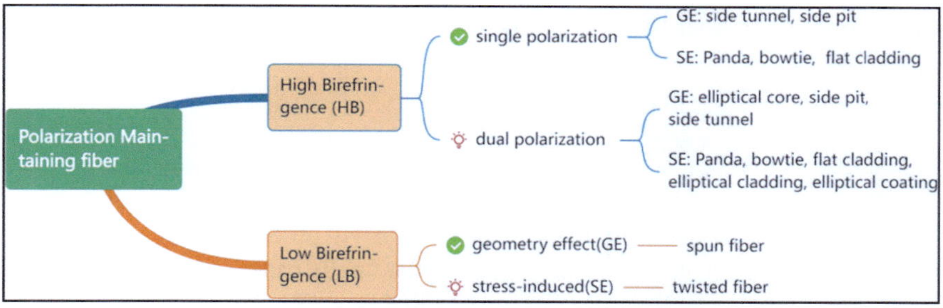

Polarization Instability in a Single-Mode Fiber

The instability of polarization state in a SMF includes internal factors of the fiber itself and external factors.

1. Internal Factors

Internal factors causing the instability of the polarization state of light in SMF are represented by internal birefringence $\delta\beta_n$, which includes two aspects: ① waveguide shape birefringence $\delta\beta_{\mathrm{GE}}$ caused by geometric distortion of the fiber cross-section; ② Stress birefringence $\delta\beta_{\mathrm{SE}}$ caused by internal stress of optical fiber. Therefore, the internal birefringence $\delta\beta_n$ is

$$\delta\beta_n = \delta\beta_{\mathrm{GE}} + \delta\beta_{\mathrm{SE}} \tag{2.84}$$

The signs of $\delta\beta_{\mathrm{GE}}$ and $\delta\beta_{\mathrm{SE}}$ may be the same or the opposite.

(1) Waveguide shape birefringence $\delta\beta_{\mathrm{GE}}$

During the process of drawing an optical fiber, the fiber core changes from round to ellipse for various reasons, and waveguide shape birefringence occurs. Let a and b be the lengths of the major and minor axes of the elliptical core respectively. The two orthogonal linearly polarized eigenstates in the optical fiber vibrate along the long and short axes accordingly. The phase difference between these two orthogonal linearly polarized lights can be expressed as

$$\delta\beta_{\mathrm{GE}} = \frac{e^2}{8a}\sqrt{(2\Delta)^3}f(V) \tag{2.85}$$

where $e = \sqrt{1 - \left(\frac{b}{a}\right)^2}$ is the ellipticity of the core, Δ is the relative refractive index difference of the optical fiber, and V is the normalized frequency. If optical fiber operates in the near-cutoff state ($V \approx 2.4$), and when $(a/b - 1) << 1$, there is

$$\delta\beta_{\mathrm{GE}} \le \frac{e^2}{8a}\sqrt{(2\Delta)^3} \tag{2.86}$$

For a single mode fiber, if $\Delta = 0.003$, $a = 2.5\mu m$, $b/a = 0.975$, we can obtain $\delta\beta_{GE} \leq 66^{\circ}/m$.

(2) Stress birefringence $\delta\beta_{SE}$

Optical fiber is composed of several layers such as core and cladding, each of which has different doping materials and thermal expansion coefficients. Therefore, even a small thermal stress asymmetry in the cross-section will produces a large stress imbalance and results in anisotropy of the core material, finally causes stress birefringence. The stress difference between the two orthogonal directions is

$$\delta\beta_{SE} \leq \frac{\pi n^3}{\lambda E}(1+\rho)(p_{12}-p_{11})\sigma \tag{2.87}$$

where E is the Young's modulus of the material, ρ is the Poisson's ratio, and p_{11}, p_{12} are the photo-elastic coefficient. For fused silica fiber: $E = 7.0 \times 10^{10}\, N/m^2$, $\rho = 0.17, p_{11} = 0.121, p_{12} = 0.270$. If the fiber operates at a wavelength of 0.6328 μm, then for a medium stress difference $\sigma = 5 \times 10^4\, N/m^2$, available $\delta\beta_{SE} \leq 109^{\circ}/m$.

2. External Factors

External factors can also affect the stability of polarization states in single-mode fibers. The expressions of external birefringence vary for numerous external factors. Due to the new anisotropy introduced by external factors, the birefringence of optical fibers varies with these conditions. For instance, optical fibers are subjected to various random external forces, including bending, twisting, vibration, compression, and other mechanical forces during the cabling and construction processes. In addition, optical fibers can operate in environments with strong electric and magnetic fields and frequently changing temperatures. When exposed to an external mechanical force, an optical fiber displays the photo-elastic effect; when subjected to an external magnetic field, the Faraday effect becomes apparent; and under the influence of an external electric field, the Kerr effect occurs. The aggregate outcome of these phenomena engenders novel anisotropic properties within the fiber, thereby inducing external birefringence.

For an optical fiber with an outer diameter of A, if its bending radius R is significantly larger than A, the stress difference caused by bending is $\sigma = A^2E/(2R)^2)$. We obtain the produced birefringence using Eq. (2.87),

$$\delta\beta \leq \frac{\pi n^3}{2\lambda}(1+\rho)(p_{12}-p_{11})\left(\frac{A}{R}\right)^2 \tag{2.88}$$

Bending also results in the deformation of the optical fiber's cross-section. According to a first-order approximation, bending causes the optical fiber's cross-section to become elliptical, and its ellipticity is

$$e = \frac{\rho a}{R} \tag{2.89}$$

By Eq. (2.86), it can be obtained that the birefringence caused by each circle of the fiber is

$$\delta\beta = \frac{\pi\sqrt{(2\Delta)^3}\rho^2 a}{4} \tag{2.90}$$

For fused silica, this value is 2.6 × 10^{-11} rad·m/circle.

When optical fiber is twisted, circular birefringence occurs due to shear stress. This phenomenon is characterized by the propagation speed difference between left-hand and right-hand circularly polarized light within the optical fiber. The magnitude of the circular birefringence can be expressed as

$$\delta\beta_t = \frac{(2\pi)^2 N}{2\lambda} n^2 (p_{12} - p_{11}) = \frac{(2\pi)^2}{2\lambda} Ng \tag{2.91}$$

where N denotes the number of fiber twists per meter and g is a constant. For fused silica fibers, the theoretical value of g is 0.18, while the experimental value is 0.14.

The polarization characteristics of light in optical fibers are of great significance for information acquisition and transmission through optical fibers. They involve the stability of many optical fiber instruments, including fiber optic gyroscopes and interferometers. People have developed optical fiber current sensors, distributed optical fiber pressure sensors, temperature sensors, and more by utilizing the variation in the polarization state of light within optical fibers due to external fields such as magnetic fields, pressure, and temperature. Polarization mode dispersion (PMD) in optical fibers significantly limits the enhancement of long-distance optical communication speeds, particularly those at or above 10 Gbit/s. The compensation of PMD for high-speed optical signals transmission in optical fiber is an important topic in the field of optical communication.

2.3.2 Perturbation Coupled Mode Theory of Weakly Guided Fiber

Perturbation Coupled Mode Equation

Under external perturbation conditions, the permittivity of a real optical fiber can be expressed as

$$\varepsilon(x, y, z) = \varepsilon(x, y) + \Delta\varepsilon(x, y, z) \tag{2.92}$$

where $\varepsilon(x, y)$ represents the dielectric properties of an ideal circular single-mode fiber, and $\Delta\varepsilon(x, y, z)$ reflects the influence of various perturbation factors, including the elliptical distortion of the fiber cross-section, bending, twisting, residual stress, and external electric, magnetic, force, and thermal fields, among others. These factors can be either scalar (such as core cross-section distortion) or tensor (such as anisotropy induced by external fields). Generally, $\Delta\varepsilon(x, y, z)$ (each component)$\ll$ $\varepsilon(x, y)$ is satisfied, so perturbation theory can be applied to analyze the polarization in optical fiber.

The light field in a perturbed optical fiber can be expressed as the superposition of the two eigen linear polarization modes in an ideal circular single-mode fiber.:

$$\begin{aligned} E_1 &= \left[J_0\left(\tfrac{u}{a}r\right)\ 0\ \tfrac{i}{\beta}J_1\left(\tfrac{u}{a}r\right)\tfrac{u}{a}\cos\phi \right] \\ E_2 &= \left[0\ J_0\left(\tfrac{u}{a}r\right)\ \tfrac{i}{\beta}J_1\left(\tfrac{u}{a}r\right)\tfrac{u}{a}\cos\phi \right] \end{aligned} \tag{2.93}$$

Among them, "a" represents the fiber core radius, while "β" denotes the propagation constant; $u = \left(k_0^2 n_1^2 - \beta^2\right)a$; k_0 is the propagation constant of light in vacuum; n_1 is the core refractive index.

The optical field in a perturbative fiber can be expressed as the superposition of two linearly polarized eigenmodes in an ideal circular single-mode fiber.

$$E = \sum_i A_i(z)E_i(x, y)\boldsymbol{e}^{-i\beta z}(i = 1, 2) \tag{2.94}$$

In eq. (2.94), $A_i(z)$ is the corresponding amplitude coefficient of mode field, which is a constant in an ideal circular SMF. In the presence of perturbation, $A_i(z)$ is a function of z. For the coupling of two modes in the same direction, $A_i(z)$ must satisfy the following coupling equation as per the coupled mode theory.

$$\frac{dA_1}{dz} = -ik_{11}A_1 - ik_{12}A_2 \tag{2.95}$$

$$\frac{dA_2}{dz} = -ik_{21}A_1 - ik_{22}A_2 \tag{2.96}$$

$$k_{vu} = \frac{\omega}{4}\int_{-\infty}^{\infty}\int_{-\infty}^{\infty} E_u^*(\Delta\varepsilon E_v)\,dxdy\ (u, v = 1, 2) \tag{2.97}$$

where, k_{vu} is the coupling coefficient. From Eqs. (2.95) and (2.96), the diagonal element k_{11}, k_{22} cause the propagation constant β of the two degenerate modes to yield an increment $\Delta\beta$, and there is

$$\Delta\beta = k_{11} - k_{22} \tag{2.98}$$

where $k_{11} - k_{22} \neq 0$, the two modes are no longer synchronized, resulting in a beat frequency with a spatial frequency of $\Delta\beta = k_{11} - k_{22}$. Off-diagonal element k_{12}, k_{21} results in power coupling between the two modes. For lossless fibers, $k_{12} = k_{21}^*$.

Eigenmodes in Perturbed Fibers

When analyzing light transmission in single-mode fiber, in principle, as long as the perturbation term $\Delta\varepsilon(x,y,z)$ is known, four coupling coefficients can be obtained from Eqs. (2.97), and (2.96) can be solved with the input light wave as initial conditions. This allows us to determine the polarization of the output light after it has transmitted through a fiber section. However, the solution to the equation is too complex under general conditions to yield an analytical result. For simplicity, we further assume that the ideal circular single-mode fiber is uniformly perturbed along the axial direction (z), which results in the coupling of HE_{x11} and HE_{y11} modes. Under this assumption, the eigenmode field solution in the perturbed fiber can be determined as follows:

$$E_1 = \begin{bmatrix} \frac{k}{\frac{\Delta\beta}{2}-s} \\ 1 \end{bmatrix} e^{-i(\overline{\beta}-s)z};\ E_2 = \begin{bmatrix} \frac{k}{\frac{\Delta\beta}{2}+s} \\ 1 \end{bmatrix} e^{-i(\overline{\beta}+s)z} \tag{2.99}$$

$$k_{12} = k_{21}^* = k \tag{2.100}$$

$$s = \sqrt{\left(\frac{\Delta\beta}{2}\right)^2 + k^2} \tag{2.101}$$

$$\overline{\beta} = \beta + \frac{k_{11} + k_{22}}{2} \tag{2.102}$$

The following conclusions can be drawn from Eq. (2.99) regarding the polarization states in perturbed fibers: the two eigenmodes are generally elliptically polarized; if k is a purely imaginary number, the eigenmode is circularly polarized light; if k is a purely real number, the eigenmode is linearly polarized light. When the eigenmodes are transmitted in the perturbed fiber, they are not coupled with each other and transmit independently, which simplifies the solution of polarization of the output light wave.

2.3.3 Perturbation Coupled Mode Theory of Birefringence Analysis in Optical Fiber

In the absence of external perturbations, the propagation constants of the two orthogonal linearly polarized modes in an ideal circular single-mode fiber are equal, $\beta_x = \beta_y$, meaning the two modes are degenerate. However, in real single-mode fiber, these two orthogonally polarized modes are no longer degenerate, meaning that $\beta_x \neq \beta_y$, due to various reasons, resulting in what is known as polarization mode birefringence. Polarization mode birefringence in single-mode fiber is quite complex, but it can generally be categorized into three types: When perturbation causes the two orthogonal linear polarization modes to no longer be degenerate, it is referred to as linear birefringence; when perturbation causes the left-hand and right-hand circular polarization modes to become non-degenerate, it is known as circular birefringence; and when perturbation results in both linear birefringence and non-degeneracy of the left-hand and right-hand circular polarization modes, it is termed elliptical birefringence.

Kerr Effect in Ideal Circular Single-Mode Fiber

The commonly employed fused silica optical fiber serves as an isotropic medium, characterized by its secondary electro-optical coefficient tensor matrix as follows:

$$[S_{vu}] = \begin{bmatrix} S_{11} & S_{12} & S_{12} & 0 & 0 & 0 \\ S_{12} & S_{11} & S_{12} & 0 & 0 & 0 \\ S_{12} & S_{12} & S_{11} & 0 & 0 & 0 \\ 0 & 0 & 0 & \frac{S_{11}-S_{12}}{2} & 0 & 0 \\ 0 & 0 & 0 & 0 & \frac{S_{11}-S_{12}}{2} & 0 \\ 0 & 0 & 0 & 0 & 0 & \frac{S_{11}-S_{12}}{2} \end{bmatrix} \tag{2.103}$$

Assuming the external electric field is directed along the x-axis, the change in each component of the relative dielectric permittivity tensor due to the electric field can be obtained as

$$\begin{aligned} \Delta b_1 &= S_{11}E^2 \\ \Delta b_2 &= S_{12}E^2 \\ \Delta b_3 &= S_{12}E^2 \\ \Delta b_4 &= \Delta b_5 = \Delta b_6 = 0 \end{aligned} \tag{2.104}$$

Therefore, the variation of each component of the corresponding dielectric coefficient tensor can be determined as

$$\begin{aligned} \Delta\varepsilon_1 &= -\varepsilon_0 n^2\left(n^2 S_{11}E^2\right) \\ \Delta\varepsilon_2 &= \Delta\varepsilon_3 = -\varepsilon_0 n^2\left(n^2 S_{12}E^2\right) \\ \Delta\varepsilon_4 &= \Delta\varepsilon_5 = \Delta\varepsilon_6 = 0 \end{aligned} \tag{2.105}$$

By substituting into Eq. (2.100), the coupling coefficient can be obtain.

$$\begin{aligned} k_{12} &= k_{21} = 0 \\ k_{11} &= -k_0 n^3\left(\frac{S_{11}+S_{12}}{2}\right)E^2 \\ k_{22} &= -k_0 n^3 (S_{12})E^2 \end{aligned} \tag{2.106}$$

Substitute the above results into Eq. (2.21), and the expression for the eigenmode in the Kerr-effect perturbed fiber is obtained as follows:

$$E_1 = \begin{bmatrix} 1 \\ 0 \end{bmatrix} e^{-i(\beta + k_{22})z}$$
$$E_2 = \begin{bmatrix} 0 \\ 1 \end{bmatrix} e^{-i(\beta + k_{11})z} \tag{2.107}$$

These are two linear polarization modes, which are the eigenmodes in an ideal fiber. The difference in propagation constants $\Delta\beta$ and the corresponding linear birefringence δ are respectively

$$\Delta\beta = k_{11} - k_{12} = -\beta\left(n^2 \frac{S_{11} - S_{12}}{2}\right)E^2 = -\beta\left(n^2 S_{44}\right)E^2$$
$$\delta = \frac{\Delta\beta}{k_0} = -n^3 S_{44} E^2 \tag{2.108}$$

The Eq. (2.108) is typically written as

$$\delta = B\lambda_0 E^2 \tag{2.109}$$

where B is the Kerr coefficient and λ_0 represents the wavelength in a vacuum. It is evident from Eq. (2.109) that the birefringence induced by the Kerr effect in the fiber increases with rising wavelength. The experimentally measured Kerr coefficient is $B = 5.4 \times 10^{-6} \text{m/V}^2$. The relationship between the Kerr coefficient and the secondary electro-optic coefficient is

$$S_{44} = -\frac{B\lambda_0}{n^3} \tag{2.110}$$

The analysis above indicates that the applied electric field causes a difference in the propagation constant of the two linear polarization modes of the optical fiber, denoted as $\Delta\beta \neq 0$, which results in linear birefringence. It should be noted that the eigenmodes, which consist of two linearly polarized modes, are obtained in the electric field perturbed fiber using the perturbation coupling theory. Compared with the eigenmodes in a non-perturbed fiber, only a perturbation term that is proportional to the applied electric field is added to the propagation constant. This is exactly the same as the electro-optic effect in an ideal cubic crystal or an isotropic medium (regardless of boundary conditions).

Faraday Effect in Ideal Round Single-Mode Fiber

Assuming that a magnetic field H is applied along the axial direction of an ideal round fiber, the dielectric tensor induced by the magnetic field has two non-zero off-diagonal elements:

$$\Delta\varepsilon_{xy} = -\Delta\varepsilon_{yx} = -i\frac{\varepsilon_0\lambda_0 nV}{\pi}H \tag{2.111}$$

where V is the Verdet constant of the material. The experimentally measured $V = 1.56 \times 10^{-2}\text{min/A}(@\lambda = 633\text{nm})$ in the optical fiber, n is the refractive index of the fiber core, and H is the axial magnetic field uniformly distributed across the fiber cross-section. For a specific medium, the V value decreases rapidly with increasing wavelength. Substitution Eq. (2.111) into (2.96), the coupling coefficient can be obtained as

$$\begin{aligned} k_{11} &= k_{22} = 0 \\ k_{12} &= -k_{21} = \frac{\omega}{4}\left(-i\frac{\varepsilon_0\lambda_0 nV}{\pi}H\right)\int\limits_0^a\int\limits_0^{2\pi} J_0\left(\frac{u}{a}r\right)rdrd\phi \\ &= -iVH \end{aligned} \tag{2.112}$$

Both cross-coupling coefficients are imaginary numbers with opposite signs, so the Faraday effect will cause circular birefringence, resulting in Faraday rotation. Substitute Eq. (2.112) into Eqs. (2.99) and (2.100), the eigenmode expression of the perturbation fiber obtains two circularly polarized modes with opposite rotation directions in the magnetic field perturbed fiber:

$$\begin{aligned} E_1(z) &= \begin{bmatrix} i \\ 1 \end{bmatrix} e^{-i(\beta+VH)z} \\ E_2(z) &= \begin{bmatrix} -i \\ 1 \end{bmatrix} e^{-i(\beta-VH)z} \end{aligned} \tag{2.113}$$

Therefore, after the incident polarized light has traveled a distance L, the polarization plane will rotate by an angle

$$\Phi(L) = VHL \tag{2.114}$$

Therefore, by measuring the polarization rotation angle of linearly polarized light, the magnitude of the applied magnetic field can be measured, or the Verdet constant of the material can be measured. It should be noted here that the eigenmodes in the magnetic field perturbed fiber obtained by the perturbation coupled mode theory are two circularly polarized modes. Compared with the eigenmodes in the non-perturbation fiber, only a perturbation term proportional to the applied magnetic field is added to the propagation constant. This is exactly the same as the optical rotation effect in cubic crystals or isotropic media (regardless of boundary conditions).

Elliptic Birefringence in an Ideal, Round, Single-Mode Fiber

When an optical fiber is subjected to the simultaneous action of an electric field and a magnetic field, the perturbation of the electromagnetic field causes the two linearly polarized modes and the left and right circularly polarized modes to become non-degenerate, resulting in elliptical birefringence. At this time, the Kerr effect and Faraday effect will cause the dielectric tensor of the optical fiber to change as follows:

$$\begin{aligned}
\Delta\varepsilon_{xx} &= -\varepsilon_0 n^2\left(n^2 S_{11} E^2\right) \\
\Delta\varepsilon_{yy} &= \Delta\varepsilon_{zz} = -\varepsilon_0 n^2\left(n^2 S_{12} E^2\right) \\
\Delta\varepsilon_{xz} &= \Delta\varepsilon_{zx} = \Delta\varepsilon_{yz} = \Delta\varepsilon_{zy} = 0 \\
\Delta\varepsilon_{xy} &= -\Delta\varepsilon_{yx} = -i\frac{\varepsilon_0\lambda_0 nV}{\pi}H
\end{aligned} \tag{2.115}$$

Substitute Eq. (2.115) into Eq. (2.97), the coupling coefficient can be obtained as,

$$\begin{aligned}
k_{11} &= -k_0 n^3\left(\frac{S_{11}+S_{12}}{2}\right)E^2 \\
k_{22} &= -k_0 n^3 S_{12} E^2 \\
k_{12} &= -k_{21} = \frac{\omega}{4}\left(-i\frac{\varepsilon_0\lambda_0 nV}{\pi}H\right)\int_0^a\int_0^{2\pi} J_0\left(\frac{u}{a}r\right) r\,dr\,d\phi \\
&= -iVH
\end{aligned} \tag{2.116}$$

None of the four coupling coefficients equals zero; substituting them into Eqs. (2.99) and (2.100), and utilizing the conclusions derived from the previous discussion in Sect. 2.3.3.2 on the Kerr effect and Faraday effect in optical fibers, the following results can be obtained:

$$\begin{aligned}
k &= -iVH \\
\Delta\beta &= k_{11} - k_{22} = 2\pi B E^2 \\
s &= \sqrt{\left(\frac{\Delta\beta}{2}\right)^2 + k^2} = \sqrt{\left(\pi B E^2\right)^2 + (VH)^2}
\end{aligned} \tag{2.117}$$

$$E_1 = \begin{bmatrix} \frac{k}{\frac{\Delta\beta}{2}-s} \\ 1 \end{bmatrix} e^{-i(\overline{\beta}-s)z};\ E_2 = \begin{bmatrix} \frac{k}{\frac{\Delta\beta}{2}+s} \\ 1 \end{bmatrix} e^{-i(\overline{\beta}+s)z} \tag{2.118}$$

At this time, the eigenmodes in the perturbed fiber are two elliptically polarized lights with opposite rotation directions, and the phase delay between the two modes is

$$\Phi = 2s = \sqrt{\left(2\pi BE^2\right)^2 + 4(VH)^2} \tag{2.119}$$

If F represents the total optical rotation effect caused by the Faraday effect and K represents the phase delay caused by the Kerr effect, the total phase delay can be expressed as:

$$\Phi = \sqrt{K^2 + 4F^2} \tag{2.120}$$

The above formula is exactly the same as the total phase delay expression derived from Maxwell's equations when both Faraday effect and linear birefringence exist. And the total phase delay is the vector superposition of the phase delay caused by Kerr effect and the optical rotation effect caused by Faraday effect. At this time, the two effects are jointly determine the eigenmodes in the fiber as well as the polarization transmission characteristics.

2.3.4 Description of Polarization of Single-Mode Fiber Using the Poincaré Sphere

General Description with Poincaré Sphere

In 1977, R. Ulrich first proposed the use of Poincaré sphere to analyze the state changes of light during transmission in optical fiber. Considering that the two orthogonal modes of polarized light in optical fiber have mutual coupling effect during transmission, the two coordinate parameters of the Poincaré sphere used on optical fibers (referring to Fig. 2.5) are

$$\begin{aligned} 2\chi &= \arctan\left(\frac{\left|\frac{A_1}{A_2}\right| - 1}{\left|\frac{A_1}{A_2}\right| + 1}\right) \\ 2\zeta &= \arg\left(\frac{A_1}{A_2}\right) \end{aligned} \tag{2.121}$$

where A_1, A_2 represent the amplitude coefficients of fiber mode field in Eq. (2.94).

In Eq. (2.121), A_1, A_2 are determined by the polarization state of the incident light.

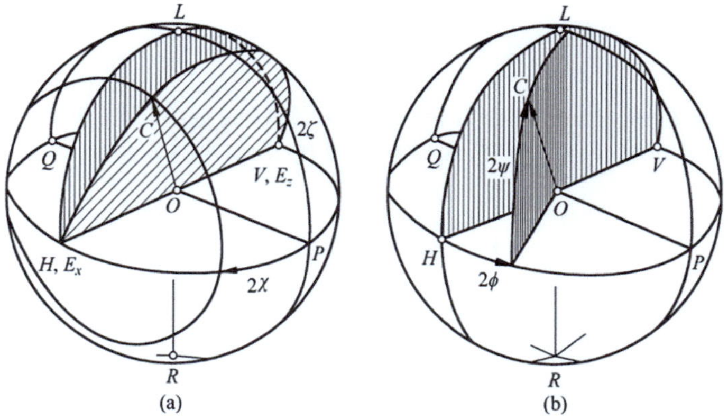

Fig. 2.5 Poincaré sphere description in optical fiber. The coordinate definition takes **a** polarized light in x and y directions as eigenstates; **b** left and right circularly polarized light as the eigenstate

$$E_{\text{in}} = A_1(z)E_1e^{-ik_1z} + A_2(z)E_2e^{-ik_2z} \tag{2.122}$$

Among them, E_1 *and* E_2 are the amplitudes of eigenstates with orthogonal polarization states. A_1, A_2 change with the polarization state of the incident light, and the corresponding point C moves on the Poincaré sphere. At this time, the movement trajectory of point C represents the evolution of the polarization state of light when it is transmitted through the optical fiber. So why is the azimuth angle determined by Eq. (2.121) selected? This can be obtained phenomenologically through the coupling equation of the two orthogonal modes in optical fiber. The coupled wave equation is as follows:

$$\begin{aligned} \frac{dA_1}{dz} &= -ik_{11}A_1(z) - ik_{12}A_2(z)e^{-i(k_2-k_1)z} \\ \frac{dA_2}{dz} &= -ik_{21}A_1(z)e^{-i(k_2-k_1)z} - ik_{22}A_2(z) \end{aligned} \tag{2.123}$$

On the Poincaré sphere, the vector $C(z) = OC$ is defined to represent the ratio $A_1(z)/A_2(z)$ of the mode field coefficients at point Z, with the coordinates of $\boldsymbol{C}(z)$ on the sphere being $(2\chi, 2\varsigma)$. When a light wave transmits through an optical fiber, the change in its polarization state is represented by the trajectory of $C(z)$ on the Poincaré sphere,

$$C'(z) = \frac{dC(z)}{dz} \tag{2.124}$$

When light travels in an optical fiber along the z direction for a distance of δz, its polarization state evolves as point C rotates around the axis of $O\Omega$ by an angle $\delta\theta$. Under the first-order approximation condition, this angle can be expressed as:

$$\delta\theta = \sqrt{(k_{11} - k_{12})^2 + 4k_{12}k_{21}}\delta z = |O\Omega|\delta z \tag{2.125}$$

Ω is a vector whose magnitude is the phase difference caused by birefringence per unit length of fiber. Its direction is expressed by the azimuth angle of the Poincaré sphere,

$$\begin{aligned} 2\zeta &= \arg(k_{12}) \\ 2\chi &= \arctan\left(\frac{k_{11} - k_{22}}{2\sqrt{k_{12}k_{21}}}\right) \end{aligned} \tag{2.126}$$

The coupling coefficient in the equation is determined by Eq. (2.97). The rotation direction of point C is clearly perpendicular to both vector $\boldsymbol{OC}$ and vector $\boldsymbol{O\Omega}$ simultaneously, indicating that it is perpendicular to the plane formed by these vectors. It is straightforward to demonstrate

$$\Delta \boldsymbol{OC} = |\boldsymbol{OC}|\delta\theta\left[\frac{\boldsymbol{O\Omega} \times \boldsymbol{OC}}{|\boldsymbol{OC}||\boldsymbol{O\Omega}|}\right] = \boldsymbol{O\Omega} \times \boldsymbol{OC}\delta z \tag{2.127}$$

For simplicity, $\boldsymbol{C}$ and $\omega(z)$ denote vector $\boldsymbol{OC}$ and vector $\boldsymbol{O\Omega}$, then there is

$$\frac{d\boldsymbol{C}}{dz} = \boldsymbol{\omega}(z) \times \boldsymbol{C} \tag{2.128}$$

Thus, the change in the polarization state over an infinitesimally short fiber segment dz is a result of the Poincaré sphere rotating by an angle ωdz around the axis $\omega(z)$. Since the coupling coefficient is known in a specific problem, in principle, $\omega(z)$ at each point of the fiber is known, and the trajectory of $C(z)$ on the Poincaré sphere can be determined for any input polarization state. Below we use this method to study birefringence in optical fibers.

Pure Linear Birefringence (Kerr Effect)

When the optical fiber is subjected to an electric field along the x-axis, the Kerr effect induces pure linear birefringence in the optical fiber. The coupling coefficient can be determined using Eq. (2.106). Substituting this coupling coefficient into Eq. (2.126), it becomes evident that the rotation axis $\omega(z)$ is located on the equatorial plane of the Poincaré sphere. This rotation corresponds to linear birefringence. In this case, the symbol $\beta(z)$ is generally used instead of $\omega(z)$. For the Kerr effect discussed herein, the position of $\beta(z)$ on the equatorial plane is as follows:

$$2\chi_\beta = \frac{\pi}{2};\quad 2\zeta_\omega = 0 \tag{2.129}$$

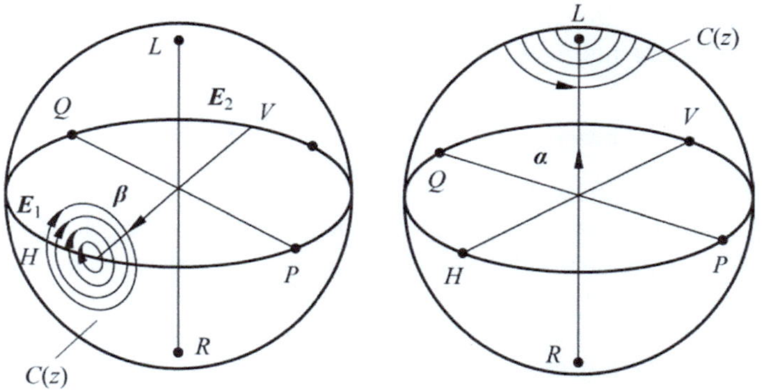

(a) Pure line birefringence; (b) circular birefringence

Fig. 2.6 Poincaré sphere description of the birefringence in an optical fiber

By substituting the coupling coefficient value into Eq. (2.98), the value of β can be found as:

$$\beta = |k_{11} - k_{22}| = 2\pi BE^2 \tag{2.130}$$

It is exactly the same as Eq. (2.109).

For an optical fiber of length L, when the polarization state of the input light is represented as $\boldsymbol{C}(0)$ on the Poincaré sphere, the polarization state of the output light is $\boldsymbol{C}(z)$. This state is the coordinate after $\boldsymbol{C}(0)$ rotating around $\beta(z)$ by an angle of βL. It is evident that as polarized light travels through the optical fiber, corresponding to various transmission lengths L (or different z), the polarization state undergoes periodic changes. The trajectory of $\boldsymbol{C}(z)$ forms concentric circles that are perpendicular to $\beta(z)$, as depicted in Fig. 2.6. At this time, the two eigenstates in the optical fiber are linearly polarized light in the x and y directions, corresponding to the H and V points, respectively. It can be observed that the description of the optical fiber under the Kerr effect on the Poincaré sphere is identical to that of general birefringent devices.

Pure Circular Birefringence (Faraday Effect)

When an optical fiber is exposed to a magnetic field along the z-axis, it will exhibit pure circular birefringence as a result of the Faraday effect. It is evident that the coupling coefficient $k_{12} = -\mathrm{i}VH$ is a pure imaginary number. By substitute into Eq. (2.126), the axis of rotation $\omega(z)$ aligns with line RL of the Poincaré sphere, which is the line connecting the north and south poles on the sphere. This rotation is circular birefringence as shown in Fig. 2.6. In this case, the symbol $\alpha(z)$ is generally used instead of ω(z). Regarding the Faraday effect discussed here, substituting the coupling coefficient into Eq. (2.127) yields the value of α

$$\alpha = 2|k_{11}k_{22}| = 2\mathrm{VH} \tag{2.131}$$

For an optical fiber of length L, when the polarization state of the input light is expressed as $C(0)$ on the Poincaré sphere, the polarization state of the output light is $C(z)$, which is the coordinate after $C(0)$ is rotated by an angle αL around $\alpha(z)$. When polarized light travels through an optical fiber, corresponding to different transmission lengths L (that is, different positions z), the polarization state changes periodically, and the trajectory of $C(z)$ is a concentric circle parallel to the equator, such as shown in Fig. 2.6.

When the incident light is linearly polarized, the trajectory of $C(z)$ is the equatorial circle, and the polarization state at any point remains linearly polarized light, with its polarization plane rotating at a rate of $\alpha/2$ per unit length of fiber. At this time, the two eigenstates in the optical fiber are left-hand and right-hand circularly polarized light corresponding to L and R points, respectively. The description of the optical fiber under the Faraday effect on the Poincaré sphere is identical to that of the optically rotating device.

Elliptical Birefringence

Elliptical birefringence can be considered the outcome of the superposition of linear birefringence and circular birefringence. For infinitesimally small rotations αdz and βdz, the total combined effect can be viewed as the sum of the two.

$$\omega dz = \alpha dz + \beta dz \tag{2.132}$$

i.e.

$$\omega = \alpha + \beta \tag{2.133}$$

When ω(z) is independent of z (as in the Faraday and Kerr effects discussed earlier), the trajectory of $C(z)$ is as shown in Fig. 2.7. At this time, there are two eigenmodes in the fiber, namely the states C_1 and C_2, whose trajectories remains unchanged with z. Generally, these two eigenstates are two orthogonal elliptical polarization modes.

Fig. 2.7 Poincaré sphere description of elliptical birefringence in a fiber

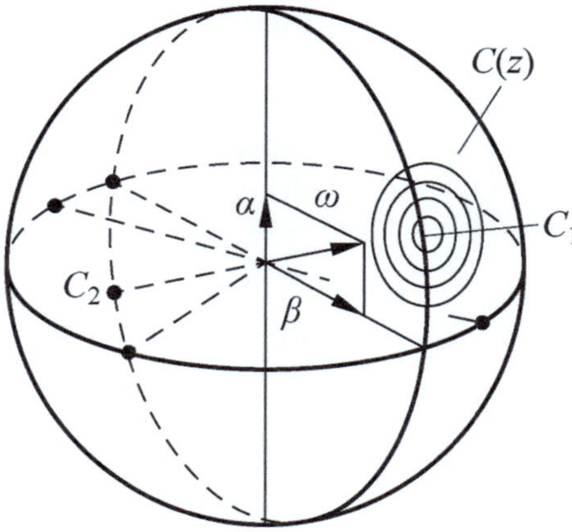

2.3.5 *Polarization Characteristics of Twisted Fiber and Spun Fiber*

Polarization is one of the most significant characteristics of single-mode fiber. The light wave propagating through an ideal, circular single-mode fiber should maintain the same polarization state. However, in practical applications, even when not influenced by external stress, ordinary round single-mode fiber will produce birefringence due to various imperfections within the optical fiber itself. This means that the input linearly polarized light becomes elliptically polarized light after traveling a certain distance, and this change occurs periodically or aperiodically as the light propagates. This instability in the polarization state is very detrimental to systems that require the maintenance of polarization.

The solution to the aforementioned issue involves the development of polarization-maintaining optical fiber. This ensures that one polarization mode within the fiber exhibits a strong birefringence effect, while the birefringence of the other mode is suppressed. Alternatively, the fiber's cross-section can be shaped to closely resemble an ideal circular cross-section, thereby making the fiber's polarization characteristics akin to those of an ideal circular single-mode optical fiber. The following discusses the polarization characteristics of a single mode low birefringence spun fiber. This kind of optical fiber rotates at a high speed during the drawing process, thus introducing an equivalent circular symmetry structure into the optical fiber, which improves the circular symmetry of the fiber cross section. This kind of optical fiber is called spun fiber. The polarization characteristics of a twisted fiber are analyzed before spun fiber.

Polarization Characteristics of a Twisted Fiber

Considering an optical fiber twisted at a velocity of ξ, the intrinsic linear birefringence caused by the core ellipse is $\Delta\beta$. Using planar wave approximation, the phase delay $R(z)$ at point Z and the optical rotation angle $\Theta(s)$ (relative to the short axis of the optical fiber ellipse at the incident point) are respectively expressed as

$$
\begin{aligned}
R(z) &= 2\arcsin\left(\frac{\rho}{\sqrt{1+\rho^2}}\sin\gamma z\right) \\
\Theta(z) &= \xi z + \arctan\left(-\frac{1}{\sqrt{1+\rho^2}}\tan\gamma z\right) \\
\Phi(z) &= \frac{\xi z - \Theta(z)}{2} \pm \frac{m\pi}{2}(m = 0, 1, 2, \cdots)
\end{aligned}
\quad (2.134)
$$

wherein,

$$
\rho = \frac{\Delta\beta}{2\xi - a}\gamma = \frac{\sqrt{(\Delta\beta)^2 + 4(\xi - a)^2}}{2} \quad (2.135)
$$

For twisted fiber, the coefficient a in Eq. (2.135) represents the optical rotation caused by the photoelastic effect, which can be expressed as:

$$
a = g\xi \quad (2.136)
$$

Here, the experimental proportional coefficient g (for quartz optical fiber) is 0.08, aligning perfectly with the theoretical value. For a spun fiber, the rotation is carried out in the molten state, and there is no torsional stress after fiber formation, so the photo-elastic effect can be considered negligible. At this time, $a = 0$.

Under the condition that the twisting rate is relatively small to the inherent linear birefringence of the fiber ($\xi \ll \Delta\beta$), it can be obtained from the Eq. (2.134) that $\Phi \approx 0$, and the direction of the main axis remains unchanged; phase delay $R \approx \Delta\beta z$, and the phase delay due to twisting is negligible. At the same time, the optical rotation angle varies linearly with z, $\Theta \approx \xi z$. In this instance, linearly polarized light that is parallel to the main axis of the optical fiber maintains its linear polarization state, and its polarization direction rotates in sync with the fiber's twist. Consequently, the optical rotation resulting from the photo-elastic effect is effectively suppressed by the strong intrinsic linear birefringence.

Under the condition that the twist rate is relatively large with respect to the intrinsic linear birefringence of the fiber ($\xi >> \Delta\beta$), the phase delay and optical rotation angle are respectively:

$$
R(z) \approx \frac{\Delta\beta}{\xi - a}\sin[(\xi - a)z]
$$

$$\Theta(z) = az \tag{2.137}$$

Under such conditions, the phase delay is significantly reduced and oscillates near zero. Linearly polarized light entering the fiber in any direction emerges as approximately linearly polarized light, with its polarization direction rotated by an angle relative to the direction of incidence. It is evident that the significant photoelastic effect resulting from large twisting effectively suppresses the intrinsic linear birefringence of the fiber. In relation to high birefringence linear polarization maintaining fibers, twisted fibers can be considered as circular birefringence polarization maintaining fibers. This approach effectively circumvents the influence of the inherent linear birefringence of the optical fiber and external factors such as microbends on the system.

Description of the Polarization Characteristics of Twisted Fiber Using Poincaré Sphere

In an optical fiber with constant linear birefringence and uniform twist, the motion of the rotation vector $\omega(z)$ can be expressed as

$$\omega'(z) = 2\xi \times \omega(z) \tag{2.138}$$

The magnitude of the twist vector ξ is $|\xi|$, and its direction aligns with the *RL* axis of the Poincaré sphere. For a right-hand twist, ξ points upward. Since the fiber is twisted, we can introduce an auxiliary local coordinate system R^0 for each cross-section of the fiber, with its X^0 axis parallel to the fast axis of the fiber (corresponding to linear birefringence), Y^0 parallel to the slow axis, and Z^0 coinciding with the Z axis of the fiber. Thus the coordinate system R^0 rotates at a speed of ξ around the Z axis of the fiber. While the Poincaré sphere S^0 corresponding to R^0, rotates around *RL* at a rate of 2ξ. All quantities related to the system R^0 and S^0 are labeled with superscript 0, where $\Delta\beta^0$ is a constant, $\alpha^0 = a$ is a constant, $\Delta\beta^0$ points to X^0. Then there is:

$$\omega^0 = \Delta\beta^0 + \alpha^0 \tag{2.139}$$

However, the polarization state of the light wave in above optical fiber does not change at a angular velocity of ω^0 on S^0, the following vector is used as the rotation vector:

$$\begin{aligned} \Omega^0 &= \boldsymbol{\alpha} + \Delta\boldsymbol{\beta}^0 - 2\xi \\ \Omega^0 &= \left|\boldsymbol{\Omega}^0\right| = \sqrt{\left|\Delta\beta^0\right|^2 + (\alpha - 2\xi)^2} \end{aligned} \tag{2.140}$$

In Eq. (2.140), the vector $\boldsymbol{\Omega}^0$ is fixed on S^0, the trajectory $C^0(z)$ of all polarization state variations are circles centered on the axis of Ω^0 in S^0.

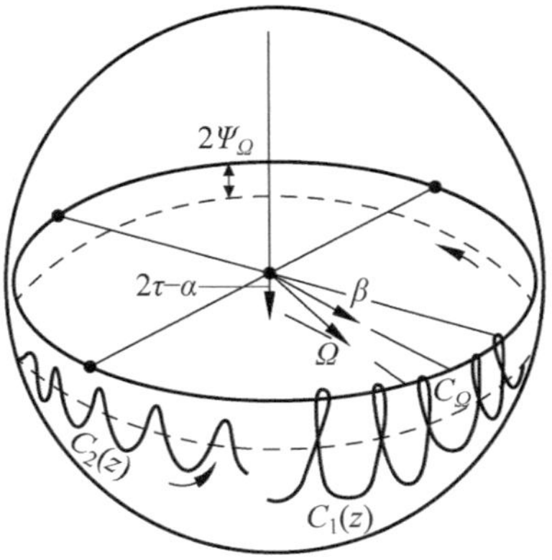

Fig. 2.8 Poincaré sphere description of the polarization of a twisted fiber

In order to find the trajectory of $C(z)$, we should obtain the trajectory from the rotating sphere S^0 back to the fixed sphere S, and $C(z)$ is moving along a constant semi-cone, whose axis is moving along a parallel circle. The latitude of the parallel circle is

$$2\Psi_\Omega = \arctan\left|\frac{\alpha - \xi}{\Delta\beta}\right| \tag{2.141}$$

Therefore, the trajectory of $C(z)$ on S is a cycloidal curve, such as shown in Fig. 2.8.

The spatial distance of points on adjacent periods of the curve on the fiber is $2\pi/\Omega$, which is equal to one rotation period on S^0. Such adjacent points is equivalent referring to the local coordinate system R^0, but in the laboratory coordinate system, the second state is rotated by an angle of $2\pi/\Omega$ relative to the previous state. Assume ξ/Ω is a rational number, $C(z)$ is a closed curve.

For a small twist, i.e., $\xi \ll \Delta\beta$, at this time there is

$$\alpha \ll \Delta\beta;\ \Omega^0 \approx \Delta\beta^0 \tag{2.142}$$

It shows that the small twist is completely concealed by the inherent linear birefringence of the fiber.

For a large twist, i.e., $\xi \gg \Delta\beta$, at this time there is

$$\alpha \gg \Delta\beta; 2\Psi \to \frac{\pi}{2}; \Omega^0 = \alpha - 2\xi = (g - 2)\xi \tag{2.143}$$

On S^0, it is equivalent to the uniform rotation around RL axis at a rate of $a - 2\xi$. Correspondingly, in the laboratory coordinate system, the polarization state of the light wave in the fiber rotates at a rate of $a = g\xi$. In real space, the rotation rate of the polarization plane is $g\xi/2$, and the linear birefringence is completely submerged by the circular birefringence. When a circularly polarized light incidents into the optical fiber, the output light maintains circularly polarized, even if there is linear birefringence. Therefore, the strongly twisted optical fiber is a polarization-maintaining optical fiber designed for circularly polarized light, also known as a "spun fiber".

Polarization Characteristics of a Spun Fiber

A high-speed twist, exceeding $2\pi \times 10^3$ rad/m, is required in certain applications to effectively suppress inherent linear birefringence. However, the optical fiber cannot withstand such a twist without breaking. Therefore, it is necessary to employ a spinning technique. The spun fiber satisfies the two conditions of $\alpha = 0$ and $\xi \gg \Delta\beta$ simultaneously, and is shown by Eq. (2.134):

$$R(z) \approx \frac{\Delta\beta}{\xi}\sin(\xi z);\ \Theta(z) \approx 0;\ \Phi(z) = \frac{\xi z}{2} \tag{2.144}$$

Compared to twisted fiber, the inherent linear birefringence and circular birefringence in spun fiber are effectively suppressed, and its polarization characteristics are consistent with those of an ideal isotropic circular single-mode fiber. For a general optical fiber, its intrinsic birefringence can be better than 1 rad/m. The same optical fiber is rotated at a rate of 1 turn/cm during the manufacturing process to make a spun fiber, and its maximum birefringence in one rotation cycle is 1.6×10^{-3} rad/cm with a residual circular birefringence of 2×10^{-4} rad/m, so the spun fiber can be treated as an isotropic round single-mode fiber.

The polarization state of light in a spun fiber on S^0 uniformly rotates around the *RL* axis at a rotation rate of -2ξ. Correspondingly, in the laboratory coordinate system, the rotation rate of the polarization state of the light wave in the optical fiber is zero. In real space, the rotation rate of the polarization plane is also zero, and the output polarization state is the same as the incident one, so the spun fiber can be regarded as a polarization-maintaining fiber.

It is important to note that the suppression of intrinsic linear birefringence in spun fiber is achieved by rendering the structure ineffective through high-speed rotation. However, this method will not suppress linear birefringence caused by external fields, where the mechanism producing the birefringence is independent of the rotational structure within the spun fiber.

2.4 Jones Matrix of Optical Fiber and Polarization Perturbation

2.4.1 Transmission Matrix of Polarized Light in Optical Fibers

In 1941, R.C.Jones expressed the x and y components of the electric field vector $\boldsymbol{E}$ using a column matrix, commonly known as the Jones vector, which generally represents elliptically polarized light,

$$\begin{bmatrix} E_x \\ E_y \end{bmatrix} = \begin{bmatrix} E_{0x}e^{i\delta_1} \\ E_{0y}e^{i\delta_2} \end{bmatrix}$$

For linearly polarized light, suppose $\boldsymbol{E}$ is in the first and third quadrants, there is $\delta_1 = \delta_2 = \delta_0$, the corresponding Jones vector is

$$\begin{bmatrix} E_x \\ E_y \end{bmatrix} = \begin{bmatrix} E_{0x} \\ E_{0y} \end{bmatrix} e^{i\delta_0} \tag{2.145}$$

Similarly, for right-hand circular polarization light, its Jones vector is,

$$\begin{bmatrix} E_x \\ E_y \end{bmatrix} = \begin{bmatrix} -i \\ 1 \end{bmatrix} E_{0x}e^{i\delta_0}$$

Since the light intensity I is given by $I = E_{\mathrm{x}}^2 + E_{\mathrm{y}}^2$ (excluding the proportionality constant), it is commonly simplified to $I = 1$. This Jones vector is then referred to as the standard Jones vector. It is calculated by normalizing each component of the Jones vector with respect to $\sqrt{I}$ (see Table 2.2). Two polarized light E_1, E_2 are orthogonal polarization states and satisfy:

$$E_1 \cdot E_2^* = \begin{bmatrix} E_{1x} & E_{1y} \end{bmatrix} \begin{bmatrix} E_{2x}^* \\ E_{2y}^* \end{bmatrix} = 0.$$

The Jones vectors of linearly polarized light are $\begin{bmatrix} E_{1x} \\ 0 \end{bmatrix}$ and $\begin{bmatrix} 0 \\ E_{2y} \end{bmatrix}$; Left-hand and right-hand circularly polarized light are also mutually orthogonally polarized light. Using Jones vector, two completely polarized coherent lights can be conveniently superimposed, polarization state transformed and so on.

Example 2.4.1 suppose that the Jones vectors of two linearly polarized lights E_1, E_2 are respectively $E_1 = \begin{bmatrix} \sqrt{3}e^{i\delta_1} \\ 0 \end{bmatrix}$, $E_2 = \begin{bmatrix} 0 \\ \sqrt{3}e^{i(\delta_1+90^\circ)} \end{bmatrix}$, then the superposition of

Table 2.2 Examples for normalized Jones vector

Example	Unnormalized	Normalized
① x-direction vibrating linearly polarized light	$\begin{bmatrix} 3e^{i\delta_x} \\ 0 \end{bmatrix}$	$\begin{bmatrix} 1 \\ 0 \end{bmatrix}$
② 45° vibrating linearly polarized light	$\begin{bmatrix} 1 \\ 1 \end{bmatrix}$	$\frac{\sqrt{2}}{2}\begin{bmatrix} 1 \\ 1 \end{bmatrix}$
③ Left-hand circularly polarized light	$\frac{1}{2}\begin{bmatrix} 1+i \\ 1-i \end{bmatrix}$	$\frac{\sqrt{2}}{2}\begin{bmatrix} 1 \\ i \end{bmatrix}$

these two polarized lights is $E = E_1 + E_2 = \begin{bmatrix} \sqrt{3}e^{i\delta_1} \\ \sqrt{3}e^{i(\delta_1+90°)} \end{bmatrix}$, leading to a right-hand circularly polarized light with an intensity of 6.

The superposition of n polarized light beams with the same frequency propagating in the same direction can be obtained from the addition of the n Jones vectors, and the amplitude and phase of the two components of the Jones vector should be considered in addition,

$$E = \begin{bmatrix} E_x \\ E_y \end{bmatrix} = \begin{bmatrix} \sum_{i=1}^{n} E_{ix} \\ \sum_{i=1}^{n} E_{iy} \end{bmatrix}$$

For another example, the polarization state of the polarized light $\begin{bmatrix} E_x \\ E_y \end{bmatrix}$ becomes $\begin{bmatrix} E'_x \\ E'_y \end{bmatrix}$ after passing through a polarization element. And the transformation can be represented by a 2 × 2 matrix $\boldsymbol{J}$:

$$\begin{bmatrix} E'_x \\ E'_y \end{bmatrix} = \begin{bmatrix} J_{11} & J_{12} \\ J_{21} & J_{22} \end{bmatrix} \begin{bmatrix} E_x \\ E_y \end{bmatrix} = \boldsymbol{J} \begin{bmatrix} E_x \\ E_y \end{bmatrix}$$

This 2 × 2 matrix $\boldsymbol{J}$ is the transmission matrix of the polarizing element, also called Jones matrix, and its elements depends only on the devices. If polarized light $\begin{bmatrix} E_x \\ E_y \end{bmatrix}$ passes through n polarization elements in sequence, their Jones matrices are $\boldsymbol{J}_i$ ($i = 1, 2, \ldots, n$), respectively. Then the Jones vectors of the light exiting from the nth polarization element is obviously

$$\begin{bmatrix} E'_x \\ E'_y \end{bmatrix} = \boldsymbol{J}_n \boldsymbol{J}_{n-1} \boldsymbol{L} \boldsymbol{J}_2 \boldsymbol{J}_1 \begin{bmatrix} E_x \\ E_y \end{bmatrix} = \boldsymbol{J} \begin{bmatrix} E_x \\ E_y \end{bmatrix}$$

Therefore, the Jones matrix characterizes the transformation performance of the device to polarized light.

Accordingly, when incident light waves travel along the optical fiber, the output optical field of the fiber equals the input optical field multiplied by a series of equivalent Jones matrices acting on the optical fiber. When the elements within the matrix are modulated with specific information, the output polarization state of light is correspondingly modulated, as seen in effects such as the Kerr effect and the Faraday effect. This allows for the detection of the information's characteristics. Since the Jones matrix is applicable only to fully polarized light, the Mueller matrix should be selected for analyzing partially polarized and incoherent light.

Assume the excitation light field at the input end of the fiber $E(0,t) = \begin{bmatrix} E_x(0,t) \\ E_y(0,t) \end{bmatrix} = \begin{bmatrix} E_{x0} \\ E_{y0} \end{bmatrix} e^{i\omega t}$, then the electric field of light wave at any position z in the optical fiber is,

$$E(z,t) = \begin{bmatrix} E_x(z,t) \\ E_y(z,t) \end{bmatrix} = \begin{bmatrix} m_{11} & m_{12} \\ m_{21} & m_{22} \end{bmatrix} \begin{bmatrix} E_{x0} \\ E_{y0} \end{bmatrix} e^{-i\beta z} e^{i\omega t}$$

where, $\boldsymbol{m}_{\text{ij}}$ is polarization transmission matrix. Please refer to relevant literature [1, 2] for the corresponding Jones transform matrices of typical optical polarizer and phase delay.

2.4.2 *Variation in Polarization Characteristics Caused by Extrusion and Twist*

External lateral pressure is a common factor that results in fiber deformation, which in turn produces stress birefringence and affects the polarization state of transmitted light waves in optical fiber. As discussed in Sect. 2.3.1, changes in the fiber's polarization characteristics due to external disturbances result in a transformation of the dielectric constant, which in turn causes changes in the fiber's transmission characteristics, i.e., alterations in birefringence. Approximately, the lateral pressure corresponds to the electric fields of the two orthogonal modes of light passing through a phase retarder with birefringence in the optical fiber. Consider an angle θ between the fast axis and the horizontal direction, and the corresponding Jones matrix is:

$$\boldsymbol{J}(\Delta\varphi) = \begin{bmatrix} e^{i\Delta\varphi}\cos^2\theta + \sin^2\theta & \left[e^{i\Delta\varphi} - 1\right]\cos\theta\sin\theta \\ \left[e^{i\Delta\varphi} - 1\right]\cos\theta\sin\theta & e^{i\Delta\varphi}\sin^2\theta + \cos^2\theta \end{bmatrix} \tag{2.146}$$

When a uniform twist is applied to the optical fiber, as discussed in Sect. 2.3.1, the transmission coefficient matrix of the optical fiber can be expressed as:

$$[m_{ij}] = \begin{bmatrix} m_{11} & m_{12} \\ m_{21} & m_{22} \end{bmatrix} = \begin{bmatrix} \cos(p\gamma z) & -\sin(p\gamma z) \\ \sin(p\gamma z) & \cos(p\gamma z) \end{bmatrix} \tag{2.147}$$

where, γ is the torsion rate (unit in rad/m), and p is a constant (generally about 0.007). When the phase difference of the eigenmodes of the twisted fiber sets to π/2, the two eigenmodes of the fiber are circularly polarized modes. Easy connection and high stability are the advantages of the circular polarization mode.

2.4.3 Thermal Effect on Polarization State

To control the refractive index of the optical fiber core and cladding, different doping materials and processes are utilized for the core and cladding during the preform manufacturing process. Given that the fiber core and cladding are drawn at high temperatures, both the geometric asymmetry of the fiber cross-section and the residual stresses from the drawing process can cause temperature-related birefringence. This is represented by a temperature-dependent rotation matrix, expressed as:

$$[m_{ij}(T)] = \begin{bmatrix} m_{11} & m_{12} \\ m_{21} & m_{22} \end{bmatrix} = \begin{bmatrix} \cos\theta(T) & -\sin\theta(T) \\ \sin\theta(T) & \cos\theta(T) \end{bmatrix} \tag{2.148}$$

Among them, θ(T) is determined based on the thermal expansion coefficient of the optical fiber, Poisson's ratio of the material, Young's modulus, as well as specific process parameters such as mechanical stress and temperature variation range.

2.5 Polarization Control in Optical Fiber and Devices

Controlling the polarization state has always been one of the challenges and high costs in optical fiber applications. Generally, the polarization control techniques for light waves transmitted through optical fibers are divided into two categories: polarization maintenance and depolarization. As the names suggest, polarization maintenance aims to preserve the original polarization state of the light waves, whether linear or circular polarization, while depolarization seeks to eliminate the distinctions between different polarization modes. Devices (such as optical fiber depolarizer) or external force intervention (like polarization scrambler) are used to have the polarization modes transmit periodically alternately, thus eliminating the differences between polarization modes caused by transmission in optical fibers.

The control methods for polarization maintenance in optical fibers typically encompass: ① Optical fiber preparation process control, which primarily involves two types of polarization-maintaining fibers: linear polarization-maintaining fibers, such as high birefringence fibers, which include processes with high doping stress regions or geometric asymmetry of the core/cladding, and low birefringence fibers, which involve twist/rotation drawing processes to ensure the circular symmetry of the optical fiber cross-section as much as possible or to eliminate parasitic stress, thus imposing stringent requirements on the optical fiber manufacturing process; ② External action control encompasses various types of optical fiber polarization controllers, including wound disc three-paddle optical fiber polarization controllers, online extrusion optical fiber polarization controllers, and Faraday rotator, among others.

2.5.1 Polarization-Maintaining Fiber, Polarizing Fiber and Spun Fiber

Polarization Maintaining Fiber

To enhance the difference in transmission constants between the two polarization modes in high birefringence optical fibers, more advanced manufacturing processes are included: ① High ovality manufacturing processes are depicted in Fig. 2.9a, which illustrates several typical polarization-maintaining fiber (PMF) structures, such as those with elliptical cores and elliptical cladding optical fibers. However, an increase in the refractive index difference between the core and cladding typically results in higher optical fiber loss. ② The high stress doping process, which includes the bow tie and panda type doping, effectively increases the difference in propagation constants between the two polarization modes by utilizing high birefringence and applying a large pre-stressing force, respectively. When a single polarization mode is incident into the optical fiber, the fiber can maintain stable transmission of that polarization state within it.

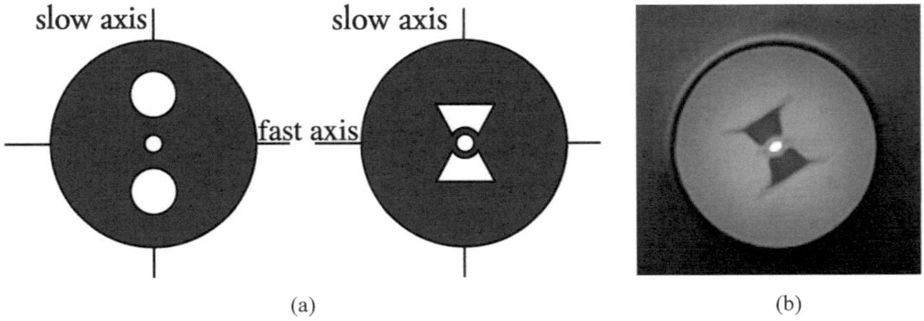

Fig. 2.9 Polarization maintaining fiber and polarizing fiber

Polarizing Fiber

Polarizing (PZ) fiber, such as the Zing™ fiber shown in Fig. 2.9b, permits only one polarization of light wave to pass, thereby polarizing the light as it propagates. PZ fiber typically employs a bow-tie structure to achieve a high birefringence effect, enabling light in a specific polarization direction to propagate within the optical fiber, whereas light with other polarization directions incurs high loss.

Unlike PMF, PZ fiber only propagates light in one polarization direction, and light polarized in other directions cannot pass through. However, PMF can preserve the polarization direction of light transmitted along the birefringence axis, allowing it to transmit light in any polarization direction. Polarization crosstalk may occur, however. Single-mode fiber can produce birefringence through pressure (as detailed in Sect. 2.5.2 Squeezed optical fiber polarization controller), which makes the fiber function more like a wave plate. Although the polarization axis is controllable, single-mode fiber does not polarize light. This single-polarization transmission of PZ fiber produces no crosstalk and achieves a polarization suppression ratio greater than 30 dB within the designed wavelength range, making it ideal for polarization-sensitive applications.

It should be noted that the polarization window and extinction ratio of PZ fiber should be adjusted by coiling (i.e. arranging: straightening, coiling, arbitrarily stacking) PZ fiber. PZ fiber boasts a wide polarization window (~100 nm), the width and center wavelength of which are contingent upon the fiber arrangement. For instance, as the diameter of the fiber coil decreases, the polarization window narrows and shifts towards the short wavelength direction. Consequently, commercial PZ fibers are typically sold with a calibrated winding diameter of Ø89 mm to prevent performance degradation. On the other hand, polarization suppression greater than 35 dB can be achieved through manual adjustment of the polarization window.

PZ fiber is an all-fiber device that offers advantages over coaxial polarizers, including lower insertion loss, a higher extinction ratio, and the absence of complicated component assembly or packaging. In summary, PZ fiber boasts a high extinction ratio (ER) and wide bandwidth, is cost-effective, functions normally as a polarizer at the operating wavelength even under pressure, and exhibits excellent temperature stability in terms of both ER and insertion loss, ensuring reliability.

Spun Fiber

A spun fiber is a unique type of high birefringence optical fiber, produced by rotating the preform of a bow-tie or panda single-mode fiber during the drawing process, rather than twisting it post-drawing. The rotation causes the bowtie or panda structure to rotate around the axis of the optical fiber, as depicted in Fig. 2.10a. Figure 2.10b presents a microscopic photograph of a spun fiber. After rotation, the stress bar of the panda PMF exhibits a "twisted pair" structure. The beat length of the Panda PMF is 10 mm, resulting in a linear birefringence of ($2\pi/10$) rad/mm, which is equivalent to

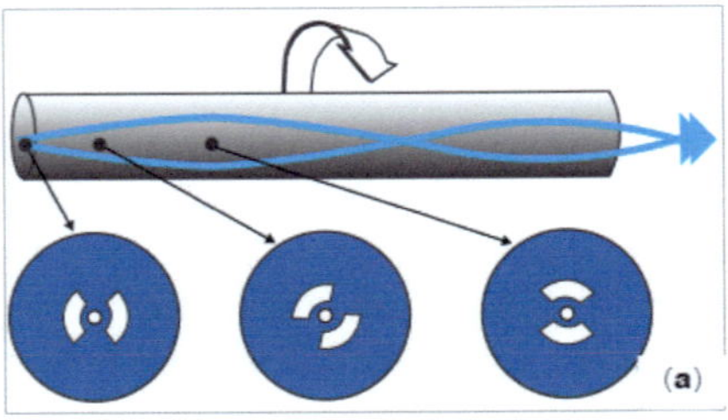

(a) Rotary bow-tie spun fiber

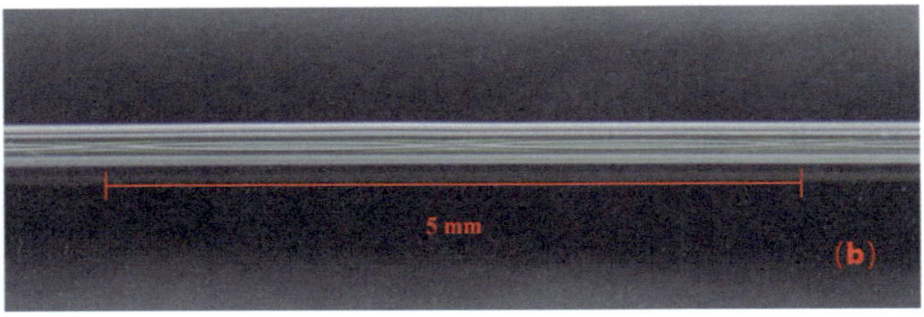

(b) Micrograph

Fig. 2.10 Rotary bow-tie spun fiber

0.2π rad/mm. The rotation speed is 0.4π rad/mm, leading to a "twisted pair" structure with a pitch of 5 mm.

Unlike traditional polarization-maintaining fibers (PMFs), spun fibers are specifically engineered to preserve circular polarization, and the output polarization remains unaffected by thermal noise, vibration noise, and drift resulting from stress birefringence. Therefore, spun fiber is suitable for high-sensitivity fiber optic current sensors (FOCS), also known as optical current transformers (OCT), which can be utilized for sensing both AC and/or DC currents. When used as FOCS or OCT, spun fiber offers several advantages over traditional methods: it enables polarization rotation within the optical fiber and remains unaffected by external electric fields.

The common theoretical analysis of a spun fiber is based on differentiation and regards the optical fiber as infinitesimal birefringent layers that are rotated and spliced together, as depicted in Fig. 2.11.

Assuming the length of the optical fiber is L and the linear birefringence per unit length is δ, measured in radians per millimeter (rad/mm), the degree of optical fiber rotation is characterized by the number of turns per unit length, also referred to as $\xi = \frac{d\theta}{dL}$, in units of rad/mm. Imagine the optical fiber is divided equally into N parts, each functioning as a combination of a linear retarder sheet and a circular retarder sheet; the length of each segment is $\frac{L}{N}$ corresponding to a phase delay of $\frac{\delta L}{N}$, and the rotation angle of each sheet relative to the preceding one is $\frac{\xi L}{N}$. Therefore, the transmission matrix of the entire optical fiber can be expressed as:

$$\boldsymbol{J} = \lim_{N\to\infty} R(-\xi L)\left[R\left(\frac{\xi L}{N}\right)J_0\left(\frac{\delta L}{N}\right)\right]^N$$

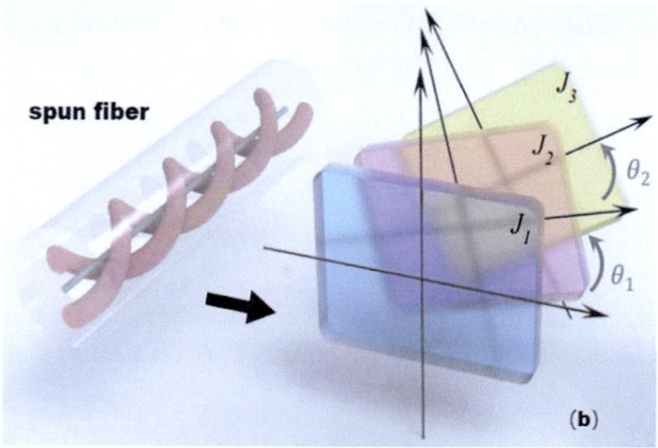

Fig. 2.11 Theoretical analysis model of a spun fiber

wherein,

$$R\left(\frac{\xi L}{N}\right) = \begin{bmatrix} \cos\left(\frac{\xi L}{N}\right) & \sin\left(\frac{\xi L}{N}\right) \\ -\sin\left(\frac{\xi L}{N}\right) & \cos\left(\frac{\xi L}{N}\right) \end{bmatrix},\quad J_0\left(\frac{\delta L}{N}\right) = \begin{bmatrix} e^{-j\frac{\delta L}{2N}} & 0 \\ 0 & e^{j\frac{\delta L}{2N}} \end{bmatrix}$$

After simplifying by Chebyshev formula, we get

$$J = \begin{bmatrix} \cos YL - j\frac{\delta}{2Y}\sin YL & \frac{\xi}{Y}\sin YL \\ -\frac{\xi}{Y}\sin YL & \cos YL + j\frac{\delta}{2Y}\sin YL \end{bmatrix}$$

where $Y = \sqrt{\xi^2 + \left(\frac{\delta}{2}\right)^2}$ denotes the elliptical birefringence per unit length caused by the rotation of the fiber geometry structure in a spun fiber, measured in rad/mm.

2.5.2 *Optical Fiber Polarization Devices*

All-fiber polarization devices are widely used due to their small size, ease of connection with fiber systems, and low connection loss. Typical all-fiber polarization devices include polarization controllers (PC), fiber polarizers, isolators, and depolarizers.

Optical Fiber Polarization Controller

Wave plates are the most commonly used devices in optical systems to alter the polarization state of light waves. In optical fiber systems, a simpler alternative involves utilizing the photoelastic effect of the optical fiber itself to change its birefringence, thereby controlling the polarization state of light waves within the optical fiber.

When an optical fiber is bent in the x–z plane, the refractive index along the x and y axes of the fiber changes due to stress, and the amplitude of this change is

$$\begin{aligned}\Delta n_x &= \frac{n^3}{4}(p_{11} - 2\mu p_{12})\left(\frac{r}{R}\right)^2 \\ \Delta n_y &= \frac{n^3}{4}(p_{11} - \mu p_{11} - \mu p_{12})\left(\frac{r}{R}\right)^2\end{aligned} \tag{2.149}$$

where r is the fiber radius; R is the radius of curvature of the fiber; p_{ij} is the photoelastic tensor of fused silica; μ is Poisson's ratio, then

$$\delta n = \Delta n_x - \Delta n_y = -0.133\left(\frac{a}{R}\right)^2 \tag{2.150}$$

From Eq. (2.150), it is evident that for the curved optical fiber depicted in Fig. 2.12a, the fast axis lies within the bending plane, while the slow axis is perpendicular to it. Therefore, a waveplate can be made by exploiting the birefringence effect of the curved optical fiber. For N-turn optical fibers with a bending radius R, if appropriate N and R are chosen so that the phase delay caused by the bending satisfies,

$$\delta = \left(\frac{2\pi}{\lambda}\right) \cdot 2\pi\alpha N\frac{r^2}{R} = \left(\frac{2\pi}{\lambda}\right) \cdot \frac{\lambda}{m} \tag{2.151}$$

Among them, α and λ are the photoelastic coefficient and operating wavelength of the optical fiber respectively, and m is a positive integer, then the optical fiber loop becomes a λ/m waveplate. For instance, consider a red light with a wavelength of 630 nm. When a fiber with a radius of 62.5 μm is wound into a single optical fiber loop with a radius (R) of 20.6 mm, it becomes a $\lambda/4$ wave plate. If the fiber is wound twice, it forms a $\lambda/2$ waveplate. This kind of fiber optic polarization controller is commonly known as a paddle type, which includes three-paddle or two-paddle configurations (refer to Fig. 2.12) [4].

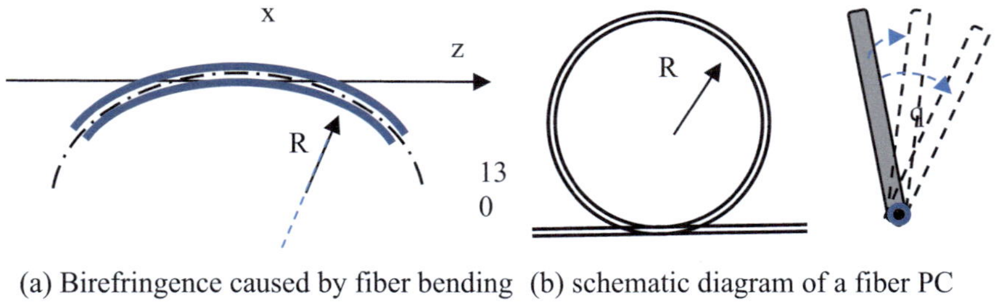

Fig. 2.12 Paddle optical fiber polarization controller

The working principle of the paddle fiber polarization controller, most commonly used in laboratories, is illustrated in Fig. 2.12b. A change in the angle of the optical fiber loop alters the orientation of the birefringence axis plane within the optical fiber, which is analogous to the rotation of the polarization axis of a wave plate. Thus, by incorporating this optical fiber loop into the optical fiber system and adjusting the angle of the fiber loop accordingly, the birefringence state of light waves can be controlled.

Commonly used polarization controllers typically consist of $\lambda/4$ and $\lambda/2$ fiber loops. Figure 2.13 illustrates a three-paddle type, while Fig. 2.14 depicts a squeeze-type device diagram of an optical fiber polarization controller. Properly adjusting the relative angle of the optical fiber loop in the paddle polarization controller or the pressure/rotation control knob of the online squeeze polarization controller functions as an adjustable wave plate, allowing for the generation of linearly polarized light in any vibration direction.

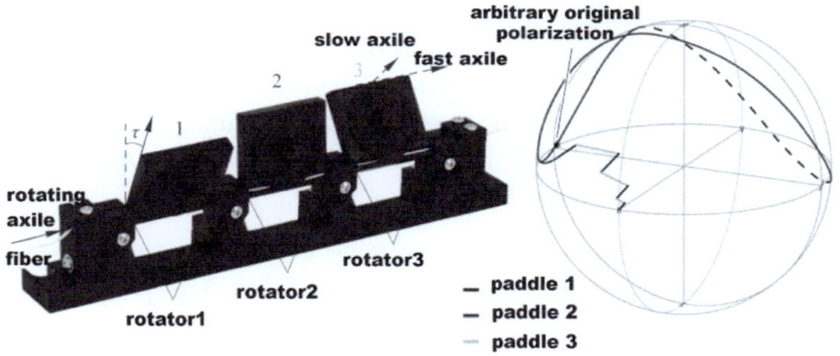

(a) Device diagram (b) Variation of polarization state with rotating paddle

Fig. 2.13 3 paddle fiber polarization controller[4]

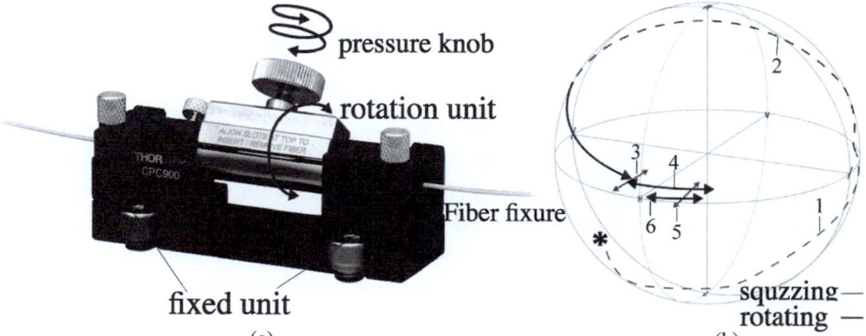

(a) (b)

Fig. 2.14 Squeezed optical fiber polarization controller [4] original arbitrary polarization:1-

Figure 2.14 illustrates the squeezed fiber polarization controller. The stress-modulated birefringence can be estimated using the following equation: Assuming the fiber diameter is d and the force is F, the phase delay generated in the quartz fiber by light with an operating wavelength of λ.

$$\delta \sim 6 \times 10^{-11} \frac{F}{\lambda d}$$

For the fiber segment in the twisted region, the angle of rotation is $\theta = \alpha\tau, \alpha = -n^2 p_{44}$. Here τ represents the rotation angle of the fiber clamp, n is the refractive index of fiber core, and p_{44} is the photoelastic coefficient of the optical fiber. The polarization extinction ratio of this device is approximately 30 dB.

Polarization Maintaining Fiber Polarizer

The design concept behind using a high birefringence fiber to create an optical fiber polarizer involves allowing one of the two polarization components to leak out from the fiber cladding of the high birefringence fiber (resulting in high loss), while the other polarization component can be transmitted with minimal loss through the fiber, thus achieving single-polarized light at the fiber's output end. A specific structural example is as follows.

Example 2.5.1 Metallization Method: Due to the high imaginary component of the metal's dielectric constant, one of the polarization components is absorbed (essentially a loss) to create an optical fiber polarizer. The device structure is shown in Fig. 2.15a. The quartz substrate features an arc-shaped groove on its top, where the PM fiber is secured along its axis and bonded. Once grounded and polished to the optical field area, the upper surface of the PM fiber is coated with a metal film, creating a dielectric-metal composite waveguide. When polarized light in the optical fiber reaches this area, the TM mode can excite a surface wave on the dielectric-metal interface, causing the optical energy to couple from the optical fiber to the dielectric-metal composite waveguide and leak out. In contrast, the TE wave does not undergo such coupling and passes through this region with minimal loss, resulting in single TE polarized light at the output end.

The key technologies involved in fabricating a PM fiber polarizer using this method encompass: fiber positioning, grinding depth monitoring, thin film evaporation, and performance testing. To guarantee an extinction ratio of 40 dB, the axis error of the polarizer must not exceed 0.5 degrees; moreover, the grinding depth requires precise control, which is achieved by measuring the leakage light power. The current performance of the PM fiber polarizer manufactured using this method achieves: ER > 35 dB, insertion loss < 0.5 dB [2].

Example 2.5.2 In example 2.5.1, the metal film within the structure is substituted with a birefringent substrate, allowing one of the polarization components to leak

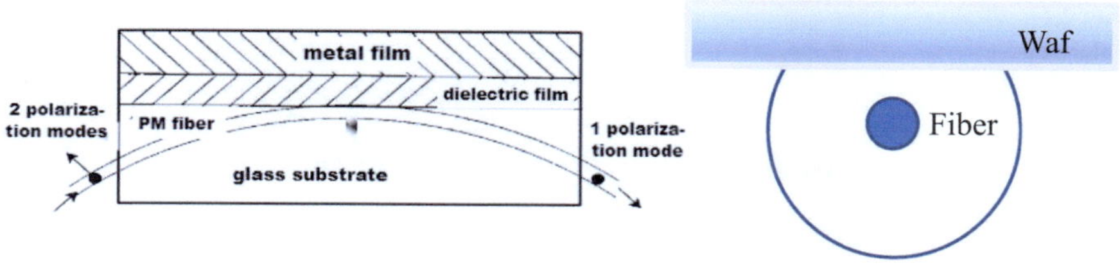

(a) metal plating film; (b) birefringent wafer

Fig. 2.15 Optical fiber polarizer composed of different materials

out. The new device structure is depicted in Fig. 2.15b, [3]. In this structure, one refractive index of the birefringent wafer is greater than that of the fiber core, while the other is less. Consequently, one polarization component of the transmitted light in the optical fiber leaks out, while the other continues to propagate within the fiber. For example, $KB_5O_3.4H_2O$ crystal is cut in the direction perpendicular to the *b*-axis, as shown in Fig. 2.15b. For red light of 0.633 μm: $n_a = 1.49$, $n_b = 1.43$, $n_o = 1.42$, while for the quartz fibers, $n_{co} = 1.456$. Thus, for light vibrating perpendicular to the crystal-fiber interface, the refractive index $n_b = 1.43$; for light vibrating parallel to the interface, its refractive index is $n_o = 1.42$. When designing a polarizer, the included angle θ of the crystal should be selected such that one of the polarization components effectively leaks from the guided mode. The calculation equation for this angle θ is as follows:

$$n_{\mathrm{co}} = \left(\sqrt{\frac{\sin^2\theta}{n_b^2} + \frac{\cos^2\theta}{n_a^2}}\right)^{-1} \tag{2.152}$$

Ultimately, the optimal extinction ratio is achieved through fine-tuning. The actual device extinction ratio can reach up to 60 dB.

Example 2.5.3 The structure of the special-shaped fiber polarizer is depicted in Fig. 2.16. It differs from standard communication optical fibers in that it features a D-shaped cavity within the cladding, which is filled with metal. The loss difference between the two orthogonal components transmitted through the fiber core can exceed 20 dB with this special-shaped metal cladding. The polarizer achieves an extinction ratio of 30 dB and an insertion loss of 1 dB across a wide spectrum ranging from 1.3 to 1.5 μm.

The in-line polarizer utilizes PZ optical fiber to permit the transmission of light in only one polarization direction—either the fast or slow axis—while blocking the transmission of light in other polarization directions. It is used to convert unpolarized light into polarized light with a high extinction ratio, thereby enhancing the polarization extinction ratio of the signal. The device operates within a wavelength range of 480 to 2000 nm.

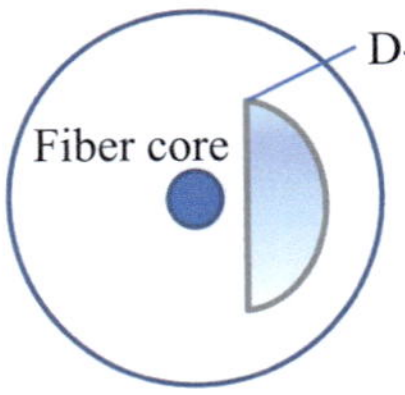

Fig. 2.16 D-shaped fiber polarizer

Fiber Optic Isolator and Phase Shifter

1. Isolator

The optical fiber isolator can be constructed using the Faraday effect of fiber materials. The polarization plane rotation angle θ of linear polarized light in an optical fiber equals the product of the fiber length L within the magnetic field, the Verdet constant of the fiber material, and the magnetic field strength H. The primary issue with fiber isolators is that the Verdet constant of current low-loss optical fiber materials (such as quartz) is very small (~ 0.0124 min/cm·Oe). Therefore, it is necessary to place a long optical fiber within a strong magnetic field to achieve a rotation angle of 45°. Taking fused silica optical fiber as an example, if H equals 10,000 Oe, then a polarization plane rotation angle of 45° can only be achieved with a fiber length in the magnetic field of approximately 2 m. Currently, optical fiber isolators are primarily made from crystalline materials with a high Verdet constant.

Figure 2.17 depicts a quartz isolator [4]. The isolator's magnetic field consists of 14 permanent magnets. Each magnet has a thickness of 0.53 cm, and the center distance between two adjacent magnets with opposite polarities is 1.65 cm, which is precisely half the beat length of the fiber (optical fiber beat length Lp = 3.30 cm). Additionally, there is a 0.03 cm wide slot, curved within the magnet, for positioning the optical fiber. The magnetic field intensity in this slot, oriented along the fiber direction, is 5 kG. The optical fiber traverses back and forth 9 times through the aforementioned magnets in series to minimize the number of permanent magnets required in the isolator. The total length of the optical fiber is 7 m, with 2 m of that length within the magnetic field region. The core diameter measures 3.3 μm, the operating wavelength is 0.633 μm, and the loss is 20 dB/km. When linearly polarized light is incident along the x direction (with the optical fiber's extinction ratio being 30 dB), the device is adjusted to achieve an optical power output of $P_x = P_y$ from the isolator, indicating that the vibration direction of the output light has rotated by 45°. At this juncture, if a polarizer is present at the incident end, it forms an isolator; in the absence of a polarizer, a circular polarizer is created. The extinction ratio of the isolator is 20 dB. Given that the beat length of optical fiber varies with temperature, it is imperative to maintain temperature control. Additionally, temperature fluctuations can be utilized for fine-tuning the device.

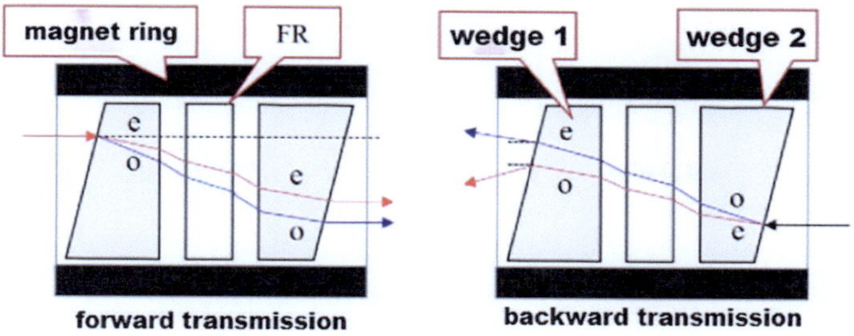

(a) Device structure diagram

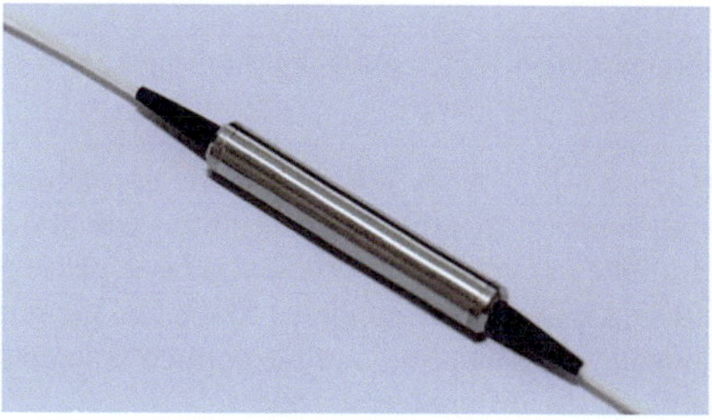

(b) Product appearance

Fig. 2.17 All-fiber isolator

2. Polarization Maintaining Fiber Phase Shifter

The polarization-maintaining phase shifter, also known as a retarder, utilizes the nonreciprocity of the forward and reverse optical rotation angles when polarized light passes through the Faraday rotator crystal. This results in a fixed optical path difference between the forward and reverse optical paths through the same device (refer to Fig. 2.18). The fiber phase shifter boasts advantages such as low insertion loss, high stability, and a high extinction ratio, making it suitable for applications in fiber lasers, fiber sensing, and other related fields.

Fiber Optic Photo-Electronic Modulator

1. All-Fiber Modulator

The Kerr effect in optical fiber can be utilized to create an optical fiber phase modulator, also known as a Kerr effect fiber phase modulator, with the structure depicted in Fig. 2.18a. Two metal electrodes are made in the cladding areas on both sides of the fiber core, and the electrode materials are indium/gallium mixture. When a voltage is applied to the electrode, birefringence occurs in the fiber core due to the Kerr effect. The magnitude of this effect is proportional to the square of the applied

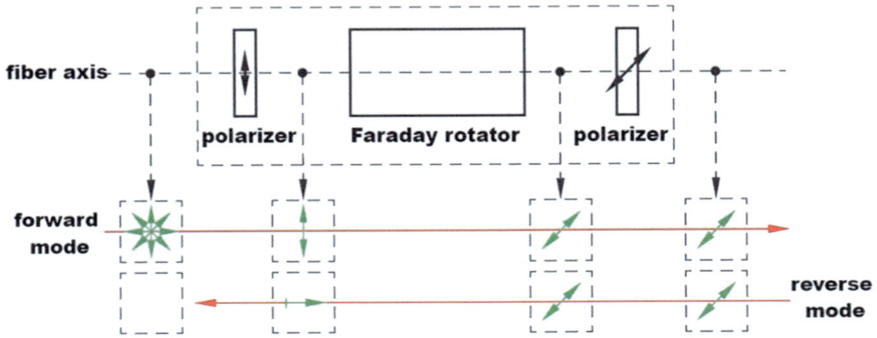

Fig. 2.18 Schematic diagram of optical fiber phase shifter principle

electric field E as $B_A = \delta\beta/\beta = KE^2$. Where K is the normalized Kerr constant of the material. Although the Kerr effect in quartz is quite weak, the extended length of optical fibers can be utilized to achieve a sufficiently large phase shift. Figure 2.19b illustrates the relationship between the applied voltage and the phase shift in a 30 m optical fiber. The frequency and amplitude of the applied voltage are 2 kHz and 5 V, respectively, to obtain a phase difference of 150°.

2. Optical Fiber Photo-Electrical Modulator Based on $LiNbO_3$ Crystal

Due to the small photo-electrical coefficient of quartz, LiNbO3 crystal is currently the most widely used optical waveguide material for fiber-coupled electro-optic modulators. Figure 2.20 depicts a LiNbO3-based electro-optic intensity modulator. Figure 2.20a illustrates the cross-sectional structure of the electro-optical modulator, while the waveguide (depicted in blue) in Fig. 2.20b is split into two paths embedded within the LiNbO3 surface (shown in green). The incident light is first modulated by a radio frequency source, the RF driver, and then by a direct current, the DC bias.

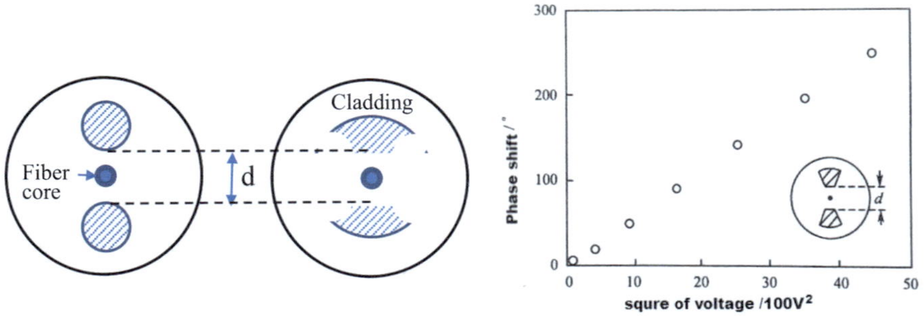

(a) transverse interface structure diagram (b) applied voltage-phase shift curve

Fig. 2.19 Kerr effect fiber modulator

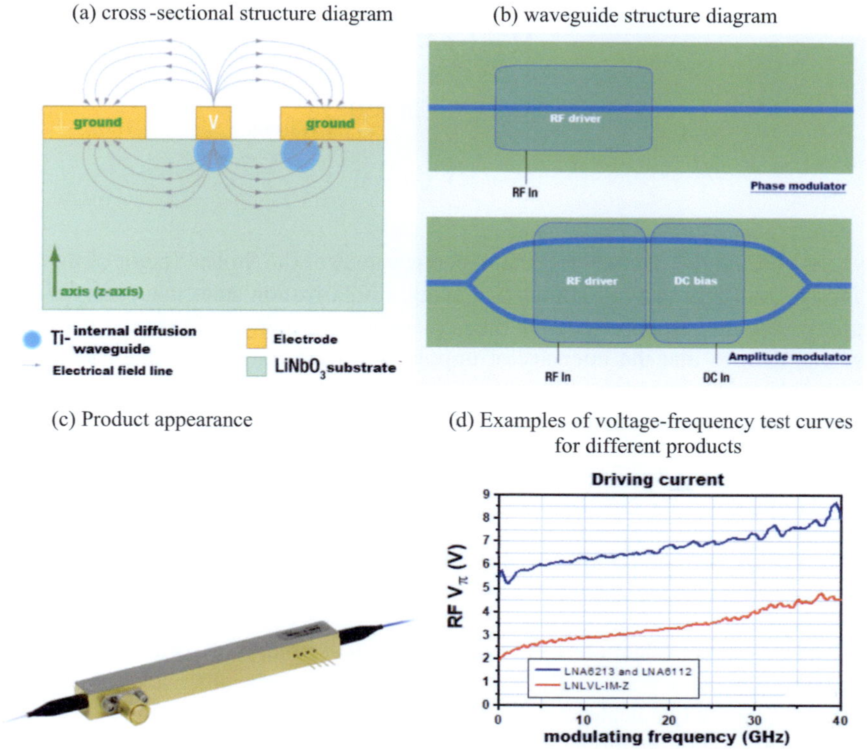

Fig. 2.20 Fiber-coupled electro-optic modulator [5]

Whether it is an intensity modulator or a phase modulator, its high-frequency performance is limited by the current waveguide technology. Currently, the modulation frequency of commercial electro-optical modulators generally ranges from 10 to 40 GHz.

Depolarizer and Scrambler

A depolarizer is a device that transforms polarized light into unpolarized light. Linearly polarized light is often unnecessary for many applications. For instance, polarization can influence the sensitivity of the detector in reflective spectrometers. Depolarizers convert polarized light into unpolarized light and are widely used in polarization-insensitive systems, the working principles of which are more complicated than those of fiber polarizers. To understand the principle of a depolarizer, we must first clarify the definition of unpolarized light.

Unpolarized light is characterized by a time average of its polarization components being zero. This concept was initially proposed by Perrin, and its mathematical expression is

$$\left.\begin{aligned} \mathrm{M} &= \frac{1}{T}\int_0^T M dt = 0 \\ \mathrm{C} &= \frac{1}{T}\int_0^T C dt = 0 \\ \mathrm{S} &= \frac{1}{T}\int_0^T S dt = 0 \end{aligned}\right\} \tag{2.153}$$

where M, C, and S denote the instantaneous values of the Stokes vector of the light. Unpolarized light can be defined as having a polarization state that changes more rapidly than the response speed of the photodetector employed. From this definition, we can deduce that the intensity of unpolarized light remains unchanged when it passes through a rotating polarizer or a wave plate. For unpolarized light of unit intensity, the Stokes vector is:

$$\begin{bmatrix} I \\ M \\ C \\ S \end{bmatrix} = \begin{bmatrix} 1 \\ 0 \\ 0 \\ 0 \end{bmatrix}$$

The depolarizer functions by dividing the input beam into two orthogonally polarized beams of equal power, delaying one beam relative to the other, and subsequently recombining the two beams using a polarization beam combiner. The relative delay between the orthogonally polarized beams exceeds the coherence length of the light source, thereby eliminating the fixed phase relationship between the beams upon recombination. Depolarizers can be categorized into two types based on their working principle: monochromatic depolarizers and broadband depolarizers.

1. Monochromatic Depolarizer

The ideal performance of a monochromatic depolarizer permits only light of the operating wavelength to pass through without attenuation. The output light, regardless of its polarization state, becomes unpolarized after passing through the depolarizer. Theoretically, two optical retarders with different delays can be connected in series to create a depolarizer. The following will demonstrate that the two retarders will form a monochromatic depolarizer as the delays of both retarders vary from 0 to 2π, and the frequency ratio of the two retarders is 2:1. The Muller matrix is employed to verify the aforementioned depolarization principle.

Suppose a voltage is applied to two electro-optical crystal (such as ADP) retarders to continuously change the delay of the light wave. At this time, the delay change over time may be linear or a sinusoidal function. The angle between the crystallographic orientations of the two crystals is 45°. Assume $\boldsymbol{D}$ is the matrix of the depolarizer, $\boldsymbol{I}_1$, $\boldsymbol{M}_1$, $\boldsymbol{C}_1$, $\boldsymbol{S}_1$ are the Stokes vectors of any polarized light incident on the depolarizer, the output lightwave of the depolarizer satisfies the following equation:

$$\boldsymbol{D}\begin{bmatrix} \boldsymbol{I}_1 \\ \boldsymbol{M}_1 \\ \boldsymbol{C}_1 \\ \boldsymbol{S}_1 \end{bmatrix} = \begin{bmatrix} \boldsymbol{I}_2 \\ \boldsymbol{M}_2 \\ \boldsymbol{C}_2 \\ \boldsymbol{S}_2 \end{bmatrix} \quad \text{or} \quad \boldsymbol{R}_2(n\delta, \beta)\boldsymbol{R}_1(\delta, 0)\begin{bmatrix} \boldsymbol{I}_1 \\ \boldsymbol{M}_1 \\ \boldsymbol{C}_1 \\ \boldsymbol{S}_1 \end{bmatrix} = \begin{bmatrix} \boldsymbol{I}_2 \\ \boldsymbol{M}_2 \\ \boldsymbol{C}_2 \\ \boldsymbol{S}_2 \end{bmatrix} \tag{2.154}$$

where $\boldsymbol{I}_2$, $\boldsymbol{M}_2$, $\boldsymbol{C}_2$, $\boldsymbol{S}_2$ are the Stokes vectors of the output light. $\boldsymbol{R}_2(n\delta, \beta)$ and $\boldsymbol{R}_1(\delta, 0)$ respectively represent the two depolarizer matrices that constitute the depolarizer, δ is the delay amount, β is the angle between the two delays, n is a positive integer, and there are $\boldsymbol{D} = \boldsymbol{R}_2(n\delta, \beta)\boldsymbol{R}_1(\delta, 0)$. Equation (2.154) presumes that the crystallographic orientation of the first retarder aligns with the system's reference direction, without loss of generality.

Depolarizers are designed to convert polarized light into unpolarized light. Consequently, an analyzer is necessary to verify that the output light from the depolarizer conforms to the definition of unpolarized light. Let A represent the matrix of the analyzer, then:

$$\boldsymbol{A}\begin{bmatrix} I_2 \\ M_2 \\ C_2 \\ S_2 \end{bmatrix} = \begin{bmatrix} I_3 \\ M_3 \\ C_3 \\ S_3 \end{bmatrix}$$

where $\boldsymbol{I}_3$, $\boldsymbol{M}_3$, $\boldsymbol{C}_3$, $\boldsymbol{S}_3$ represent the Stokes vector of the output light after passing through the analyzer. As defined by unpolarized light, the output light $\boldsymbol{I}_3$ should be independent of the parameters $\boldsymbol{I}$, $\boldsymbol{M}$, $\boldsymbol{C}$ and $\boldsymbol{S}$ of the incident light, and should also be independent of A. Therefore, the time average of $\boldsymbol{M}_2$, $\boldsymbol{C}_2$, $\boldsymbol{S}_2$ is zero:

$$\frac{1}{\tau}\int_0^\tau M dt = \frac{1}{\tau}\int_0^\tau D_{21} dt + \cdots = 0 \tag{2.155}$$

where D_{ij} is the instantaneous value, and the observation time is t. This Eq. (2.155) shows that the characteristic parameters of D should vary with time and satisfy Eq. (2.155) to function as a depolarizer. Consequently, many methods can theoretically construct a depolarizer. Among them, the simplest one consists of two wave plates connected in series, with their crystallographic orientations forming a specific included angle β, and the retardations being d and $n\delta$ respectively. At this juncture, the matrix expression for the combination of the two wave plates is

$$\begin{bmatrix} 1 & 0 & 0 & 0 \\ 0 & \cos^2 2\beta + \sin^2 2\beta \cos n\delta & \begin{array}{l} \sin 2\beta \cos 2\beta(1 - \cos n\delta)\cos\delta \\ + \sin 2\beta \sin n\delta \sin\delta \end{array} & \begin{array}{l} -\sin 2\beta \cos 2\beta(1 - \cos n\delta)\sin\delta \\ + \sin 2\beta \sin n\delta \cos\delta \end{array} \\ 0 & \sin 2\beta \cos 2\beta(1 - \cos n\delta) & \begin{array}{l} \cos\delta\left[\sin^2 2\beta + \cos^2 2\beta \cos n\delta\right] \\ + \sin\delta(-\cos 2\beta \sin\delta) \end{array} & \begin{array}{l} -\sin\delta\left[\sin^2 2\beta + \cos^2 2\beta \cos n\delta\right] \\ + \cos\delta(-\cos 2\beta \sin n\delta) \end{array} \\ 0 & -\sin 2\beta \sin n\delta & \cos\delta \cos 2\beta \sin n\delta - \sin\delta \cos n\delta & -\sin\delta \cos 2\beta \sin n\delta + \cos\delta \cos n\delta \end{bmatrix}$$

Where δ is a periodic function of time. From Eq. (2.155), there is

$$\overline{D}_{22} = \frac{1}{T}\int\limits_{0}^{T}\left(\cos^2 2\beta + \sin^2 2\beta \cos n\delta\right)dt = 0$$

$$\text{or}\,\overline{D}_{22} = \cos^2 2\beta + \frac{1}{T}\int\limits_{0}^{T}\left(\sin^2 2\beta \cos n\delta\right)dt = 0$$

It is evident that β equals 45°, meaning the angle between the crystallographic orientations of the two retarders is 45°. When δ varies linearly over time, such that $\delta = Qt$ (where $0 \le t \le \tau$ and Q is a constant), there is $\overline{D}_{22} = \frac{1}{2\pi}\int_0^{2\pi} \cos n\delta d\delta = 0$. Similarly,

$$\overline{D}_{23} = \frac{1}{2\pi(n-1)}\sin\pi(n-1) + \frac{1}{2\pi(n+1)}\sin\pi(n+1)$$

Clearly, D_{23} is zero when $n = 2$. It can be proven that the time-averaged values of the remaining components of D are zero at this juncture, indicating that the delays of the retarders are in a 2:1 ratio and vary linearly with time, thus forming a monochromatic depolarizer.

2. Broadband Depolarizer

For broadband light, Lyot depolarizer is currently a more suitable type[5]. The optical Lyot depolarizer consists of two quartz wave plates connected in series. The thickness ratio of the two wave plates is 2:1, and the angle between the crystallographic axis is 45°(as Fig. 2.21 shown). The Lyot depolarizer, also known as a broadband depolarizer, differs from a monochromatic depolarizer in two key aspects: it is designed for wide spectral ranges rather than monochromatic light, and it features two fixed quartz plates whose retardation remains constant over time.

The proof procedure for the Lyot depolarizer is similar to that of the monochromatic depolarizer, except that the time average is replaced by a wavelength average. Likewise, it is proven that two relatively fixed retarders can form a broadband depolarizer, with the angle between the crystallographic orientations of the two retarders being 45°. The difference between the retardations is an integer multiple of d (typically $n = 2$), and d is sufficiently large. This forms the fundamental principle of the Lyot depolarizer ($n = 2$). The error of such a depolarizer is dependent on the error in the ratio n of the retardations of the two delays. It is evident that an all-fiber Lyot depolarizer can also be constructed using a high birefringent optical fiber [6].

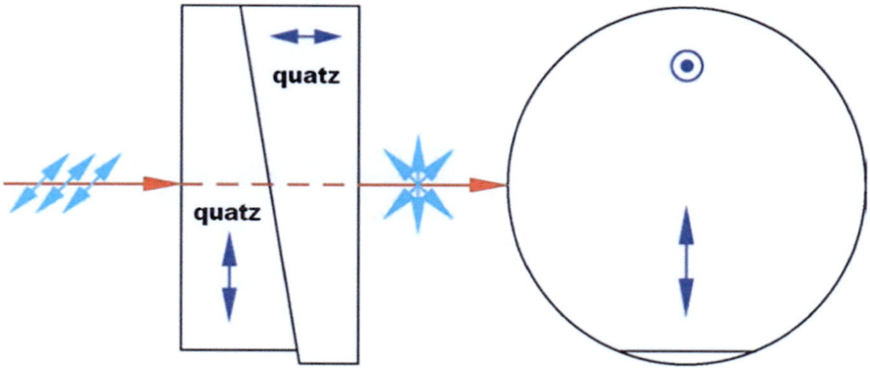

Fig. 2.21 Schematic diagram of Lyot depolarizer

The all-fiber Lyot depolarizer, as depicted in Fig. 2.22, generates a pseudo-random polarization state output. Any input polarization state (SOP) results in unpolarized light with a low degree of polarization (DOP), effectively eliminating the polarization of lasers with a coherence length of up to 10 m. A passive device that utilizes the coherence of a light source to generate random polarization states for depolarization purposes is commonly employed in applications with high depolarization requirements, such as ASE broadband light sources and high-power pump amplifier lasers.

3. Fiber Optic Scrambler

The fiber optic scrambler depicted in Fig. 2.23a operates in a manner analogous to the online squeeze polarization controller detailed in Sect. 2.5.2.1, as shown in Fig. 2.14b. This is based on the principle that the polarization state of fiber changes under the influence of strain. It consists of a squeeze polarization controller and a high-speed scrambler driver, capable of randomly homogenizing any input polarization state. The polarization controller consists of four electro-mechanical optical fiber extruders, and the scrambling circuit generates multi-frequency control signals. The scrambler is essentially equivalent to a cascade of several birefringent modulators operating at different frequencies, thereby achieving the homogenization of the

Fig. 2.22 Fiber optic Lyot depolarizer

Fig. 2.23 Typical scrambler types

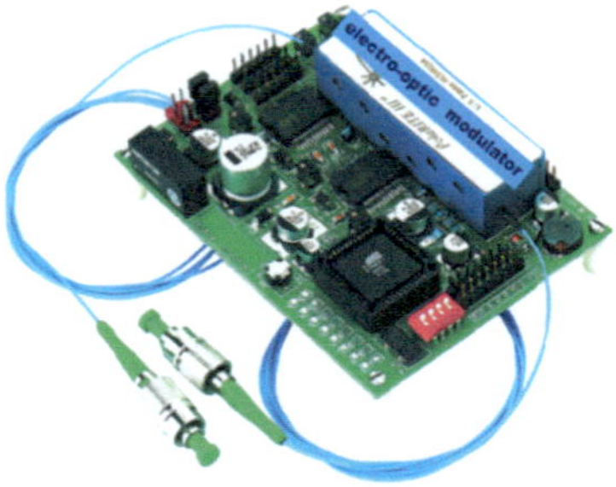

(a) mechanically squeezed scrambler

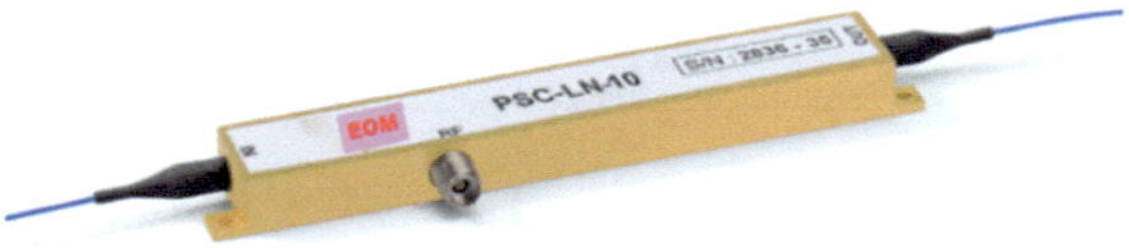

(b) high-speed $LiNiO_3$ Electro-optical scrambler

output polarization state. This structure boasts the advantages of low insertion loss, low return loss, and a wide operating spectrum. The scrambling frequency typically ranges in the several hundred kHz.

Another integrated optical device, the high-speed $LiNiO_3$ electro-optical scrambler (Fig. 2.23b), is manufactured to further reduce the loss and fiber deterioration caused by the vertical extrusion of the fiber. The scrambler features a low-loss single-mode waveguide capable of modulating the polarization state within a frequency range of DC to 10 GHz, boasting a bandwidth of 100 nm. Its working principle is akin to that of an electro-optic modulator. The linear polarization mode is decomposed into TE and TM modes at the input end of the fiber, and is applied with a sawtooth wave or sinusoidal voltage to generate the phase delay between the two modes through the linear electro-optic effect. The device is capable of stable depolarization while maintaining a residual degree of polarization.

The multimode fiber scrambler is primarily utilized to homogenize the intensity distribution of the far-field spot of multimode fiber, as depicted in Fig. 2.24.

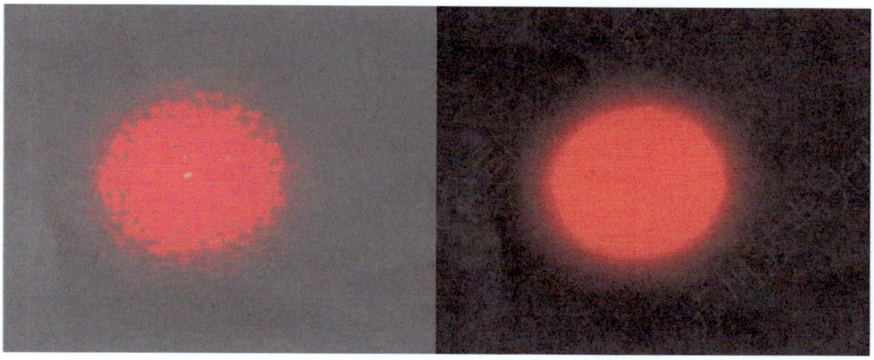

Fig. 2.24 Comparison of output spot homogenization of multimode fiber before (left) and after (right) loading scrambler

References

1. Lee K-J (2010) Handbook of optics. Shaanxi Science and Technology Press, Xian
2. Libo Y, Yang J (2016) Principles and applications of optical fiber white light interference. Science Press, Beijing
3. Hu Y (1999) Research on polarization-maintaining fiber polarizer. PhD thesis, Tsinghua University
4. Turner EH et al (1981) Opt Lett 6(7):322–323
5. http://www.thorlabschina.cn/
6. Fengping Y et al (1995) J Northern Jiaotong Univ 19(3):314–319

Chapter 3
Types and Optical Characteristics of Optical Fiber Interferometers

3.1 Optical Fiber Interferometer: Phase Modulation Optical Fiber Sensor

A fiber optical sensor that operates based on the phase change of light waves due to external physical quantities is known as a phase modulation optical fiber sensor. In this type of sensor, phase modulation is primarily achieved through the use of a fiber interferometer, which is a method of interferometry. Optical fiber interferometry is a measurement method characterized by high sensitivity and a very large dynamic range. In principle, optical fiber interferometers can be categorized into dual-beam interferometers and multi-beam interferometers. Common interference structures include the Mach-Zehnder interferometer (MZI), Michelson interferometer, Fabry-Perot interferometer (FPI), ring resonator interferometer, and Sagnac interferometer, among others, all of which have been utilized in optical fiber sensing applications. The main features of phase modulation fiber optic sensors include:

- High sensitivity

Optical interferometry is one of the most sensitive detection techniques known. Fiber optic interferometers are more sensitive than ordinary optical interferometers because they utilize optical fibers that are several meters or even hundreds of meters in length, eliminating the need for moving parts and providing a stable, solid structure.

- Flexible structures

Since the sensitive unit of a fiber optic sensor is the fiber itself, the geometry of the probe can be flexibly designed to meet various needs.

- Diverse measurands

Any physical quantity that influences the optical path of the interferometer can be utilized for sensing purposes. Currently, fiber optic sensors have been developed to measure various physical quantities, such as pressure (including underwater sound),

Y. Liao et al., *Fiber Optic Interferometer*, Springer-TUP Physics Series,
https://doi.org/10.1007/978-981-95-2558-4_3

temperature, acceleration, current, magnetic field, and liquid composition, using various types of fiber optic interferometers. Furthermore, a single interferometer can frequently sense multiple physical quantities.

- Special requirements for optical fiber

To achieve interference, firstly, the light of the same mode must be superimposed, for which a single-mode fiber should be utilized. Admittedly, a certain interference pattern can also be achieved using multimode fiber, but performance significantly diminishes and signal detection becomes more challenging. Secondly, to obtain the optimal interference effect, the vibration directions of the two coherent lights must be aligned. It is preferable to use "high birefringence" single-mode fiber in fiber interferometers. Research indicates that the materials of the fiber, particularly the coating and outer cladding, significantly impact the sensitivity of the optical fiber interferometer. Consequently, to enhance the sensitivity of the fiber interferometer to the measured physical quantities and reduce sensitivity to those not being measured, single-mode optical fiber must undergo a special process to meet the requirements for various physical quantities. When studying fiber optical interferometers, particular attention should be given to the performance of the optical fiber employed.

The following text introduces the optical characteristics of several typical fiber optic interferometers, which are categorized into two groups: double-beam interferometers and multi-beam interferometers, based on the number of beams involved in the interference process.

3.2 Double Beam Interferometer

3.2.1 Mach-Zehnder Interferometer

Both the Mach-Zehnder interferometer (hereinafter referred to as MZI) and the Michelson interferometer are two-beam interferometers. The structure of the MZI is depicted in Fig. 3.1. The monochromatic light emitted from the light source S passes through the coupler C_1, where it is divided into two beams of light. Arm a carries the signal light, while arm b carries the reference light. After traversing their respective optical paths, a and b, they are recombined at the coupler C_2, where interference occurs and is detected by photodetectors D_1 and D_2. The incorporation of optical fiber has effectively resolved the discrete issues of traditional optical paths, enabling the realization of a fully integrated optical fiber interferometer structure.

Assuming the polarization states of the optical signals in the two arms remain constant and parallel, the amplitudes of the two output lights from the second coupler C_2 are respectively, [1]:

$$\begin{aligned} E_1 &= k_{2c} \exp(i\varphi_b) k_{1c} E_0(\tau_b) + k_{2t} \exp(i\varphi_a) k_{1t} E_0(\tau_a) \\ E_2 &= k_{2t} \exp(i\varphi_b) k_{1c} E_0(\tau_b) + k_{2c} \exp(i\varphi_a) k_{1t} E_0(\tau_a) \end{aligned} \tag{3.1}$$

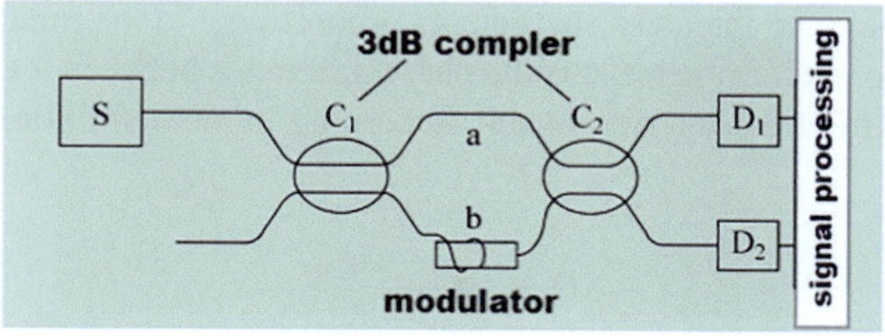

Fig. 3.1 Optical fiber structure diagram of MZ interferometer

where k_{ij} is the beam splitting ratio of the coupler, $i = 1, 2$ is the number of the couplers, $j = t, c$ represents the through arm and the coupling arm; $E_0(\tau_\mathrm{a})$ and $E_0(\tau_\mathrm{b})$ are the incident light amplitudes, ϕ_a and ϕ_b are the phase corresponding to the optical paths of the two arms. Since the cross arm of the coupler has an additional $\pi/2$ phase difference relative to the through arm, generally the beam splitting ratio of the through arm is expressed as a real number, while the beam splitting ratio of the coupling arm is expressed as an imaginary number, then $k_\mathrm{c} = \mathrm{i}k'_\mathrm{c}$, k'_c is a real number.

Because $I = \langle E \cdot E^* \rangle$ (I is the light intensity), it is obtained

$$\begin{aligned} I_1 =& k_{1c}^{\prime 2} k_{2c}^{\prime 2} \langle E_0^2(\tau_b) \rangle + k_{1t}^2 k_{2t}^2 \langle E_0^2(\tau_a) \rangle \\ &+ 2\,Re\left[-k_{1c}' k_{2c}' k_{1t} k_{2t} exp(i(\phi_a - \phi_b))\right] \langle E_0(\tau_b) \cdot E_0^*(\tau_a) \rangle \end{aligned} \tag{3.2}$$

The coherence intensity of the light source is defined as [1]:

$$\gamma(\tau_a - \tau_b) = \frac{\langle E_0(\tau_b) \cdot E_0^*(\tau_a) \rangle}{I_0}$$

Then

$$I_1 = I_0\left[k_{1c}^{\prime 2} k_{2c}^{\prime 2} + k_{1t}^2 k_{2t}^2 - 2k_{1c}' k_{2c}' k_{1t} k_{2t} \gamma \cos(\varphi_a - \varphi_b)\right] \tag{3.3}$$

In a similar way

$$I_2 = I_0\left[k_{1c}^{\prime 2} k_{2c}^{\prime 2} + k_{1t}^2 k_{2t}^2 + 2k_{1c}' k_{2c}' k_{1t} k_{2t} \gamma \cos(\varphi_a - \varphi_b)\right] \tag{3.4}$$

i.e.,

$$I_1 = A - B\cos\phi \tag{3.5}$$

$$I_2 = A + B\cos\phi \tag{3.6}$$

where A and B are the intensity coefficients proportional to the input light intensity of the fiber optic interferometer; ϕ is the phase difference between the signal arm and the reference arm of the interferometer. According to the definition of interference contrast [1]:

$$Q = \frac{I_{\max} - I_{\min}}{I_{\max} + I_{\min}} \tag{3.7}$$

The contrast ratio of the two outputs can be found as follows:

$$Q_1 = Q_2 = \frac{B}{A} = \frac{2k'_{1c}k'_{2c}k_{1t}k_{2t}}{k'^2_{1c}k'^2_{2c} + k^2_{1t}k^2_{2t}}\gamma \tag{3.8}$$

Then, the contrast of the output light from the MZI is related to the beam splitting ratio and the coherence intensity of the coupler. According to Eq. (3.8), the calculation indicates that when the insertion loss of the coupler is 3dB (corresponding to a beam splitting ratio of 1:1), the interference contrast reaches its maximum.

Additionally, based on the phase difference of the output light wave of the MZ interferometer, which is $\phi = 2\pi nlv/c$, the phase difference change can be found as follows:

$$\Delta\varphi = \frac{2\pi nlv}{c}\left(\frac{\Delta n}{n} + \frac{\Delta l}{l} + \frac{\Delta v}{v}\right) \tag{3.9}$$

where c is the velocity of light in vacuum, n is the refractive index of the optical fiber, l is the length difference between the two arms of the interferometer, and v is the optical frequency.

3.2.2 Michelson Fiber Interferometer

A Michelson fiber interferometer comprises a fiber coupler C, a reference arm a, a signal arm b and fiber coated end faces D_a and D_b. Its structure and principle are shown in Fig. 3.2 as follows.

The light beam emitted by the monochromatic laser source LD is divided into two orthogonal light beams by a coupler C: a reference light a and a signal light b, which can be expressed as $E_a e^{j(\omega t + 2kl_0)}$ and $E_b e^{j(\omega t + 2kl)}$, respectively. Wherein

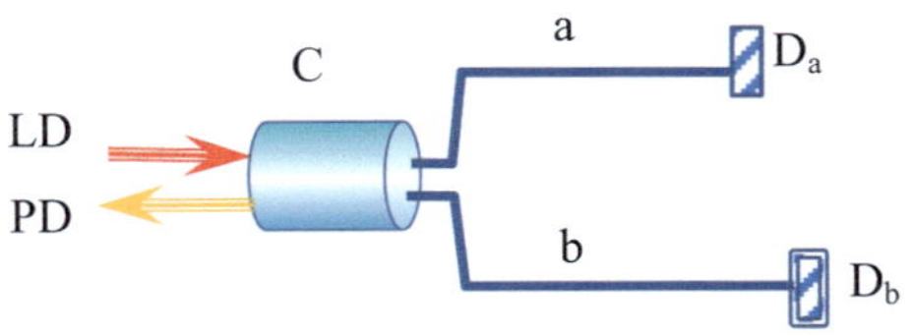

Fig. 3.2 Optical fiber schematic diagram of Michelson interferometer

E_a, E_b represent the amplitudes of the two optical signals respectively; l and l_0 are the optical path; k is the medium propagation constant, and $k = \frac{2\pi n}{\lambda}$; ω is the optical frequency. After reaching the reflectors D_a and D_b along their respective paths, the two light beams are reflected back to the coupler C (which acts as a light combiner at this time) to be recombined and interfere. The amplitudes of the two light beams are:

$$\begin{aligned} E_1 &= k_c r_b exp(i\phi_b) k_c E_0(\tau_b) + k_t r_a exp(i\phi_a) k_t E_0(\tau_a) \\ E_2 &= k_t r_b exp(i\phi_b) k_c E_0(\tau_b) + k_c r_a exp(i\phi_a) k_t E_0(\tau_a) \end{aligned} \tag{3.10}$$

where r_a and r_b are the reflectivity of the reflective surfaces D_a and D_b, respectively. To prevent the reflected light back from the coupler returns to the light source, an optical isolator must be added after the light source to avoid echo interference or even damage to the light source. Therefore, the output of Michelson type only has the second signal, which is one of the differences between Michelson and MZI.

Using a similar analysis to MZ interferometer, the second output light intensity signal and interference contrast can be obtained as follows:

$$\begin{aligned} I_2 &= I_0\left[1 - Qcos(\phi_a - \phi_b)\right] \\ Q &= \frac{2r_a r_b \gamma}{r_a + r_b} \end{aligned} \tag{3.11}$$

It can be seen from Eq. (3.11) that the contrast of the Michelson interferometer is also related to the coherence intensity through the reflectivity of the reflector, but unlike the MZI, it is not related to the beam splitting ratio of the coupler.

Such as shown in Fig. 3.2, the output light signal is converted into photocurrent $i_d = \overline{i_d} + \tilde{i}_d$ by the photodetector PD. Wherein $\overline{i_d}$ is a DC component, $\tilde{i}_d$ is an AC component, and $\tilde{i}_d = Q\cos\phi(t)$. Q is a composite coefficient; $\varphi(t) = \phi_a - \phi_b = 2k|l - l_0|$ is the phase difference of the two light beams. Thus, the phase change of the output arm can be measured by detecting the AC component of the output light, allowing the measurand to be obtained.

According to the principle of double-beam interference, external factors such as temperature and pressure directly affect the sensing arm of the interferometer, causing changes in the fiber length l (corresponding to the elastic deformation of the optical fiber) and the refractive index n (corresponding to the photoelastic effect of the optical fiber). From the phase equation: $\phi = \beta L$, we can deduce that:

$$\Delta\phi = \beta\Delta L + L\Delta\beta = \beta L\frac{\Delta L}{L} + L\frac{\delta\beta}{\delta n}\Delta n + L\frac{\delta\beta}{\delta D}\Delta D \tag{3.12}$$

where, β is the propagation constant of the optical fiber; L is the length of the optical fiber; n is the refractive index of the fiber material. The variation in the fiber diameter ΔD corresponds to the waveguide effect, by which the phase shift caused is 2 to 3

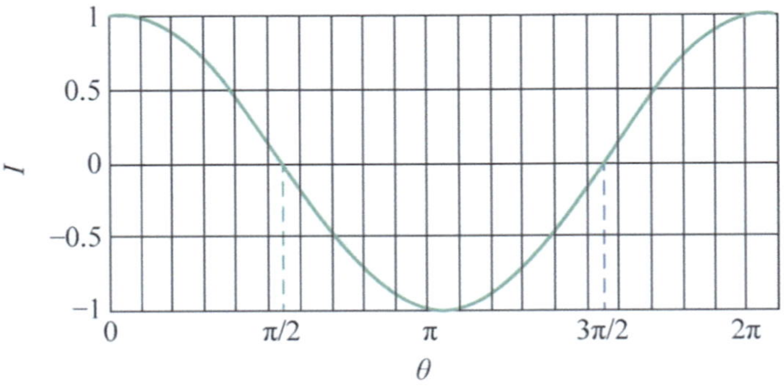

Fig. 3.3 $I \sim \theta$ relationship diagram of double beam interference

orders of magnitude smaller than the first two items and can be ignored. Formula (3.12) is a general expression of phase change caused by external factors of a MZI.

Both the Michelson and MZ interferometer are types of double-beam interference. For example, vector $\boldsymbol{e}_1$ and $\boldsymbol{e}_2$ represent the electric field components of the two light beams respectively. Suppose $e^2(\theta)$ is the square of the modulus of the vector sum of the two electric field, I is the combined light intensity. Theoretical analysis proves that under the condition that the two beams are linearly polarized in the same direction with the same amplitude, I is proportional to $e^2(\theta)$, where θ is the relative phase angle of the two vectors in time and space. Thus, the $I \sim \theta$ relationship curve is Shown in Fig. 3.3. It can be seen that I change with cos θ. There is a peak at $\theta = 2n\pi(n = 0, \pm 1, \pm 2, \ldots)$; and the lowest point is at $\theta = (2n + 1)\pi(n = 0, \pm 1, \pm 2, \ldots)$; the change is the most rapid when $\theta = n\pi + \frac{\pi}{2}(n = 0, \pm 1, \pm 2, \ldots)$, indicating that the interferometer has the highest sensitivity at this time.

As previously mentioned, both Michelson and MZ interferometers should operate at a point where the optical path difference (OPD) between the two arms is $\frac{\pi}{2}$, maximizing the interferometer's sensitivity.

In the Michelson interferometer, light travels along two optical paths to their respective reflecting surfaces and then returns along the same path after reflection. Consequently, the light traverses the same section of fiber twice, resulting in a phase shift sensitivity per unit length of fiber that is double that of the MZ interferometer (MZI).

Michelson and MZ interference structures each have their own set of advantages and disadvantages. Structurally, the Michelson interferometer has one less coupler, and its sensitivity for the same fiber length is twice that of the MZI, with the interference contrast being independent of the coupler's beam splitting ratio. However, during implementation, the Michelson structure requires measures to prevent reflected light from returning to the light source, necessitating the addition of an isolator at the light source end and imposing strict requirements on the reflecting end face.

3.2.3 *Sagnac Fiber Interferometer*

Fundamental Principles and Features

Two light waves transmitted in opposite directions within an optical fiber coil, which is wound from the same optical fiber, produce phase shifts of equal magnitude but in opposite directions when acted upon by external factors. The detection of these phase shifts through the interference effect is the fundamental principle behind the Sagnac optical fiber interferometer. The most typical application of this principle is in rotational sensing, specifically in fiber optic gyroscopes. The Sagnac interferometer, which has no moving parts, exhibits no nonlinear effects, and lacks the locking zone characteristic of laser gyroscopes at low speeds, is well-suited for high-performance, low-cost rotation detection devices. Figure 3.4 depicts the schematic diagram of the Sagnac fiber interferometer. An optical fiber with a length of L is wound into a loop with a radius of R. A laser beam is split into two by a coupler, with each beam entering the optical fiber ring from opposite ends and exiting at the other end. The superposition of the two output beams results in an interference effect, the intensity of which is detected by a photodiode (PD).

When the annular fiber loop rotates at Ω relative to the inertial space (assuming Ω is perpendicular to the loop plane), a non-reciprocal OPD will be produced for the lights traveling clockwise and counterclockwise

$$\Delta L = \frac{4A}{c}\Omega \tag{3.13}$$

where A is the area of the annular optical path; c is the speed of light in vacuum. When the annular optical path is composed of N turns of single-mode optical fibers, the phase difference between the clockwise and counterclockwise optical paths is

$$\Delta\phi = \frac{8\pi NA}{\lambda c}\Omega \tag{3.14}$$

where λ is the wavelength in vacuum.

Compared with traditional mechanical gyroscopes, fiber optic gyroscopes have several advantages:

Fig. 3.4 Schematic diagram of fiber Sagnac interferometer

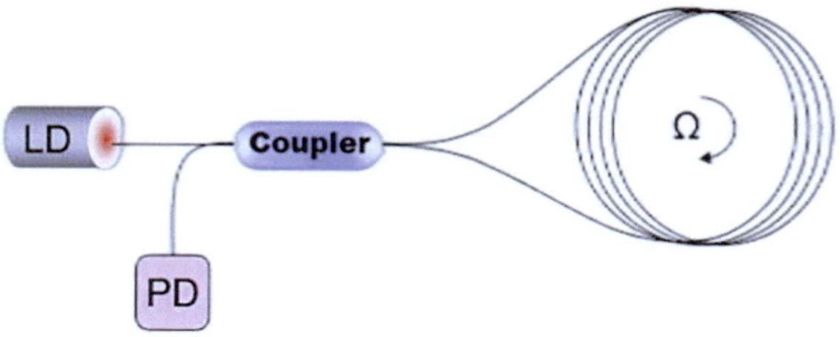

- High sensitivity

Since the fiber optic gyroscope can easily increase the loop area by winding multiple turns of optical fibers, the area changes from A to AN (where N represents the number of turns of optical fiber), significantly enhancing the sensitivity of phase shift detection without increasing the instrument's volume.

- No rotating part

Since the fiber optic gyroscope is mounted on the rotating component being measured, its application field is significantly expanded.

- Small size

The challenge in applying fiber optic gyroscopes for measurement lies in the extremely stringent requirements for their components, parts, and systems. For example, in order to detect a rotational speed of 0.01°/h, assuming that the fiber length L is 1 km, the operating wavelength is 1 μm, and the fiber ring diameter is 10 cm, the phase shift $\Delta\phi$ caused by the Sagnac effect is 10^{-7} rad. Meanwhile the phase shift after a 1 km fiber is 9×10^9 rad, resulting in a relative phase shift of $\frac{\Delta\varphi}{\varphi} \approx 10^{-17}$. It demonstrates the high detection accuracy. Since the Sagnac optical fiber interferometer encompasses all the issues that should be considered in general fiber interferometers, the discussions below are also relevant for other optical fiber interferometers.

4 Key Questions

(1) Reciprocity and polarization

For precise measurements, it is essential that the two coherent beams traveling in opposite directions within the fiber experience only non-reciprocal phase shifts due to rotation, while phase shifts caused by all other factors should be reciprocal. Only then can the corresponding phase shifts be canceled out. Typically, three identical measurements of the same optical path, mode, and polarization are employed.

- Same optical path

Only one coupler is utilized in the primary optical path (Fig. 3.4), causing one beam of light to pass through the coupler twice, while the other beam is reflected by the coupler twice, resulting in an additional OPD between the two beams. Should one coupler be substituted with two couplers, the two beams of light traveling clockwise and counterclockwise will each undergo two transmissions and reflections between the light source and the detector, without introducing any additional OPD.

- Same mode

If a multimode fiber is used in the interferometer, the intput light of a certain mode will generally be transmitted to another output mode at the other end of the fiber.

The phase shift caused by the coupling and interference of these two different modes of light will be non-reciprocal and very unstable. Therefore, single-mode optical fiber and single-mode filter should be used to ensure that the detected light is the superposition of the same mode.

- Same polarization state

Since single-mode fiber typically exhibits birefringence, this can also result in a non-reciprocal phase shift. Additionally, the energy coupling between the two polarization states can diminish the contrast of interference fringes. The birefringence effect, induced by mechanical stress and the shape of the optical fiber, is also unstable. To ensure that the polarization states of the two light beams remain identical, a polarization state compensation and/or control system is commonly employed in the optical path, along with high birefringence/polarization-maintaining fiber. A more effective solution to this issue is the use of a single polarization (PZ) fiber, which supports only one polarization state.

(2) Bias and phase modulation

The optical power detected by the interferometer is

$$P_D = \frac{P_0}{2}(1 + \cos\Delta\phi) \tag{3.15}$$

where P_0 is the input optical power; $\Delta\phi$ is the phase difference introduced by the non-reciprocal measurand.It is evident that for slow rotations (i.e., small $\Delta\Phi$), the detection sensitivity is quite low. To address this, a phase difference bias $\Delta\Phi_b$ must be added to the detection signal, which falls between the maximum and minimum values of PD, as depicted in Fig. 3.5.

There are two types of bias states: 45° bias and dynamic bias. There is $P_D \propto \sin\Delta\phi$ at the 45 ° bias, which has the advantage of zero output without rotation. The main issue is that the instability of the bias point itself will introduce a significant error to the measurement results. There is $P_D(t) \propto P_0 \sin(\Delta\phi)\sin(\omega_m t)$ at dynamic bias, the output remains zero without rotation; however, the stability of the bias point is significantly enhanced. The phase shift bias is typically achieved through phase modulation. This can be achieved by incorporating a phase modulator into the optical path to utilize additional rotational capabilities, or by employing magneto-optic and acousto-optical modulation techniques, as well as heterodyne modulation technology. The scheme for using a magneto-optic modulator involves generating a 45° phase bias through an external magnetic field, ensuring it operates at the highest sensitivity, along with a sinusoidal dynamic modulation of ΔB. The scheme for acousto-optical modulation involves using two counter-propagating light beams to create a frequency difference Δf that compensates for the phase shift induced by the rotation within the acousto-optical modulator. Consequently, the rotation can be quantified by detecting this frequency.

Fig. 3.5 Variation of optical power with phase difference

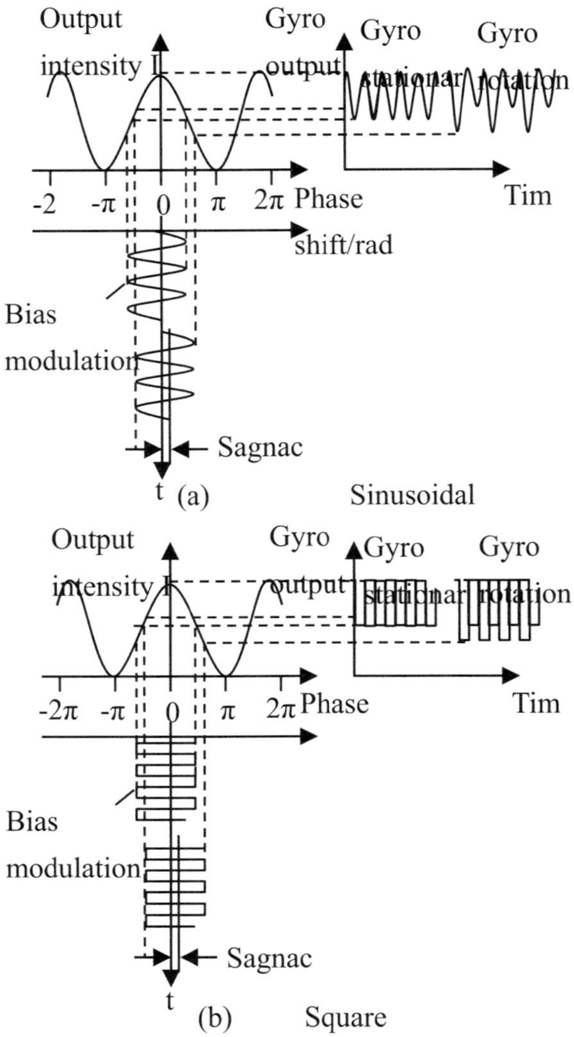

(3) Photon noise

There are various types of noise present in Sagnac fiber optic gyroscopes, significantly impacting the system's signal-to-noise ratio (SNR). Consequently, this issue merits serious attention, as photon noise represents the fundamental limitation. The intensity of the noise correlates with the light power incident upon the detector, a value determined by DC bias. The average number of photons received on the detector within the integration time T is

$$\bar{N} = \frac{P_0 T}{2hv} \tag{3.16}$$

Its standard deviation (by Poisson distribution) $\sigma = \sqrt{\bar{N}}$. Therefore, the root mean square value of phase noise is

$$\Delta\phi_{\text{rms}} = \frac{\sigma}{\bar{N}} = \sqrt{\frac{h\nu}{\frac{1}{2}P_0}}\sqrt{B} \tag{3.17}$$

where $B = 1/T$ is the receiver bandwidth. If $P_0 = 200\,\mu\text{W}$, $\nu = 3 \times 10^{14}\,\text{Hz}$ ($\lambda = 1.0\,\mu\text{m}$), then

$$\frac{\Delta\phi_{\text{rms}}}{\sqrt{B}} \approx \frac{10^{-7}}{\sqrt{Hz}} \tag{3.18}$$

Corresponding ($\lambda = 1\,\mu\text{m}$, $L = 1\,\text{km}$, $D = 10\,\text{cm}$)

$$\frac{\Omega_{\text{rms}}}{\sqrt{B}} = \frac{\lambda c}{2\pi LD}\frac{\Delta\phi_{\text{rms}}}{\sqrt{B}} = 10^{-2}deg/\left(\frac{h}{\sqrt{\text{Hz}}}\right)$$

(4) Impact of parasitic effects and methods of mitigation

- Direct dynamic effect

The temperature and mechanical stress acting on the optical fiber can cause changes in the propagation constant and the size of the optical fiber, resulting in phase noise at the receiver. The reciprocity theorem applies only to time-invariant systems. If the disturbance source is symmetric about the center point of the system, the total effect will cancel out. Therefore, it is advisable to avoid having a single disturbance source close to one fiber end and to pay attention to the winding techniques of the fiber loop.

- Reflection and Rayleigh backscattering

Secondary waves a_1, a_2 generated by Rayleigh backscattering in optical fibers and reflections from each end face, when superimposed with primary waves A_1 and A_2, result in a coherent superposition, as depicted in Fig. 3.6, leading to noise at the receiver.

Rayleigh scattering in optical fibers occurs due to the inhomogeneity of the medium within the fiber. The scattered wave is omnidirectional and has a constant frequency, with the light intensity being proportional to $\frac{1}{\lambda^4}$. Set

$$\left.\begin{aligned} a_1 &= a_1\exp[i\phi_1] \\ a_2 &= a_2\exp[i\phi_2] \\ A_1 &= A_1\exp\left[i\left(\phi_0 + \tfrac{\Delta\phi_s}{2}\right)\right] \\ A_2 &= A_2\exp\left[i\left(\phi_0 - \tfrac{\Delta\phi_s}{2}\right)\right] \end{aligned}\right\} \tag{3.19}$$

Considering only the coherence term, the output signal is,

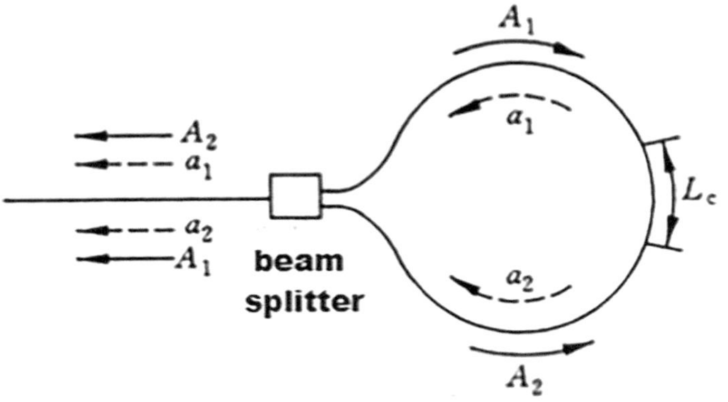

Fig. 3.6 Schematic diagram of signal and reflected waves in the loop

$$\begin{aligned} I =& I_0 + \gamma_1\left(a_1 a_2^* + a_1^* a_2\right) + \gamma_2\left[\left(a_1 A_2^* + a_1^* A_2\right) + \left(a_2 A_1^* + a_2^* A_1\right)\right] \\ &+ \left(A_1 A_2^* + A_1^* A_2\right) = I_0 + 2\gamma_1 |a_1 a_2| \cos(\phi_1 - \phi_2) \\ &+ 2\gamma_2\left[|a_1 A_2|\cos\left(\phi_1 - \left(\phi_0 - \frac{\phi_s}{2}\right)\right)\right. \\ &\left. + |a_2 A_1| \cos\left(\phi_2 - \left(\phi_0 + \frac{\phi_s}{2}\right)\right)\right] + 2|A_1 A_2|\cos(\Delta\phi_s) \end{aligned} \tag{3.20}$$

where, I_0 is the average light intensity; γ_1 is the coherence degree of a_1 and a_2; γ_2 is the degree of coherence of a_1A_2 or a_2A_1; the coherence degree of A_1A_2 is approximately 1; The coherence of a_1A_1 or a_2A_2 is zero; $\Delta\phi_s$ is the Sagnac phase difference.

Due to the non-reciprocity between a_1 and a_2, a_1 and A_2 and a_2 and A_1, their interference terms $\Delta\phi_s$ contribute different to the measurand, thereby introducing measurement errors of phase or angular velocity. For a 1 km fiber, the maximum phase error caused by Rayleigh backscattering is 10^{-2} rad; for a fiber optic gyro with a diameter of $D = 10$ cm and a wavelength of $\lambda = 1$ μm, the corresponding angular velocity error is on the order of 10^3 rad/h.

To minimize phase errors resulting from Rayleigh backscattering, one approach is to employ a light source with a short coherence length. The superluminescent diode (SLD) is acknowledged as an ideal light source, distinguished by its excellent spatial coherence and poor temporal coherence. Another method involves implementing alternating phase modulation at one end of the optical fiber loop and choosing an appropriate modulation frequency. This ensures that the backscattered light near the center point $\pm L/2$ of the optical fiber loop contributes nothing to the output signal. It is evident that modulation technology can stabilize the bias point and separate the signal from backscattering noise. These measures can reduce Rayleigh backscattering noise to levels below quantum noise.

- Faraday effect

The optical fiber loop in a magnetic field induces noise in the fiber gyroscope due to the Faraday effect, which introduces non-reciprocal circular birefringence (the rotation direction of light vibration depends on the direction of light propagation). This is superimposed on the original reciprocal birefringence. The magnitude of this effect is contingent upon the strength and orientation of the magnetic field. For instance, in the geomagnetic field, the Faraday effect causes a rotation of the light direction by 10°/h. A more effective method to mitigate this effect is to encase the fiber system within a magnetic shielding box.

- Optical Kerr effect

The optical Kerr effect is characterized by a change in the refractive index of materials due to an optical field. This implies that the propagation constant of the guided wave is dependent on the power of the light wave within a single-mode fiber. For instance, in the context of a fiber optic gyroscope and a linear material such as fused silica, a power difference of just 10 nW between the light waves traveling in opposite directions can result in a significant error, particularly for inertial navigation systems. Therefore, for the general case of a total power of 100 μW, this requires a power stability better than 10^{-4}. To reduce the error caused by this effect, there are currently three methods: ① Adjust the splitting ratio K of the beam splitter to 1/2 to nullify the phase error from the Kerr effect; ② Employ a broadband light source, such as an SLD with a duty cycle of 1/2, and sum the contributions from all wavelength components to achieve an average phase error from the Kerr effect that is zero; ③ Implementing intensity modulation of the light source can reduce the error by 1 to 2 orders of magnitude.

Examples of Error Analysis

The analysis of various error sources has been conducted above, and the subsequent section provides a calculation example for error analysis. The calculation principle is based on the distribution of the total error to each error source with equal precision, offering reference data for system design. Consider four error sources: Rayleigh backscattering, the Faraday effect, the Kerr effect, and the polarization effect. Assuming that the total accuracy of the fiber optic gyroscope is $\Delta\Omega = 0.01$ °/h, the allocated accuracy for each error source is $\delta\Omega = \frac{\Delta\Omega}{\sqrt{4}} = 0.0025$ °/h (since there are 4 error sources).

(1) Rayleigh backscatter error

To simplify the problem, we will ignore the interference terms of a_1, a_2 in Eq. (3.19), and consider the phase error caused by Rayleigh backscattering in the worst case scenario,

$$\Delta\varphi_{\text{Rmax}} = 2\Delta\varphi = 2\sqrt{\frac{I_s}{I_P}} \tag{3.21}$$

where $\Delta\varphi_{\text{Rmax}}$ is the maximum phase error; I_s is the intensity of backscattered light; I_P is the CW or CCW wave intensity. Additionally, assuming the presence of a scattering center within the optical fiber, where the backscattered light constitutes P portion of the total scattered light, then

$$P = \pi \frac{(NA)^2}{4\pi} = \frac{1}{4}(NA)^2 \tag{3.22}$$

where NA represents the numerical aperture of the optical fiber. For single-mode fibers, NA ranges from 0.1 to 0.2.

$$\frac{I_S}{I_P} = P\alpha_s \frac{L}{2} = \frac{1}{8}(NA)^2 \alpha_s L \tag{3.23}$$

where α_s is the attenuation coefficient of light; L is the fiber length. Thus

$$\Delta\phi_{\text{Rmax}} = (NA)\sqrt{\frac{1}{2}\alpha_s L} \tag{3.24}$$

Let $\alpha_s = 1.15 \times 10^{-4}$/m (0.5 dB/km), L=1 km, NA=0.1. According to the Eq. (3.24), $\Delta\phi_{Rmax} = 0.024$ rad, which is equivalent to an angular velocity error of the fiber gyro of 2363°/h with D = 10 cm, λ = 1 μm.

In practical applications, there are two measures to reduce the phase error caused by Rayleigh backscattering.

The first measure involves utilizing a light source with a short coherence length. For instance, the $\Delta\phi_R$ can be diminished by three orders of magnitude when employing a laser with a coherence length of 0.5 mm, $\Delta\phi \leq 3 \times 10^{-5}$ rad. To achieve an objective accuracy of 0.005°/h for $\Delta\phi_{\text{Rmax}}$, the coherence length must be reduced by 12 orders of magnitude to decrease the error by 6 orders of magnitude. Even if a super-radiant light-emitting diode (SLD) with a coherence length of only 3 μm is used, $\Delta\Omega_R$ is expected to reduce to about 0.1 °/h, nearly two orders higher than the design accuracy.

The second measure involves installing an alternating phase modulator (PM) at one end of the fiber coil. To enhance sensitivity, a fixed phase bias of π/2 is introduced at the opposite end. This setup allows for the coherent wave and the backscattered wave to be respectively.

$$A_1 = A_1 \exp\left\{ i \left[\phi_0 + \phi(t-\tau) + \frac{\Delta\phi_s}{2} + \frac{\pi}{2} \right] \right\}$$
$$A_2 = A_2 \exp\left\{ i \left[\phi_0 + \phi(t) - \frac{\Delta\phi_s}{2} + \frac{\pi}{2} \right] \right\}$$
$$a_1 = a_1 exp\{i[\phi_1 + \phi(t-\tau) + \phi(t)]\}$$
$$a_2 = a_2 \exp\{i[\phi_2 + \pi]\}$$

where; $\phi(t) = \phi_m \sin(\omega_m t)$; τ is the time difference between A_1 and A_2 arriving at PM; ϕ_0, ϕ_1 and ϕ_s are constant phase related to the path; $\Delta\phi_s$ is Sagnac phase shift. The interference terms in Eq. (3.19) are

$$\left.\begin{aligned}\left(a_1^* a_2 + a_1 a_2^*\right) &= 2|a_1 a_2|\cos[\phi(t-\tau) + \phi(t) + \phi_1 - \phi_2] \\ \left(a_1^* A_2 + a_1 A_2^*\right) &= 2|a_1 A_2|\sin\left[\phi(t-\tau) + \phi_1 - \phi_0 + \tfrac{\Delta\phi_s}{2}\right] \\ \left(a_2^* A_1 + a_2 A_1^*\right) &= 2|a_2 A_1|\sin\left[\phi(t-\tau) + \phi_0 + \tfrac{\Delta\phi_s}{2} - \phi_2\right] \\ \left(A_1^* A_2 + A_1 A_2^*\right) &= 2|A_1 A_2|\cos[\phi(t-\tau) - \phi(t) + \Delta\phi_s]\end{aligned}\right\} \tag{3.25}$$

where,

$$\begin{aligned}\phi(t-\tau) + \phi(t) &= 2\phi_m \cos\left(\omega_m \frac{\tau}{2}\right)\sin\left[\omega_m\left(t - \frac{\tau}{2}\right)\right] \\ &= \psi_m \sin\left[\omega_m\left(t - \frac{\tau}{2}\right)\right]\end{aligned},$$

$$\begin{aligned}\phi(t-\tau) - \phi(t) &= 2\phi_m \sin\left(\omega_m \frac{\tau}{2}\right)\sin\left[\omega_m\left(t - \frac{\tau}{2}\right) - \frac{\pi}{2}\right] \\ &= \theta_m \sin\left[\omega_m\left(t - \frac{\tau}{2}\right) - \frac{\pi}{2}\right]\end{aligned}$$

Where θ_m and ψ_m represent the modulation depth, and θ_m is set to 1.8 rad so that the first-order Bessel function, which includes the Sagnac phase difference, reaches its maximum value for the highest sensitivity. On the other hand, the interference signal $a_1 a_2$ is orthogonal to the interference signal $A_1 A_2$. By decomposing the signal in the direction of maximum interference signal $A_1 A_2$, the contribution of the scattered light $a_1 a_2$ to the interference term $\Delta\phi_R$ can be nullified.

Theoretical analysis indicates that the backscattering error can be nullified when the optimal modulation frequency, such as $f_m = 1/(1.36\tau)$, is chosen. Typically, the scattering error peaks when $f_m = 1/(2\tau)$ is selected. Furthermore, by maintaining the stability and modulation depth error of the modulation frequency within the range of 10^{-5} to 10^{-3}, the coherence length of the light source can be extended to tens of centimeters, thus achieving inertial navigation-level gyro performance. In summary, with the SLD light source and appropriate phase modulation, a precision of $\Delta\Omega_R < 0.005$ °/h can be attained.

(2) Faraday effect error

The angular velocity drift resulting from the geomagnetic Faraday effect is typically 10 °/h. Theoretical analysis indicates that controlling the angular velocity drift within the desired accuracy is not possible by solely employing high birefringence fiber; magnetic shielding of the fiber loop is also necessary. With this additional measure, the angular velocity drift can be reduced by at least two orders of magnitude, thereby fully meeting the distribution requirements.

(3) Kerr effect error

Kerr effect is a third-order nonlinear optical effect. In single-mode fiber, the propagation constants of CW and CCW light waves are, respectively:

$$\begin{cases} \beta_{\mathrm{cw}} = \beta + \Delta\beta_{\mathrm{cw}} = \gamma[I_{\mathrm{cw}} + 2I_{\mathrm{ccw}}] \\ \beta_{\mathrm{ccw}} = \beta + \Delta\beta_{\mathrm{ccw}} = \gamma[I_{\mathrm{ccw}} + 2I_{\mathrm{cw}}] \end{cases} \tag{3.26}$$

where, γ is a constant. If $I_{\mathrm{cw}} \neq I_{\mathrm{ccw}}$, then the error of propagation constant is:

$$\Delta\beta = \beta_{\mathrm{cw}} - \beta_{\mathrm{ccw}} = \gamma[I_{\mathrm{ccw}} - I_{cw}] \tag{3.27}$$

$\Delta\beta$ will inevitably lead to phase error or angular rate error $\Delta\Omega_K$:

$$\begin{aligned} &\Delta\Omega_K \\ &= \frac{C}{R}\eta n_a \delta(1-2K)\left|\frac{\langle I_0^2(t)\rangle - 2\langle I_0(t)\rangle^2}{\langle I_0(t)\rangle}\right| \\ &= \frac{C}{R}\eta n_a \delta\langle I_0(t)\rangle(1-2K)\left|\frac{\langle I_0^2(t)\rangle}{\langle I_0(t)\rangle^2} - 2\right| \end{aligned} \tag{3.28}$$

where, η is the dielectric impedance; n_{a} is the Kerr dielectric constant; δ is a factor related to the lateral distribution of the mode, and K is the splitting ratio of DC. Assuming $\Delta\Omega \leq 0.005$ °/h, $C/R = 6 \times 10^9\ \mathrm{s}^{-1}$, $\eta n_a \delta = 10^{-14}\ \mu\mathrm{m}$, $< I_0(t) >$ H1 $\mu\mathrm{W/mm}^2$, using a continuous wave light source with stable intensity $\langle I_0^2(t)\rangle / \langle I_0(t)\rangle^2 \approx 1$, then the splitting ratio $K = 0.5 \pm (2 \times 10^{-4})$. The requirement for the splitting ratio K is too high to be realized. However, some thermal light source, such as SLD, which have a duty cycle of 1/2, meet $\langle I_0^2(t)\rangle = 2\langle I_0(t)\rangle^2$. According to Eq. (3.28), $\Delta\Omega_K \rightarrow 0$. Therefore, SLD is an ideal light source in fiber optic gyroscope.

Utilizing multi-longitudinal mode semiconductor lasers, $\Delta\Omega_K$ can similarly be minimized because

$$\left|\frac{\langle I_0^2(t)\rangle - 2\langle I_0(t)\rangle^2}{\langle I_0(t)\rangle^2}\right| = \frac{1}{N} \tag{3.29}$$

where N is the number of modes. when $N = 10$, $\Delta\Omega_K$ will be reduced by an order of magnitude. Indeed, $\Delta\Omega_K$ can be reduced by 1~2 orders magnitude at $N \geq 50$ when utilizing a semiconductor laser. Consequently, the accuracy requirement for K can be relaxed to 10^{-2}. Nonetheless, there remains challenges to achieving $|K_{1/2}| \leq 10^{-2}$.

(4) Polarization error

The poor polarizer, energy coupling between orthogonal polarization modes, and other factors in the fiber optic gyroscope can introduce polarization errors, which manifest as a phase difference $\Delta\phi_b$. Considering the first-order approximation of the amplitude ratio, $\Delta\phi_{\mathrm{bmax}}$ is

$$\Delta\phi_{\mathrm{bmax}} = \varepsilon\left|\frac{a_B}{a_A}\right|\left|\frac{t_{\mathrm{AB}} + t_{\mathrm{BA}}}{t_{\mathrm{AA}}}\right| = 2\varepsilon\left|\frac{a_B}{a_A}\right|\frac{|t_{\mathrm{AB}}|}{|t_{\mathrm{AA}}|} \tag{3.30}$$

In the formula, a_B and a_A represent the amplitude components of the incident light that are respectively orthogonal and parallel to the transmission axis of the polarizer; $|t_{AB}| = |t_{BA}|$ is the amplitude coupling coefficient between the two orthogonal polarization modes; $|t_{AA}|$ is the amplitude transmission coefficient of the polarization state a_A. If $\Delta\Omega_b \le 0.005$ °/h, i.e., $\Delta\phi_{bmax} = 2.5 \times 10^{-8}$ rad, then a high birefringence fiber should be selected first, and the h-parameter currently reaches 10^{-6} m^{-1}. At this time, $\left|\frac{t_{AB}}{t_{AA}}\right| \approx 0.33(L = 1$ km). Secondly, choose $\left|\frac{a_B}{a_A}\right| \le 10^{-2}$, and $\varepsilon = 8.3 \times 10^{-5}$ is available via Eq. (3.30), corresponding to the polarizer with an extinction ratio of 80 dB, which remains technically challenging to achieve at present. The typical extinction ratio of commercial polarizers is about 60dB. However, $\Delta\phi_{bmax}$ represents the worst result, where the requirements for polarizers can actually be less stringent.

The previous section discussed the four fundamental sources of error in fiber optic gyroscopes and the methods to limit these errors to a specific range. Additional steps must be implemented to address the harsh operating conditions and other angular velocity errors associated with fiber optic gyroscopes. For instance, the operating temperature range for fiber optic gyroscopes is typically −40 to 50 degrees Celsius, and temperature fluctuations can significantly affect the fiber coil, phase modulator, and fiber coupler. The measurement results indicate that a temperature change of 1°C causes a scale factor variation of 5%, necessitating precise control of the fiber's temperature. Furthermore, stress induces additional phase errors, thereby demanding more stringent assembly processes for the fiber gyroscope, particularly in the ring winding technology. Ultimately, the accuracy of the fiber gyroscope is limited by quantum noise [2].

3.3 Fiber FP Interferometer-Double Beam and Multi-beam Interferometry

3.3.1 Introduction

An optical fiber Fabry-Perot Sensor is an FP interferometer composed of an optical fiber. Currently, there are three typical FP cavities: intrinsic, extrinsic, and linear composite. The intrinsic F-P cavity is composed of optical fibers, whereas the extrinsic type utilizes the air gap between the end faces of two fibers (with or without a highly reflective film on both ends) to form a microcavity of length L (see Fig. 3.7). Among these, the extrinsic type exhibits the best performance and is currently the most widely used.

When a coherent beam is incident into the microcavity along the fiber, the reflected beams from both ends of the microcavity return along the original path and interfere, producing an output signal that is dependent on the microcavity's length. When external parameters such as force, deformation, displacement, temperature, voltage, current, and magnetic field act upon the microcavity in a specific manner, the change

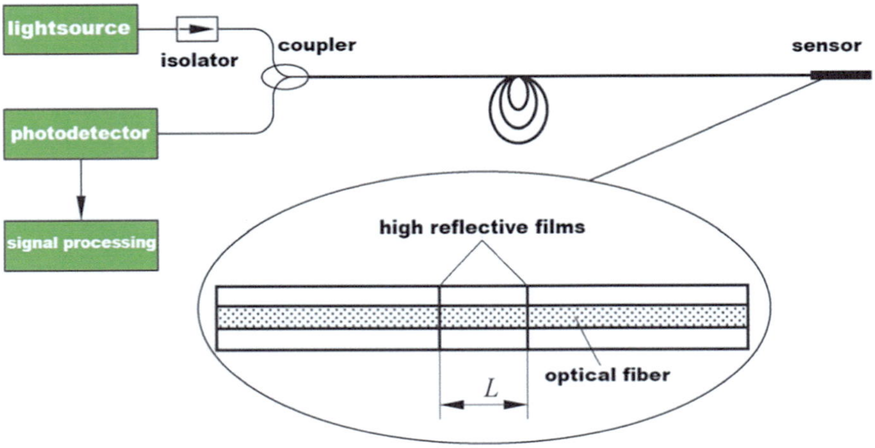

Fig. 3.7 Schematic diagram of optical fiber FP sensor

in cavity length L results in corresponding changes to the interference output signal. Based on this principle, the microcavity length can be determined from the variation of the interference signal, and subsequently, the change in external parameters can be ascertained, thereby achieving the sensing of these parameters. For instance, when the fiber FP cavity is directly affixed to a deformed object, any slight deformation of the object is directly imparted to the FP cavity, causing a variation in the output interference light. This results in the creation of an optical fiber FP strain/stress/pressure/vibration sensor. An optical fiber FP temperature sensor is produced by securing the optical fiber FP cavity to a material with a thermal expansion coefficient that exhibits good linearity, allowing the cavity length to vary with the thermal expansion and contraction of the material. By attaching the optical fiber FP cavity to magnetostrictive material, an optical fiber FP magnetic field sensor is formed. Similarly, an optical fiber FP voltage sensor is created by fixing the optical fiber FP cavity onto electrostrictive material.

In an optical fiber Fabry-Pérot (FP) sensor system, as depicted in Fig. 3.7, the fiber FP cavity serves as the sensing unit to acquire measurand information. To detect various parameters, there are diverse fiber FP cavity structures, each with unique sensing characteristics. Furthermore, the signal from the fiber FP cavity must undergo processing to achieve the desired result, a process known as signal demodulation for fiber FP sensors. Demodulation techniques for fiber FP sensors encompass both intensity demodulation and phase demodulation. Despite the challenges, phase demodulation stands out due to its extensive research, numerous implementation strategies, and it is currently the most widely adopted demodulation method.

3.3.2 Fundamentals

The optical FP interferometer depicted in Fig. 3.8 has been developed. An optical FP interferometer is an optical resonant cavity, commonly known as an FP cavity, consisting of two optically parallel flat plates with high-reflectivity films on their ends and a spacing of L, which are held strictly parallel to one another. If the reflectivities of both end mirrors are R, and the wavelength and intensity of the incident light are λ and I_0 respectively, the reflection and transmission output of the optical FP cavity can be expressed as I_R and I_T, respectively, based on the principle of multi-beam interference:

$$\begin{cases} I_R = \frac{2R(1-\cos\Phi)}{1+R^2-2R\cos\Phi} I_0 \\ I_T = \frac{(1-R)^2}{1+R^2-2R\cos\Phi} I_0 \end{cases} \tag{3.31}$$

where Φ is the optical phase, and there is

$$\Phi = \frac{4\pi}{\lambda} n_0 L \tag{3.32}$$

where n_0 is the refractive index of the material within the cavity. When the cavity is filled with air, $n_0 \approx 1$. When two optical fiber end faces are employed in place of the two mirrors of the FP interferometer, the FP interferometer depicted in Fig. 3.8 transforms into the Fiber Optic FP Sensor shown in Fig. 3.7. Considering the optical FP interferometer in Fig. 3.8, both the reflected and transmitted output signals can be utilized via Eq. (3.31). However, the fiber FP sensor in Fig. 3.7 can only utilize reflected light.

When the specular reflectance R decreases, double-beam interference can be utilized in place of multi-beam interference. Consequently, I_R in Eq. (3.31) can be approximately simplified as:

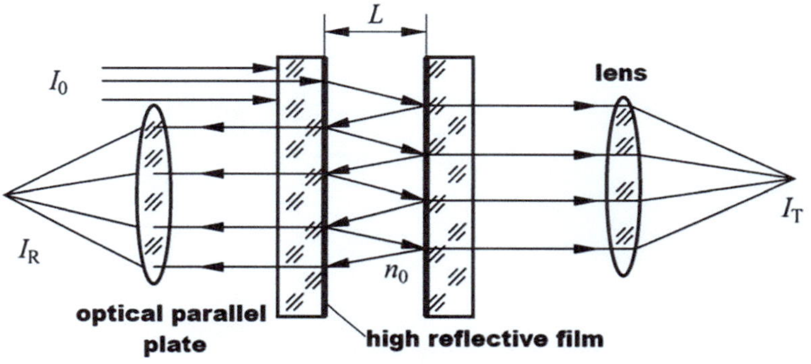

Fig. 3.8 Principle diagram of optical FP interferometer

$$I_R = 2R(1 - \cos\Phi)I_0 = D + C\cos\Phi \tag{3.33}$$

Due to the output interference signals from both Eqs. (3.31) and (3.33), the incident light beam to the fiber FP sensor must be coherent. This necessitates that not only the light source in Fig. 3.7 is coherent but also that the fiber is single-mode.

3.3.3 *Classification and Characteristics*

Based on the structure of the fiber FP cavity, fiber FP sensors can be categorized into three types: the intrinsic Fabry-Perot interferometer (IFPI), the extrinsic Fabry-Perot interferometer (EFPI), and the linear composite cavity FP sensor, also known as the in-line Fabry-Perot (ILFP).

Intrinsic Optical Fiber FP Sensor

The intrinsic fiber FP sensor is the earliest type of optical fiber FP sensor. It involves cutting the optical fiber into three sections labeled A, B, and C. A high reflective film is deposited on the end faces of sections A and C, which are adjacent to section B, before fusing them together, as illustrated in Fig. 3.9. At this point, the length L of segment B represents the FP cavity length. Clearly, this constitutes an intrinsic optical fiber FP sensor. Given that the cavity length L of a fiber FP sensor is typically only tens of microns, the processing of section B is particularly challenging.

Furthermore, it can be seen from Eq. (3.31) that the length L and refractive index n of the B-section fiber, which serves as the resonant cavity, are affected by external parameters. Consequently, the output becomes a two-parameter function of L and n. Hence, in practice, it is difficult to distinguish the influences of these two parameters.

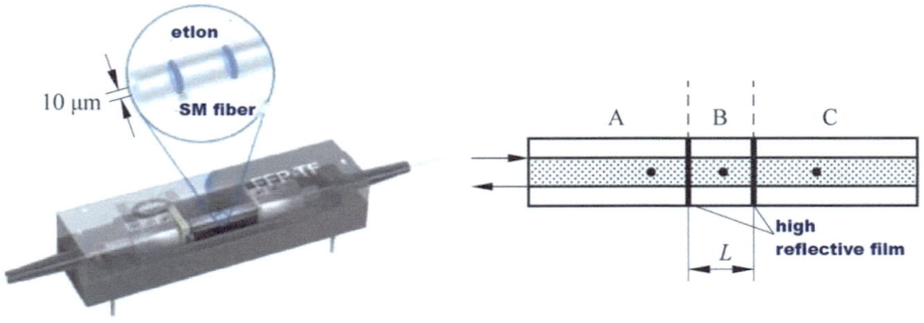

Fig. 3.9 Schematic diagram of intrinsic optical fiber FP sensor

Extrinsic Optical Fiber FP Sensor

EFPI is the most widely used fiber optic pressure sensor currently available. It comprises two single-mode optical fibers with coated end faces that are precisely aligned parallel and coaxial, and are encased within a specialized tube of length D and inner diameter d ($d \geqslant 2a$, $2a$ is the outer diameter of the optical fiber) (Fig. 3.10). Owing to its structural characteristics, EFPI offers several advantages:

- Easy control of cavity length: During the assembly process of fiber FP cavity, it is easy to adjust and accurately control the cavity length L by a special fine-tuning mechanism.
- Tunable sensitivity: Given that the catheter length D is greater than he cavity length L and the actually functional length of the sensor, the sensitivity of the sensor can be adjusted by varying the length of D.
- FP cavity is a single value function of L: Composed of an air gap and its refractive index $\boldsymbol{n}_0 \approx 1$, the FP cavity can be approximately considered as a single-parameter function of L.
- Excellent temperature characteristics: When the thermal expansion coefficient of the catheter material matches that of the optical fiber, the thermal elongation of the catheter aligns with that of the optical fiber. This essentially negates the change in cavity length L resulting from the thermal expansion or contraction of the materials. Consequently, the temperature characteristics of the EFP sensor are significantly superior to those of the IFP sensor, allowing the influence of temperature to be disregarded.

If the sensor experiences excessive tension during transportation or installation, the distance between the two fiber end surfaces, known as the FP cavity length L, may become too great for them to remain parallel. This could cause the light beam to reflect only a few times between the end surfaces, making it even less likely to return to the original optical fiber, thus resulting in sensor failure. An improved structure, which includes a transitional buffer gap [3] (Fig. 3.11), could be employed to address this issue.

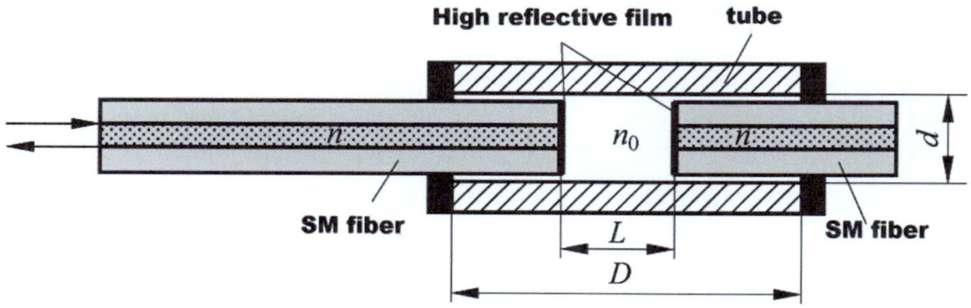

Fig. 3.10 Schematic diagram of extrinsic fiber FP sensor

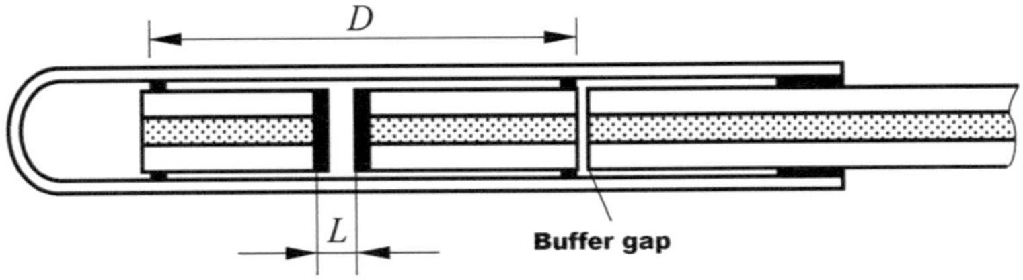

Fig. 3.11 An improved EFPI sensor schematic

In Eq. (3.31), it is assumed that parallel light emitted from the SM fiber is reflected multiple times within the FP cavity and completely returns to the SM fiber. However, the actual light beam exiting a SM fiber is divergent (Fig. 3.12), meaning only a portion of the light energy returns to the incident optical fiber, resulting in coupling loss upon reflection. This loss is related to the core diameter $2a$ of the SM fiber, the receiving angle θ_c, and the length L of the FP cavity. According to the principles of wave optics, the loss $\alpha(L)$ for monochromatic light is approximately:

$$\alpha(L) = \left(\frac{\alpha}{\alpha + L \cdot \tan\theta_c}\right)^2 \tag{3.34}$$

Due to the influence of $\alpha(L)$, Eq. (3.31) becomes

$$I_R = \frac{R\left[1 + \alpha^2(L) - 2\alpha(L)\cos\Phi\right]}{1 + R^2\alpha^2(L) - 2R\alpha(L)\cos\Phi} I_0 \tag{3.35}$$

There are significant differences between Eqs. (3.31) and (3.35). The corresponding output curves are shown below in Fig. 3.13. The output intensity of the actual EFPI sensor varies with the cavity length L, which introduces certain difficulties in subsequent signal processing. In contrast, the IFPI sensor does not have this issue because the light beam always propagates within the optical fiber.

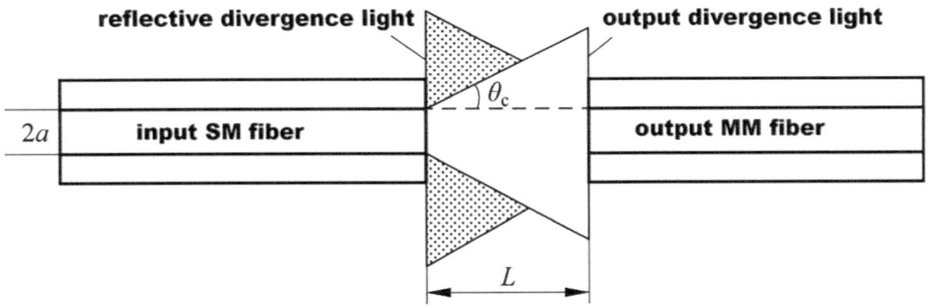

Fig. 3.12 Relations of insertion loss with cavity length L and fiber divergence angle θ_c of the FP Cavity

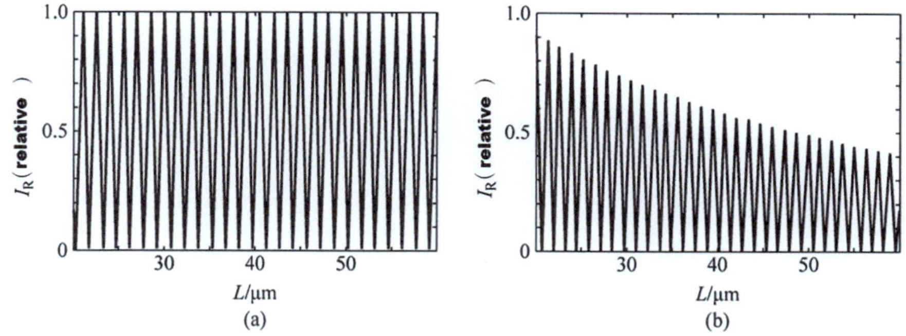

(a) Ideal output without cavity loss (b) actual output with cavity loss

Fig. 3.13 Output interference fringe variation with fiber FP cavity length

Linear Composite Cavity Fiber FP Sensor

The principle diagram of the ILFP sensor is schematically illustrated below in Fig. 3.14. It is formed by replacing the B-section fiber in Fig. 3.9 with a conduit that has the same outer diameter as the optical fiber. Thus, it is a composite structure of intrinsic and extrinsic types, exhibiting characteristics of both [3]. However, due to the processing technology challenges of the IFPI sensor, the ILFP sensor requires the processing accuracy of the microtube length L to be on the order of microns. Consequently, this type of sensor has been rarely practically studied and reported.

3.3.4 *Resonance Modes and Enhancement of Short Cavity FP Structures*

The cavity length of conventional Fabry-Pérot (FP) cavities typically ranges from tens to hundreds of micrometers (μm). However, it has been observed that short cavity FP sensors, which are only micrometers (μm) or even sub-micrometers in length and consist of highly reflective interfaces, exhibit greater coupling efficiency and a higher signal-to-noise ratio (SNR). This, in turn, effectively enhances their sensitivity.

Fig. 3.14 Schematic diagram of a ILFP sensor

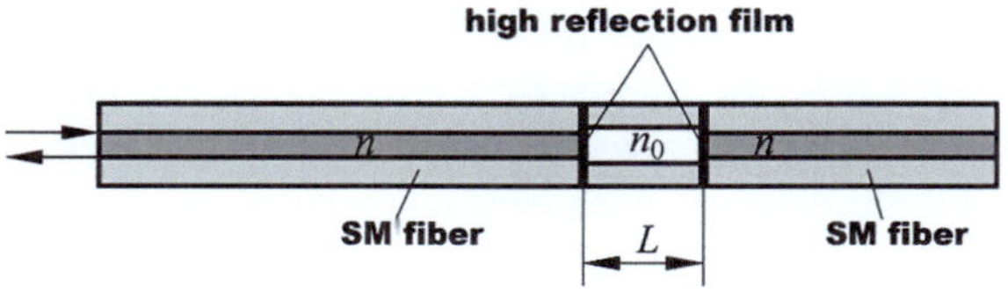

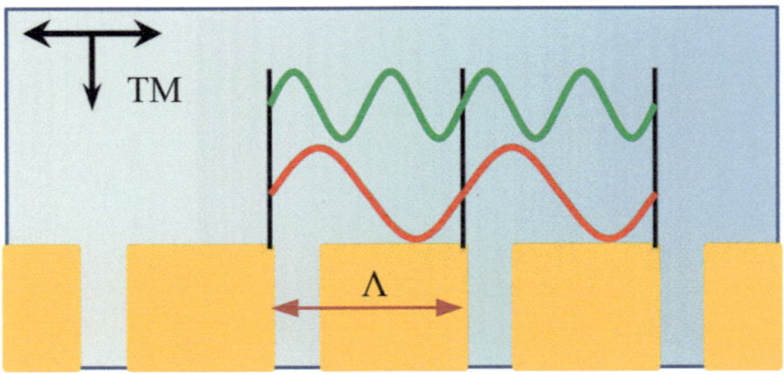

Fig. 3.15 Schematic diagram of Bloch mode of metal grating

The sandwich resonant structure of Au-grating/dielectric/Au-film is fabricated on the fiber end face, capable of exciting a variety of resonant modes distinct from the conventional FP cavity. These include the localized surface plasmon resonance (LSPR) mode, surface plasmon-Bloch waves (SPP-BW), TM0, and FP mode. This section primarily discusses the impact of grating period and dielectric layer thickness on the reflection spectrum of the micro-nano resonator, and presents the electric field distributions of different resonance modes, as calculated by the finite-difference time-domain (FDTD) method.

When an electromagnetic wave strikes the grating surface, its wave vector changes. The grating structure excites a special type of surface plasmon known as SPP-BW, whose wavelength is related to the grating period and the incident medium (as illustrated in Fig. 3.15). The wavelength of the Bloch mode is determined by the grating's array period, denoted by Λ.

When a dielectric layer with a permittivity of ε_1 sandwiched between two semi-infinite gold films with permittivities of ε_2 that are very close to each other, a metal-insulator-metal (MIM) waveguide structure is formed. By analyzing the conduction mode supported by MIM waveguide which involves solving the eigen Eq. (3.36) of metal-coated plate dielectric waveguide, the dispersion curve of the MIM waveguide mode is available in Fig. 3.16.

$$\kappa_1 d = m\pi + 2\phi (m = 0, 1, 2, \ldots) \tag{3.36}$$

where $\varphi^{\mathrm{TE}} = arctan\left(\frac{\alpha_2}{\kappa_1}\right)$, $\varphi^{\mathrm{TM}} = arctan\left(\frac{\varepsilon_1}{\varepsilon_2}\frac{\alpha_2}{\kappa_1}\right)$, $\kappa_1 = \sqrt{k_0^2\varepsilon_1 - \beta^2}$, $\alpha_2 = \sqrt{\beta^2 - k_0^2\varepsilon_2}$.

As can be seen in Fig. 3.16, TM_0 and TM_1 are two special modes of the MIM structure. When the thickness (d) of the intermediate dielectric layer in the MIM structure approaches infinity, the structure effectively becomes two independent metal-dielectric interfaces. Consequently, the propagation constants of these two modes converge towards that of SPR (Surface Plasmonic Resonance). Hence, the SPR wave can be considered a special instance of the guided modes TM_0 and TM_1. Furthermore, it can be observed that the TM_0 mode lacks a cut-off thickness; in other words, TM_0 persists irrespective of the spacing between the two gold films.

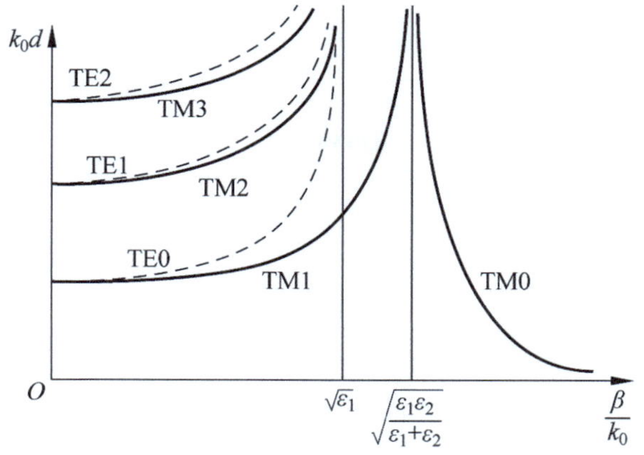

Fig. 3.16 Dispersion curves of gold-dielectric-gold waveguide structures

To replace the gold film at the incident end with a gold grating, the gold film at the incident layer is truncated, and the TM_0 mode is reflected. Assuming that the reflection coefficients of the two interfaces of the resonant cavity are identical, spectral resonance will occur if Eq. (3.36) is satisfied.

$$\pi k = Re(n_{\text{eff}})k_0L + Angle(r),\ \ k \in Z \tag{3.37}$$

Equivalent refractive index $n_{\text{eff}} = \beta/k_0$, β is the propagation constant of the TM_0 mode; the Angle function serves to extract the argument of a complex number.

To enhance the contrast of interference fringes in the FP cavity formed by a fiber end face and a gold film, it is necessary to coat the fiber end to match the high reflectivity of the gold film.

When both reflective surfaces possess the same high reflectivity, as per the relation (Eq. 3.38) between the coupling coefficient and cavity length (illustrated in Fig. 3.17), a shorter cavity length results in a higher energy coupling coefficient. This indicates that light divergence has a reduced impact on reflectivity. Consequently, a short cavity length is advantageous for achieving a high contrast of the fiber end face FP resonance mode.

$$I = \frac{R_1 + \xi R_2 - 2\sqrt{\xi R_1 R_2}\cos\delta}{1 + \xi R_1 R_2 - 2\sqrt{\xi R_1 R_2}\cos\delta} I_0;$$
$$\xi = \frac{4\left[1 + \left(\frac{2\lambda L}{\pi N_0 w_0^2}\right)^2\right]}{\left[2 + \left(\frac{2\lambda L}{\pi N_0 w_0^2}\right)^2\right]^2} \tag{3.38}$$

where I_0 is the intensity of the incident light, R_1 and R_2 are the reflectivity of the fiber end face and the gold film, respectively. ξ is the coupling coefficient of light transmission in the FP cavity, δ is the phase difference of two coherent light beams,

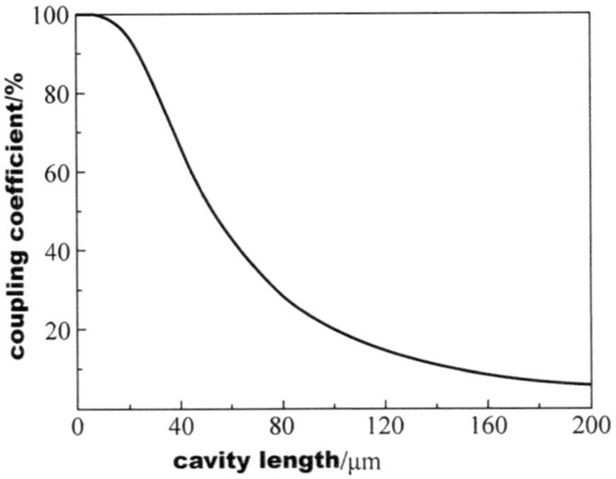

Fig. 3.17 The coupling coefficient-cavity length curve of FP cavity

θ is incident angle, λ is incident light wavelength, and N_0 is the refractive coefficient of the medium in FP cavity, ω_0 is the radius of the mode field. Assuming air medium and SM fiber used, N_0 and ω_0 are 1 μm and 4.9 μm, respectively.

From our previous discussion, it is evident that the MIM structure with a cavity length on the order of the wavelength meets the criteria for both FP resonance mode and SPR mode simultaneously. Figure 3.18 depicts the variation in the reflection spectrum of a grating-dielectric-gold film resonator as the thickness of the dielectric layer changes, for both TM mode (a) and TE mode (b) of the incident light. The gold grating in Fig. 3.18 has a period of $P = 900$ nm, a grating height $t_g = 120$ nm, a grating width $W = 800$ nm, and the gold film mirror has a thickness of $t_m = 100$ nm.

In Fig. 3.18, three resonant modes, TM_0-SP waves, Bloch surface wave, and FP resonance waves are observed in the resonant cavity. The first-order FP resonance mode and TM_0-SPR mode intersects near the thickness of the dielectric layer of 400 nm. This phenomenon suggests a new approach to enhancing the resonance effect in short FP cavities. In Fig. 3.18c, d, as the grating period P increases, the electric field energy scatters, leading to less energy being coupled into the guided mode and a subsequent decrease in resonance mode intensity.

3.3.5 Signal Demodulation of Fiber FP Sensor

As previously mentioned, external physical quantities influence the fiber FP cavity, thereby affecting its output optical signal I_R by altering the cavity length L of the sensor. Consequently, the cavity length L of the fiber FP sensor is a critical parameter for the measurand, and the signal demodulation process of the fiber FP sensor involves deducing the cavity length L from the output signal I_R. Depending on the optical parameters utilized during demodulation, the signal demodulation of the fiber FP sensor primarily consists of intensity demodulation and phase demodulation.

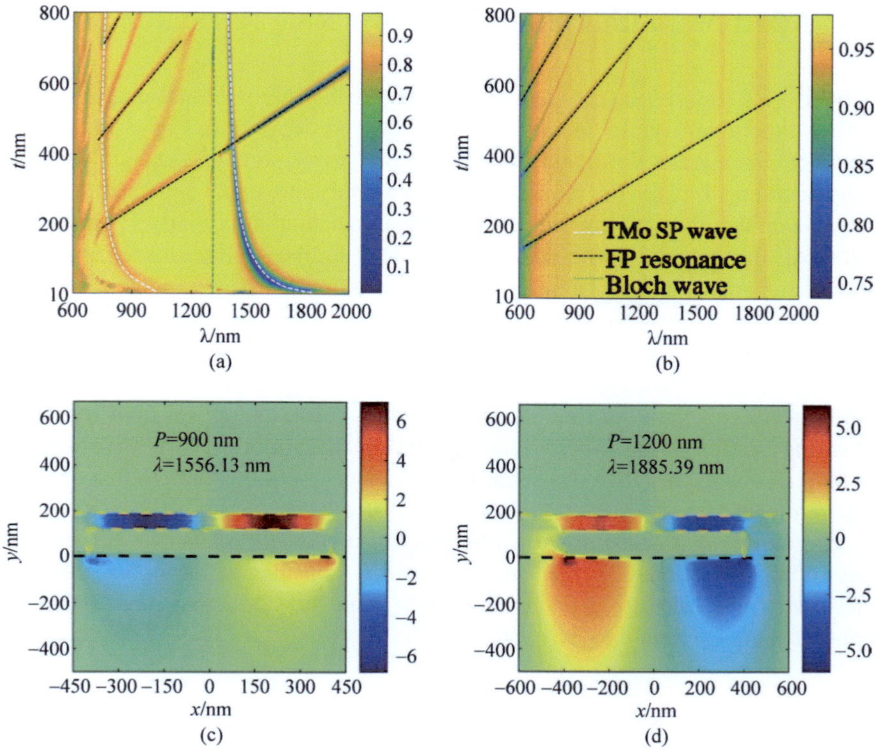

Fig. 3.18 Reflectance spectra of grating-dielectric-gold film resonator with dielectric layer thickness

Intensity demodulation typically employs a monochromatic light source (with a fixed wavelength) to directly solve for L in Eq. (3.35). In contrast, phase demodulation utilizes a broadband or wavelength-tunable light source, determining L by solving Eqs. (3.32) and (3.35) as the output signal I_R varies with wavelength λ, ϕ_R and L. The intensity demodulation method is straightforward, yet its accuracy is relatively low, making it a commonly used technique in early fiber FP sensor research. Phase demodulation is more complex but offers significantly higher accuracy, and is therefore widely used today. Chapter 4 will delve into the demodulation of optical fiber F-P sensor signals in greater detail.

Due to their slender and fragile nature, bare optical fibers are seldom used in practice as fiber FP cavities. Typically, protective structures are added to create fiber FP sensors tailored to the characteristics of the application, such as stress/strain, pressure, temperature, and vibration sensors. By making specific structural modifications, fiber FP strain sensors can be adapted to develop a range of sensors with varying parameters to suit different measurands. Currently, the most prominent application of fiber FP sensors is in the safety monitoring of large structures, such as bridges.

3.4 Multi-beam Interferometer: Fiber Ring Resonator

A fiber optic ring resonator is created by connecting the input and output ends of a fiber directional coupler to form a closed loop, as depicted in Fig. 3.19. When a laser beam is injected from end 1 of the annular cavity, some of the light energy is coupled to end 4, while the remainder passes directly into end 3 and enters the fiber ring. Given that the coupling ratio of the directional coupler is nearly 1, most of the light exits through end 4 when the fiber ring does not meet the resonance condition, and the transmission light intensity is nearly equal to the input light intensity. Upon satisfying the resonance condition, the optical field within the ring is amplified by resonance, and the optical field passing through end 2→4 interferes destructively with the optical field coupled from end 1→4, resulting in a decrease in the output light intensity of the fiber ring. After numerous cycles, the optical field within the fiber ring establishes multi-beam interference, and the output signal at end 4 becomes a sharp resonance negative peak near the resonance condition, akin to that of an FP interferometer.

The output characteristics of a fiber ring resonator are related to the coupling ratio of the directional coupler and the transmission loss of the optical fiber. Its intracavity light intensity, I_3, and output light intensity, I_4, are expressed as relative light intensities within the cavity [4],

$$I_3 = \frac{|E_3|^2}{|E_1|^2} = (1-\gamma)\frac{1-K}{\left(1-\sqrt{KT}\right)^2 + 4\sqrt{KT}\sin^2\left(\beta L + \frac{\pi}{2}\right)} \tag{3.39}$$

The relative light intensity output of the annular cavity is,

$$I_4 = \frac{|E_4|^2}{|E_1|^2} = (1-\gamma)\frac{\left(\sqrt{K}-\sqrt{T}\right)^2 + 4\sqrt{KT}\sin^2\left[\frac{1}{2}\left(\beta L + \frac{\pi}{2}\right)\right]}{\left(1-\sqrt{KT}\right)^2 + 4\sqrt{KT}\sin^2\left[\frac{1}{2}\left(\beta L + \frac{\pi}{2}\right)\right]} \tag{3.40}$$

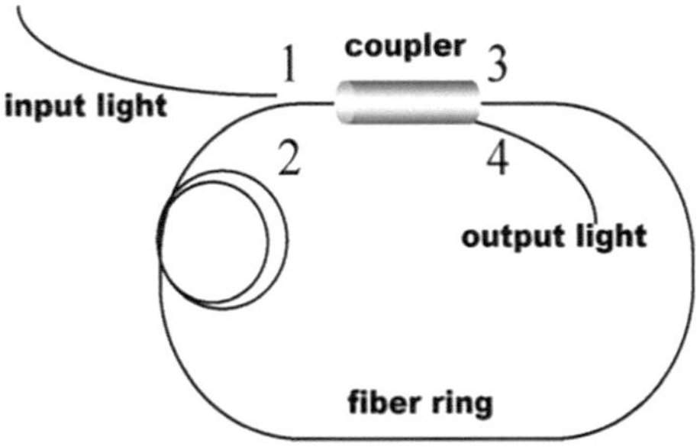

Fig. 3.19 Optical fiber ring cavity interferometer

Where, $\begin{cases} E_2 = exp(-\alpha L)exp(i\beta L)E_3 \\ E_3 = \sqrt{1-\gamma}\left(\sqrt{1-K}E_1 + \sqrt{K}E_2\right) \\ E_4 = \sqrt{1-\gamma}\left(\sqrt{K}E_1 + \sqrt{1-K}E_2\right) \end{cases}$, E_i is the optical amplitude at the directional coupler end i; K and γ are the coupling ratio and insertion loss of the coupler, respectively; α is the amplitude attenuation factor of fiber; β is the propagation constant of light waves in the optical fiber; L is the length of the fiber ring; T is the light intensity transmission factor of the annular cavity loop, and its value is determined by equation $T = (1-\gamma)e^{-2\alpha L}$. T represents the ratio of the light intensity to the initial light intensity after a single cycle of transmission through the optical fiber loop.

From Eqs. (3.39), (3.40), it is evident that the light intensity within the fiber ring is a periodic function of βL. When the phase condition $\beta L = 2m\pi - \frac{\pi}{2}$ (m=1, 2, 3 ...) is met, the relative light intensity of the ring resonator output reaches the minimum, while the relative light intensity in the ring is at the maximum:

$$\left.\begin{array}{r} I_{4min} = \frac{(\sqrt{K}-\sqrt{T})^2}{(1-\sqrt{KT})^2} \\ I_{3max} = (1-\gamma)\frac{1-K}{(1-\sqrt{KT})^2} \end{array}\right\} \tag{3.41}$$

On the contrary, when $\sin^2\left(\frac{1}{2}\beta L + \frac{\pi}{4}\right) = 1$, there are

$$\left.\begin{array}{r} I_{4max} = (1-\gamma)\frac{(\sqrt{K}-\sqrt{T})^2+4\sqrt{KT}}{(1-\sqrt{KT})^2+4\sqrt{KT}} \\ I_{3min} = (1-\gamma)\frac{1-K}{(1-\sqrt{KT})^2+4\sqrt{KT}} \end{array}\right\} \tag{3.42}$$

Figure 3.20 depicts the characteristic curve of the relative intracavity light intensity (I_3) and the relative output light intensity (I_4) of the fiber ring resonator as they vary with the phase βL, given $K = T = 0.95$. As these results stem from multi-beam interference, the interference peaks are notably sharp, yielding an output spectrum characterized by dark peaks against a bright background.

The fineness of the fiber ring resonator is defined as the ratio of the free spectral range of the ring resonator to the width at half maximum of the resonant peak. From the output characteristics of the annular cavity, it can be obtained that the width at the half peak value $\Delta\nu$ is

$$\Delta\nu = \left|\nu_{+\frac{1}{2}} - \nu_{-\frac{1}{2}}\right| = \frac{2c}{n\pi L}\arcsin\left[\frac{1-\sqrt{KT}}{\sqrt{2(1+KT)}}\right] \tag{3.43}$$

And because the width of the free spectral region of the fiber ring is:

$$\text{FSR} = |\nu_{n+1} - \nu_n| = \frac{c}{nL}$$

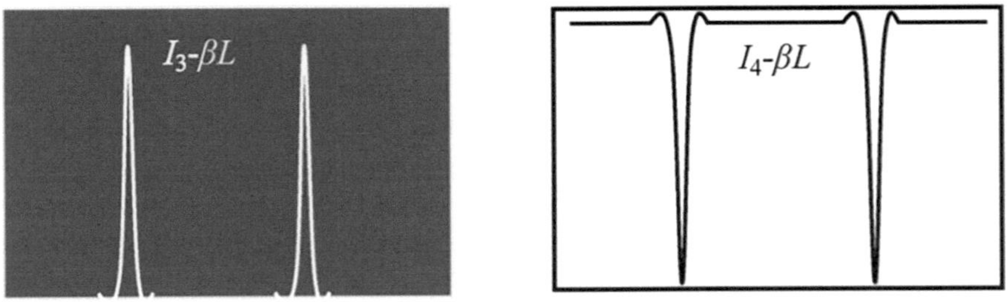

(a) Curve $I_3 - \beta L$; (b) Curve $I_4 - \beta L$

Fig. 3.20 The relative light intensity in fiber ring and output relative light intensity with phase change

The expression for interference fineness can be obtained as follows:

$$F = \frac{\text{FSR}}{\Delta\nu} = \frac{\pi}{2\,arcsin\left[\frac{1-\sqrt{KT}}{\sqrt{2\left(1-\sqrt{KT}\right)}}\right]} \tag{3.44}$$

When $K \approx 1, T \approx 1$, the above equation simplifies to

$$F = \frac{\pi\sqrt{1+KT}}{\sqrt{2}\left(1-\sqrt{KT}\right)} \tag{3.45}$$

3.5 Fiber Optic White Light Interferometer

The phase modulation optical fiber sensor boasts a significant advantage in high sensitivity. However, one drawback is that it can only conduct relative measurements, suitable for gauging changes rather than absolute state quantities. In recent years, interferometers utilizing broadband light sources, such as white light, have been developed for absolute measurements and have consequently garnered increasing interest. Currently, market offerings include low-cost interferometric demodulation devices based on white light interferometric technology.

3.5.1 Principles and Characteristics

An optical fiber white light interference sensor comprises two fiber interferometers. One is positioned at the measurement site as the sensing unit, the FP fiber

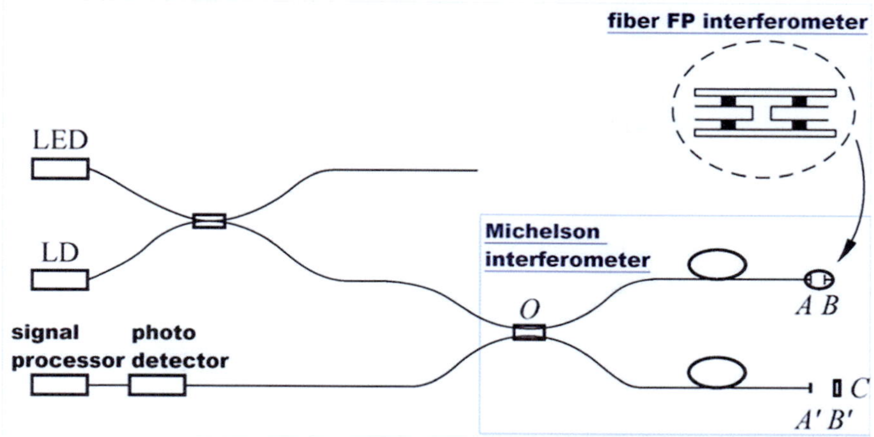

Fig. 3.21 Optical path diagram of fiber white light interferometer

interferometer in Fig. 3.21, and also functions as the sensing arm of the second interferometer.

The other arm of the second interferometer, the Michelson interferometer depicted in Fig. 3.21, acts as a reference arm, offering phase compensation outside the field in the control room. The OPD of each interferometer exceeds the coherence length of the light source. Suppose that position A' in Fig. 3.21 is an equal optical path of point O to A, and B' is an equal optical path of point O to B. At this time, as the reflector C passes through position A' from left to right, the zero-order interference fringes of the white light will appear at the receiving end of the Michelson interferometer. Similarly, when the reflector C passes through position B', the zero-order interference fringes of white light will reappear. The displacement between the positions of A' and B' corresponding to the two zero-order interference fringes represents the optical path of the FP cavity. Consequently, the absolute value of the optical path of the FP cavity can be determined by measuring the spacing between A' and B' using appropriate methods.

In Fig. 3.21, $OA = L_1$, $OB = L_2$, $OC = L$. When adjusting the optical path, ensure $L_2 - L_1 > 2L_C$, L_C represents the coherence length of the light source. The case of interference at plane A is considered below. At this juncture, plane A and reflector C form a Michelson interferometer. According to two-beam interference theory, for monochromatic light with a wavelength of λ, the light intensity detected is

$$\begin{aligned} I_0 &= I_1 + I_2 + 2\sqrt{I_1 I_2}\cos\left[\frac{2\pi}{\lambda}(L - L_1)\right] \\ &= I_\lambda a\left\{1 + \gamma\cos\left[\frac{2\pi}{\lambda}(L - L_1)\right]\right\} \end{aligned} \tag{3.46}$$

where, $a = a_1^2 R_A + a_2^2 R_C$, $\gamma = \frac{2a_1 a_2 \sqrt{R_A R_C}}{a_1^2 R_A + a_2^2 R_C}$. I_λ is the output light intensity of the monochromatic light source; R_A, R_C are the reflectivity of plane A and plane C, respectively; a_1, a_2 are the transmittance of the two arms of the Michelson interferometer; γ is the contrast of the double-beam interference fringes. For a wide spectrum LED, its spectrum exhibits a Gaussian distribution:

$$I_\lambda d\lambda = I_m A \exp\left[-B^2(\nu - \nu_0)^2\right] d\nu \tag{3.47}$$

where, $A = \frac{2}{\Delta\nu_D}\sqrt{\frac{ln2}{\pi}}$, $B^2 = \frac{4ln2}{\Delta\nu_D^2}$. At this time, the light intensity detected by the interferometer is:

$$I_0 = \int I_{out} d\nu = \int I_m A\, exp\left[-B^2(\nu - \nu_0)^2\right] d\nu$$

After substituting the aforementioned conditions and integrating, we obtain:

$$I_0 = I_m a\left[1 + \gamma\, exp\left(-\frac{\pi^2}{4\, ln\, 2}\frac{\Delta L^2}{L_C^2}\right)\cos\left(\frac{2\pi}{\lambda_0}\Delta L\right)\right] \tag{3.48}$$

In actual detection, only the AC component of the output signal is generally taken into consideration, that is:

$$\begin{aligned} I_{OAC} &= I_m a\gamma\, exp\left(-\frac{\pi^2}{4\, ln\, 2}\frac{\Delta L^2}{L_c^2}\right)\cos\left(\frac{2\pi}{\lambda_0}\Delta L\right) \\ &= I_m 2a_1 a_2\sqrt{R_A R_C}exp\left(-\frac{\pi^2}{4\, ln\, 2}\frac{\Delta L^2}{L_c^2}\right)\cos\left(\frac{2\pi}{\lambda_0}\Delta L\right) \end{aligned} \tag{3.49}$$

From the above equation, we can deduce the following conclusion:

- When there is $\boldsymbol{\Delta L = L - L_1 = 0}$, and the two reflecting surfaces have equal optical paths, zero-order interference fringes appear, which is independent of external interference factors.
- The amplitude of the interference signal is dependent on the output power of the light source, the transmission loss of the fiber, and the reflectivity of each mirror.
- External disturbances will affect the amplitude of interference fringes, but will not alter the position of the zero-order interference fringe.

The performance of a white light interferometer largely depends on the scanning accuracy and speed of the scanning mirror, as well as the identification accuracy of the zero-order central fringe. With high-speed mechanical scanning technology, the scanning speed can reach hundreds of meters per second. Electronic scanning technology is more compact, precise, and faster than mechanical scanning methods, and it eliminates the need for any moving parts. Another method to enhance the accuracy of central fringe recognition is a multi-stage squaring signal processing scheme.

3.5.2 *Advantages and Difficulties*

Advantages

(1) Absolute measurement. The ability to measure the absolute optical path.
(2) Strong anti-interference ability. The system exhibits robust anti-interference capabilities, with its resolution being unaffected by the wavelength stability and power fluctuations of the light source or fiber disturbances. The measurement accuracy is solely determined by the positional accuracy of the interference fringe center and the reference reflector.
(3) Versatility. The sensing fiber utilized is a standard single-mode fiber. Typically, no additional special processing of the optical fiber is necessary. Its small size and excellent compatibility make it suitable for embedding within fiber-reinforced composite materials, concrete, or affixing to structural surfaces without significantly affecting its mechanical properties.
(4) Flexible design. The sensor size can be selected according to specific needs. The length of the sensing fiber can range from tens of microns to tens of kilometers, depending on the specific application. A tiny strain sensor is suitable for placement at critical points where high strain is expected in the structure to detect the local strain state of the material. For large constructions, such as suspension bridges, it is crucial to monitor their spatial stability, including deformation. The sensor length must be on the order of meters or more.
(5) Simple to multiplex. Fiber-optic white light interferometers can measure the required information over long distances and are easily multiplexed to achieve quasi-distributed measurements. Multiple sensing optical fibers, or segments (i.e., multiple sensors), are connected to one or more optical fiber buses. Only a single scanning demodulation interferometer is required to interrogate all sensors, and the necessary signal processing circuitry is relatively simple.

Difficulties

The acquisition of a low coherence light source and the detection of zero-order interference fringes are two key issues in white light interference technology. Theoretical analysis indicates that to accurately determine the position of the zero-order interference fringe, it is essential to minimize the coherence length of the light source as much as possible and to select appropriate instruments and methods to enhance the precision of locating the center of the zero-order interference fringe.

As fiber white light interferometric sensing technology advances, an increasing number of application fields are being explored. For instance, the fiber low-coherence large-scale structure sensor, developed for the Swiss industrial construction industry, has achieved a resolution of several microstrains within a measurement range of several thousand microstrains. It is utilized to monitor temperature and one or two-dimensional strains within concrete samples.

Based on the concept of distributed sensors, a quasi-distributed fiber white light interferometric system has been further developed. The experimental system reported by Lecot [5] contains over 100 multiplexed temperature sensors for monitoring the heat generation of a AC stators in a nuclear power plants. It can be expected that such absolute strain sensors, based on white light interferometry will play an increasingly important role in smart structures and materials [6].

3.5.3 Strain and Temperature Measurement Technology Based on Optical Fiber White Light Interferometry

Characterization of Optical Path Variables for Strain and Temperature Measurement

The fundamental parameter of a fiber white light interference sensor is the sensing optical path. Under the assumption of homogeneity, it can be expressed as:

$$S = nL \tag{3.50}$$

where n is the refractive index of the fiber core and L is the nominal length of the fiber sensor.

Generally, the optical path is a function of applied strain and temperature (ε, T), which can be expressed as follows:

$$\mathrm{S} = \mathrm{S}(\varepsilon, T)$$

The increment resulting from the change is

$$dS = \left[\frac{\partial S}{\partial \varepsilon}\right]_T d\varepsilon + \left[\frac{\partial S}{\partial T}\right]_\varepsilon dT$$

where, $d\varepsilon$ and dT represent the changes caused by local strain and temperature, respectively. $\left[\frac{\partial S}{\partial \varepsilon}\right]_T$ 和 $\left[\frac{\partial S}{\partial T}\right]_\varepsilon$ are the derivatives of S with respect to ε and T, respectively. Using Eq. (3.50), the optical path change can be further expanded into

$$dS = \left\{n\left[\frac{\partial L}{\partial \varepsilon}\right]_T + L\left[\frac{\partial n}{\partial \varepsilon}\right]_T\right\}d\varepsilon + \left\{n\left[\frac{\partial L}{\partial T}\right]_\varepsilon + L\left[\frac{\partial n}{\partial T}\right]_\varepsilon\right\}dT \tag{3.51}$$

Using $d\varepsilon = \frac{dL}{L}$ with the thermal expansion coefficient of optical fiber $\alpha_f = \frac{1}{L}\left(\frac{\partial L}{\partial T}\right)$, Eq. (3.47) is rewritten as:

$$dS = nL\left\{\left(1 + \frac{1}{n}\left[\frac{\partial n}{\partial \varepsilon}\right]_T\right)d\varepsilon + \left(\alpha_f + \frac{1}{n}\left[\frac{\partial n}{\partial T}\right]_\varepsilon\right)dT\right\} \tag{3.52}$$

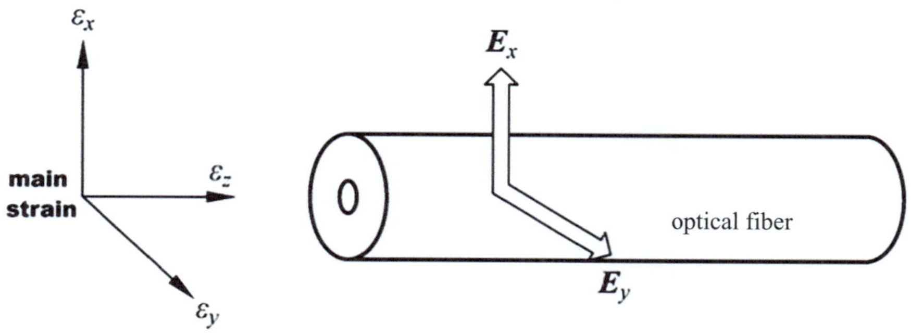

Fig. 3.22 Relationship of the orientation of strain ε_x and ε_y and ε_z with the orientation of $\boldsymbol{E}_{\mathrm{x}}$, $\boldsymbol{E}_{\mathrm{y}}$

Reapplying Eqs. (3.50), (3.52) can be simplified as

$$\frac{dS}{S} = (1 + C_\varepsilon)d\varepsilon + \left(\alpha_f + C_T\right)dT \tag{3.53}$$

where the definitions of coefficients C_ε and $\mathrm{C_T}$ are $C_\varepsilon = \frac{1}{n}\left[\frac{\partial n}{\partial \varepsilon}\right]_T$ and $C_T = \frac{1}{n}\left[\frac{\partial n}{\partial T}\right]_\varepsilon$, respectively.

Under conditions of homogeneous, isotropic, constant temperature and only axial strain ε_z, according to the assumptions of Butter and Hocker [7],

$$C_\varepsilon = -\frac{n^2}{2}[P_{12} - \mu(P_{11} + P_{12})]\varepsilon_z$$

where, P_{11} and P_{12} are the photo-elastic coefficient and μ is the fiber's Poissons ratio.

For more general strain fields, consider the three strain components $\{\varepsilon_x, \varepsilon_y, \varepsilon_z\}$ shown in Fig. 3.22. For light polarized in different directions ($\boldsymbol{E}_{\mathrm{x}}$, $\boldsymbol{E}_{\mathrm{y}}$), the values C_ε will be respectively as follows,

$$\boldsymbol{C}_\varepsilon = -\frac{\boldsymbol{n}^2}{2}\left[\boldsymbol{\varepsilon}_x\boldsymbol{P}_{11} + (\boldsymbol{\varepsilon}_z + \boldsymbol{\varepsilon}_x)\boldsymbol{P}_{12}\right] \quad \text{and} \quad \boldsymbol{C}_\varepsilon = -\frac{\boldsymbol{n}^2}{2}\left[\boldsymbol{\varepsilon}_y\boldsymbol{P}_{11} + (\boldsymbol{\varepsilon}_z + \boldsymbol{\varepsilon}_x)\boldsymbol{P}_{12}\right]$$

White Light Measurement System

There are two reflected signals in a fiber optic white light Michelson interferometer used for actual measurements (as shown in Fig. 3.21): one from the front end face and the other from the rear end face of the sensing fiber. When the reflector in the reference arm scans, two sets of white light interference fringes emerge. The two sets of fringes correspond to the reference optical path, which equals the optical path of the two reflected signals. However, the difference between the two positions of the reflective scanner corresponding to the center fringes of the two groups of white light interference ($X = X_2 - X_1$) depends on the sensor length.

$$X = X_2 - X_1 = nL_0 \tag{3.54}$$

When strain or temperature acts on the sensor, the center fringe position of the white light interference shifts. Equation (3.54) becomes:

$$X' = X_2' - X_1' = (nL_0)'$$

where "'" denotes the position after loading,

$$\Delta X' = X' - X = (nL_0)' - (nL_0) = \Delta(nL_0) = \Delta S \tag{3.55}$$

According to the definition of $S = nL_0$, and the relationship between ΔS and temperature/strain is obtained from Eqs. (3.52) or (3.53).

(1) Strain measurement

Assuming the temperature remains constant and only axial strain $\Delta\varepsilon_z = \varepsilon_z - \varepsilon_{z0}$ is applied over the sensing length, from Eqs. (3.53) and (3.55), we obtain,

$$\Delta X = \Delta S = (1 + C_\varepsilon)nL_0\Delta\varepsilon_z = n_{\text{eq}}L_0\Delta\varepsilon_z \tag{3.56}$$

where, the equivalent refractive index $n_{eq} = n\{1 + C_\varepsilon\}$. For a SMF, the refractive index $n = 1.46(@\lambda = 1300)$ nm, the Poissons ratio $\mu = 0.25$, and the strain coefficient P_{11}=0.12 and P_{12}=0.27, $C_\varepsilon = -0.1332 \times 10^{-6}/\mu\varepsilon$ then $n_{\text{eq}} = 1.19$. The strain acting on the fiber can be given by:

$$\Delta\varepsilon_z = \frac{\Delta X}{n_{\text{eq}}L_0} \tag{3.57}$$

(2) Temperature Measurement

In the absence of strain, if the ambient temperature changes from T_0 to T, the sensitivity of the fiber temperature sensor is calculated by Eq. (3.53) as follows,

$$\begin{aligned} &\Delta X = \Delta S = L_0 n\big(\alpha_f + C_T\big)(T - T_0) = \xi L_0(T_0)(T - T_0) \\ &\text{Or,}\quad (T - T_0) = \frac{\Delta X}{\xi L_0(T_0)},\ \xi = \big(\alpha_f + C_T\big)n(\lambda, T_0) \end{aligned} \tag{3.58}$$

where, ξ is the sensitivity coefficient. For standard SMF, the typical values of relevant parameters are: When $\lambda = 1310$ nm, $n = 1.4681$, $\alpha_f = 5.5 \times 10^{-7}$/°C, the C_{T} value is 0.762×10^{-5}/°C. When $\lambda = 1550$ nm, $n = 1.4675$, $\alpha_f = 5.5 \times 10^{-7}$/°C, the C_{T} value is 0.811×10^{-5}/°C. Utilizing these data, the sensitivity coefficient ξ of the fiber temperature sensor is 11.99 (m·°C) for the unit length of sensing fiber at 1310 nm and 12.71 (m·°C) at 1550 nm, respectively.

3.6 Optical Fiber Polarization Interferometer

A significant drawback of a fiber MZI is the interference occurs between the two arms. External disturbances to the reference arm frequently cause substantial interference or can even disrupt the normal operation of the instrument. To mitigate this issue, sensors can utilize the phase shift difference between the two orthogonal polarization modes in a single high-birefringence SM fiber to detect applied external influence quantities. Figure 3.23 depicts a schematic diagram of a fiber temperature sensor constructed using this method, which is a fiber optic polarization interferometer.

After passing through a fiber polarizer and a $\lambda/4$ wavelet, the laser beam becomes circularly polarized light, which uniformly excites two orthogonal polarization states of the high birefringence SM fiber for sensing. Due to the different phase shifts, the synthetic polarization state of the output light can vary between left-hand circularly polarization light, 45° linearly polarization, right-hand polarization, and 135° linearly polarization. If only the 45° linear polarization component is detected at the output end, the intensity is

$$I = \frac{1}{2} I_0 (1 + \cos \phi)$$

where, φ is the phase change affected by external factors. In order to reduce the instability of the light source itself, Wollaston prism can be used to simultaneously detects the output of the two orthogonal components I_1 and I_2, after data processing there is

$$P = \frac{I_1 - I_2}{I_1 + I_2} = \cos \phi$$

Experiments indicate that the sensitivity of a temperature sensor employing a high birefringence fiber (with a beat length Λ of 3.2 mm) is approximately 2.5 rad/(°C·m). It has a sensitivity of approximately 1/40 that of the MZ double-arm interferometer,

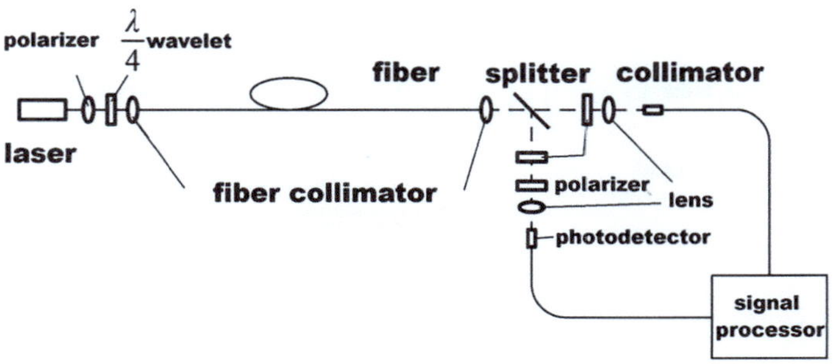

Fig. 3.23 Single fiber polarization interferometer

which is about 100 rad/(°C·m). However, the device is significantly simpler, and its pressure sensitivity is just 1/7300 of the MZ interferometer's, making it effectively pressure desensitized.

3.7 Optical Fiber Long-Range Interferometer

Fiber interferometric sensors and related digital signal processing technologies can be utilized to construct a fiber optic distributed event monitoring and positioning system, which is suitable for long-distance or large-scale regional security monitoring. Currently, the commonly used long-range interferometer structures primarily include high-sensitivity long-range Mach-Zehnder (MZ) interferometers, as well as distributed sensing based on optical time domain reflectometry (OTDR) and/or optical frequency domain reflectometry (OFDR). This section primarily presents the long-range interference structure utilizing the MZ interferometer. The structure boasts high sensitivity and rapid response times, yet it suffers from low positioning accuracy. The system is capable of real-time determination of any dangerous events within the monitored area, with a false alarm rate that is nearly zero. Furthermore, the system can discern the type and urgency of the event. The reported real-time positioning error is under 20 meters and can be achieved within 1 second. The precision and velocity of event monitoring and identification satisfy the security system's requirements. The OTDR and OFDR-based schemes are detailed in Chap. 4.

3.7.1 Principle of Event Monitoring with MZ Interferometer

The basic principle of the system is illustrated in Fig. 3.24. After passing through coupler 1, the light is split and transmitted through two SM fibers, and then combined at coupler 2 where they interfere with each other.

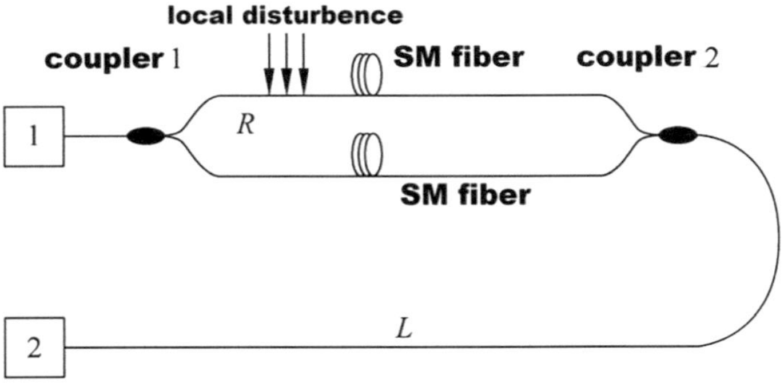

Fig. 3.24 Schematic diagram of distributed MZ interferometer

When the sensor remains undisturbed, the light emitted by light source 1 will generate stable interference fringes at photo detector 2. Suppose a disturbance occurs at a distance R from end 1 of the sensing SM fiber, and the length of the disturbance action is l, the resulting phase change is:

$$\Delta\varphi = \beta\Delta l + l\Delta\beta = \beta l\left(\frac{\Delta l}{l}\right) + l\left(\frac{\partial\beta}{\partial n}\right)\Delta n + l\left(\frac{\partial\beta}{\partial r}\right)\Delta r \tag{3.59}$$

where, $\Delta\varphi$ is phase change, β is the propagation constant of fiber, r is fiber core radius, and n is the refractive index of the fiber core.

MZI is a highly sensitive phase detection solution. Due to its double-arm interferometric structure, changes in the external environment, such as temperature fluctuations, can cause corresponding variations in the fiber length and refractive index. These changes often lead to random phase noise in the transmission process, which can easily reach 2π, resulting in a low signal-to-noise ratio (SNR) of the monitoring signal and the inability to demodulate the event signal. In practical applications, event recognition is typically not achieved through signal demodulation, but rather by using a neural network algorithm to reduce the requirements for the signal-to-noise ratio (SNR) of the optical system.

3.7.2 The Principle of Event Location Using Time Delay Estimation

Localization Principle of Time Delay Estimation

To realize the event monitoring function alone, a single interferometer suffices. When an event affects one arm of the interferometer, the optical path difference (OPD) between the two arms varies, causing a corresponding change in the interference output signal. Event information can be obtained by monitoring these signal changes. However, when it is necessary to both locate and monitor the event, two sets of interferometers with opposite light transmission directions and reciprocal structures are required.

The optical fundamentals of event localization are depicted in Fig. 3.24. Light emitted from end 1 interferes after passing through couplers 1 and 2. Due to the symmetry of the optical path, light emitted from end 2 may also interfere at coupler 1. In the absence of disturbance, the light emitted from end 1 will produce stable interference fringes at 2. Simultaneously, the light emitted from end 2 will also produce stable interference fringes at 1.

Assuming that the phase change is caused by the narrow linewidth laser emitted from end 1 and modulated by the disturbance signal at R, the interference signal from this event reaches 2 after the elapsed time of $t_1 = \frac{(2L-R)n}{c}$, causing changes in the interference fringes, thus altering the light intensity monitored by a PIN photodetector

(PD) at 2. Similarly, the narrow-linewidth laser emitted from end 2 is also modulated by the disturbance signal at R, altering its phase. After a certain elapsed time $t_2 = \frac{Rn}{c}$, it reaches the PIN PD at 1, thus changing the light intensity monitored by the PIN PD at 1. Idealistically, if the two interferometers are perfectly reciprocal, the signal waveforms detected by the PIN PDs at both ends should be identical, except for a brief time delay of ΔT resulting from the varying transmission path lengths.

$$\Delta T = t_1 - t_2 = \frac{(L - R)2n}{c} \tag{3.60}$$

Among them, L represents the length of the sensing arm of the interferometer, n is the refractive index of the fiber core, c is the speed of light, and R is the distance from the event to end 1, which is the position where the event occurred. Given that L, n, and R are known, the disturbance position R can be determined as long as the signal delay $\Delta T = \tau_d$ is measured.

$$R = L - \frac{c \cdot \tau_d}{2n} \tag{3.61}$$

Simultaneously, since the total optical path of the optical circuit is $2L$, the following relationship also holds:

$$t_1 + t_2 = \frac{2Ln}{c} \tag{3.62}$$

As shown in Fig. 3.25, consider the output interference signal of light transmitted counterclockwise, which is received by detector PD1 at point 1, denoted as S1, and the light transmitted clockwise, received by detector PD2 at point 2, denoted as S2.

This structure necessitates the simultaneous injection of light into both ends of the fiber. If the two light sources are not perfectly matched, interference fringes may not be achieved. To address the coherence issue of the light source, coupler A is employed to divide the light from a single laser into two beams. These beams then serve as the light sources for the two interferometers respectively.

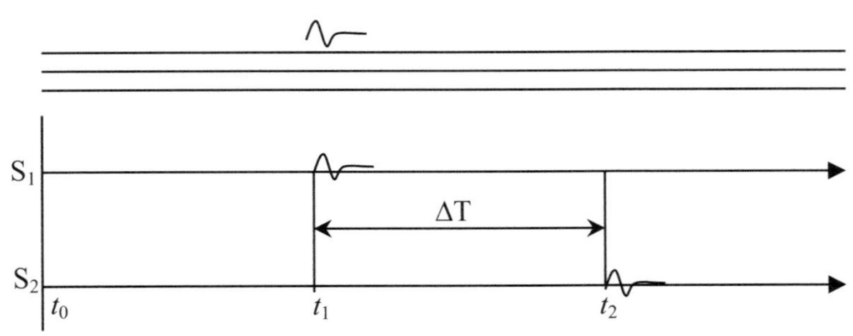

Fig. 3.25 Event location scheme

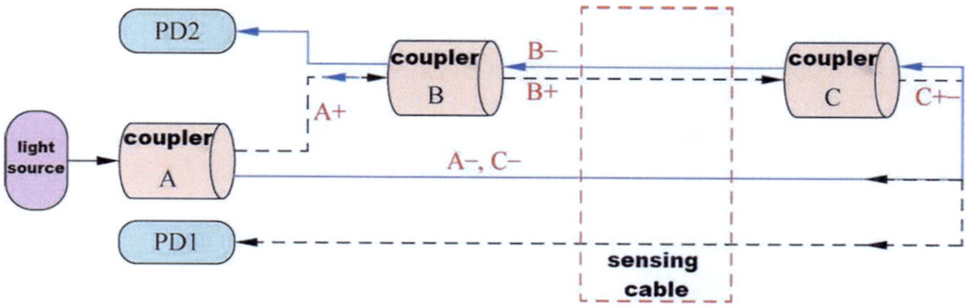

Fig. 3.26 Principal block diagram of optical module of the sensor

The optical path of the system, as depicted in Fig. 3.26, can essentially fulfill the requirements for locating. However, as the sensing cable length continues to increase, the output consistency of the two interferometers gradually deteriorates, impacting the accuracy of the location. Based on the fundamental principles of interferometry, if the two reverse interferometers within the system are reciprocal, the outputs of the optical system should be consistent. Consequently, the residual non-reciprocity in the two interferometers emerges as the primary factor that diminishes the interference consistency of the system. Thus, the factors leading to non-reciprocity must be identified and rectified.

Non-reciprocal Solutions

The light path through the two interferometers in the optical system described above can be characterized as follows:

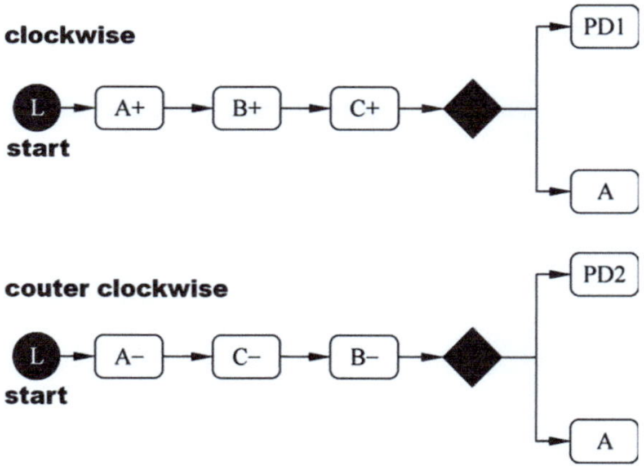

In an ideal scenario, the phase difference between the output light of the two arms after passing through the coupler in an interferometer should be π, meaning the alternating current (AC) terms of the two output light intensities should have opposite signs. Since the error originates from the interference optical path and coupler A has minimal impact on the performance of the interference optical path, we will temporarily disregard the influence of coupler A. Let the outputs of the two arms of coupler B be B_0 and B_π respectively, the absolute value of the disturbance caused by coupler B to the two arms being δ_{B0} and $\delta_{B\pi}$. Similarly, the absolute values of disturbance carried by the sensing arm of coupler C can be set to δ_{C0} and $\delta_{C\pi}$ respectively. Additionally, let the absolute value of the accumulated disturbance between the two interference fiber arms, between B and C, be δ_1 and δ_2.

Due to the phase difference of π between the outputs of the two arms of the coupler, we assume that the phase of one path is 0, and the phase of the other path is π. Consequently, the interference δS accumulated by the clockwise interferometer on the interference optical path can be expressed as:

$$\begin{aligned}\delta_S &= \delta_{B_0} + \delta_1 + \delta_{C_0} + \left(-\delta_{B_\pi}\right) + (-\delta_2) + \left(-\delta_{C_\pi}\right) \\ &= \left(\delta_{B_0} - \delta_{B_\pi}\right) + (\delta_1 - \delta_2) + \left(\delta_{C_0} - \delta_{C_\pi}\right)\end{aligned} \tag{3.63}$$

Similarly, the interference δ_N accumulated by an interferometer in the counter-clockwise direction on the interference optical path can be expressed as:

$$\begin{aligned}\delta_N &= \delta_{C_0} + \delta_2 + \delta_{B_0} + \left(-\delta_{C_\pi}\right) + (-\delta_1) + \left(-\delta_{B_\pi}\right) \\ &= \left(\delta_{B_0} - \delta_{B_\pi}\right) + (\delta_2 - \delta_1) + \left(\delta_{C_0} - \delta_{C_\pi}\right)\end{aligned} \tag{3.64}$$

From the two equations presented, it is evident that for the system's interferometers to be reciprocal, meaning that $\delta_S = \delta_N$, it is necessary for $\delta_1 - \delta_2 = \delta_2 - \delta_1$, which implies that $\delta_2 = \delta_1$. Achieving this with two fibers spanning several kilometers, or even tens of kilometers, is challenging. Moreover, the difference between δ_1 and δ_2 typically varies with the fiber length and the environment, resulting in poor consistency and output instability of the two interferometers. The optimized optical system is shown as follows: the fiber connection mode of the coupler C is made into a crossover (Fig. 3.27).

The primary difference from the previous scheme is that the light path through the two interferometers has changed.

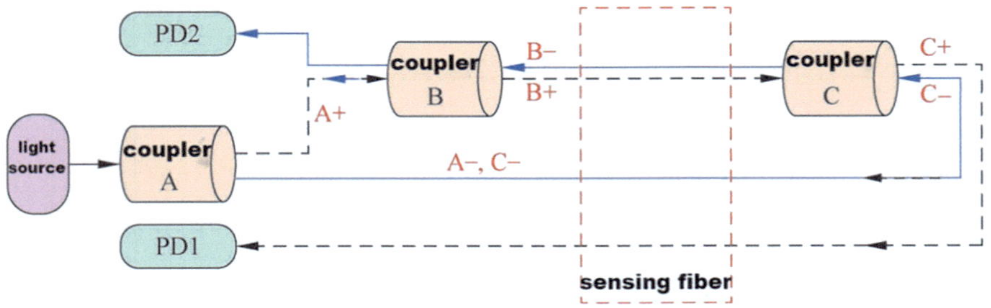

Fig. 3.27 Principal block diagram of the improved sensor optics part

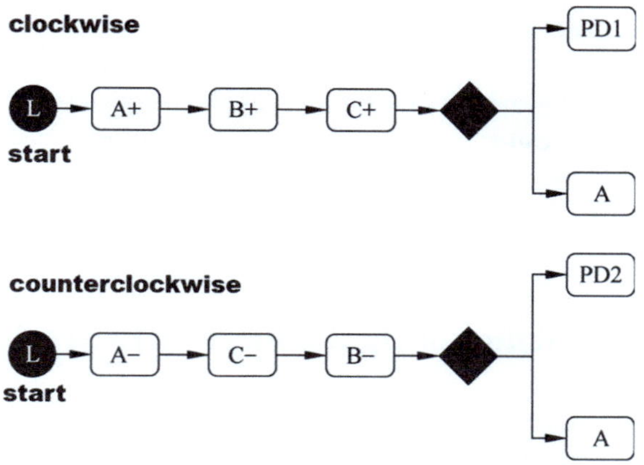

For the time being the influence of coupler A is not taken into account. Let the outputs of the two arms of coupler B be B_0 and B_π respectively, and the absolute value of the disturbance caused by the coupler of the two arms is set to δ_{B0} and $\delta_{B\pi}$ respectively. Similarly, the absolute values of interference carried by the output of the sensor arm of coupler C can be assumed to be δ_{C0} and $\delta_{C\pi}$ respectively. Additionally, let the absolute value of the accumulated interference between the two optical fibers used to form the interference between B and C be δ_1 and δ_2.

Due to the phase difference of π between the outputs of the two arms of the coupler, we assume that the phase of one path is 0, and the phase of the other path is π. The interference δ_S accumulated by the clockwise interferometer on the interference optical path can be expressed as:

$$\begin{aligned}\delta_S &= \delta_{B_0} + \delta_1 + \delta_{C_\pi} + (-\delta_{B_\pi}) + (-\delta_2) + (-\delta_{C_0}) \\ &= (\delta_{B_0} - \delta_{B_\pi}) + (\delta_1 - \delta_2) + (\delta_{C_\pi} - \delta_{C_0})\end{aligned} \tag{3.65}$$

Similarly, the interference δ_N accumulated by an interferometer in the counterclockwise direction on the interference optical path can be expressed as:

$$\begin{aligned}\delta_N &= \delta_{C_0} + \delta_1 + \delta_{B_\pi} + (-\delta_{C_\pi}) + (-\delta_2) + (-\delta_{B_0}) \\ &= (\delta_{B_\pi} - \delta_{B_0}) + (\delta_1 - \delta_2) + (\delta_{C_0} - \delta_{C_\pi})\end{aligned} \tag{3.66}$$

From the above two equations, it is evident that if the two interferometers within the system of the aforementioned structure are to be reciprocal—that is, if $\delta_S = \delta_N$—then δ must satisfy $\delta_{B_0} - \delta_{B_\pi} = \delta_{C_0} - \delta_{C_\pi}$. In other words, the performance of couplers B and C must be consistent. This is relatively much easier and can be achieved by enhancing the coupler manufacturing process and device calibration. The experiment demonstrates that following the improvement of the scheme, the output signals of the two interferometers are essentially consistent, and the locating accuracy is significantly improved.

Moreover, coupler A can also influence the experimental results in a real-world setting. Should coupler A experience considerable interference, the polarization state of the laser might change rapidly, resulting in a significant contrast difference between the two signals. This can also impact the accuracy of location and sensing performance. A potential solution is to incorporate a polarization-independent coupler into the optical path.

3.8 Phase Modulation of Optical Fiber by External Stress

When the external factor is stress, the relationship between $\Delta L/L$ and Δn with pressure necessitates a rewrite of Eq. (3.9). Upon transverse compression of the fiber optic interferometer, the relative phase change can be derived from Eq. (3.12). Based on the principles of elastic mechanics, assuming the relationship between the change in refractive index and strain for isotropic materials is ε_i, the following can be deduced:

$$\begin{bmatrix} \Delta B_1 \\ \Delta B_2 \\ \Delta B_3 \\ \Delta B_4 \\ \Delta B_5 \\ \Delta B_6 \end{bmatrix} = \begin{bmatrix} B_1 - B_0 \\ B_2 - B_0 \\ B_3 - B_0 \\ B_4 \\ B_5 \\ B_6 \end{bmatrix} = \begin{bmatrix} P_{11} & P_{12} & P_{12} & 0 & 0 & 0 \\ P_{12} & P_{11} & P_{12} & 0 & 0 & 0 \\ P_{12} & P_{12} & P_{11} & 0 & 0 & 0 \\ 0 & 0 & 0 & P_{44} & 0 & 0 \\ 0 & 0 & 0 & 0 & P_{44} & 0 \\ 0 & 0 & 0 & 0 & 0 & P_{44} \end{bmatrix} \times \begin{bmatrix} \varepsilon_1 \\ \varepsilon_2 \\ \varepsilon_3 \\ 0 \\ 0 \\ 0 \end{bmatrix}$$

where

$$P_{44} = \frac{1}{2}(P_{11} - P_{12}); \quad B_1 = \frac{1}{n_1^2}; \quad \Delta B_1 = -\frac{2}{n_1^3}\Delta n_1$$

In general, an approximation $n_1 \approx n^0$ can be taken, thus we get,

$$\Delta n_1 = -\frac{1}{2}\left(n^0\right)^3 \Delta B_1 = -\frac{1}{2}\left(n^0\right)^3 (P_{11}\varepsilon_1 + P_{12}\varepsilon_2 + P_{13}\varepsilon_3)$$

Similarly, there are

$$\Delta n_2 = -\frac{1}{2}\left(n^0\right)^3 (P_{12}\varepsilon_1 + P_{11}\varepsilon_2 + P_{12}\varepsilon_3)$$

$$\Delta n_3 = -\frac{1}{2}\left(n^0\right)^3 (P_{12}\varepsilon_1 + P_{12}\varepsilon_2 + P_{11}\varepsilon_3)$$

And there are $B_4 = B_5 = B_6 = 0$; Considering $\beta \approx nk_0$, $\frac{d\beta}{dn} \approx k_0$, and omitting the phase change caused by ΔD, then Eq. (3.12) can be rewritten as

$$\Delta\phi = \beta L\varepsilon_3 + Lk_0\Delta n_i \quad (i = 1,\ 2,\ 3)$$

Or expressed in terms of relative change as

$$\frac{\Delta\phi}{PL} = \frac{\beta}{P}\varepsilon_3 + \frac{k_0}{P}\Delta n_i \quad (i = 1,\ 2,\ 3) \tag{3.67}$$

where, P is the pressure acting on optical fiber, $\varepsilon_1, \varepsilon_2$ are the transverse strains of fiber; $\varepsilon_3 = \frac{\Delta L}{L}$ is the longitudinal strain of fiber; P_{11}, P_{12} are the elastic-optical coefficients of fiber material; n is the refractive index of fiber material. Understanding the strain of the optical fiber after compression, the relative phase change in the sensing arm of the optical fiber interferometer can be determined using the aforementioned formula. For instance, the calculation results for the simplest scenario, where the optical fiber comprises solely a core and cladding, are provided below.

According to the principles of elastic mechanics, the relationship between stress σ and strain ε is

$$\begin{bmatrix} \varepsilon_1 \\ \varepsilon_2 \\ \varepsilon_3 \\ \varepsilon_4 \\ \varepsilon_5 \\ \varepsilon_6 \end{bmatrix} = \begin{bmatrix} \frac{1}{E} & -\frac{\mu}{E} & -\frac{\mu}{E} & 0 & 0 & 0 \\ -\frac{\mu}{E} & \frac{1}{E} & -\frac{\mu}{E} & 0 & 0 & 0 \\ -\frac{\mu}{E} & -\frac{\mu}{E} & \frac{1}{E} & 0 & 0 & 0 \\ 0 & 0 & 0 & \frac{1}{G} & 0 & 0 \\ 0 & 0 & 0 & 0 & \frac{1}{G} & 0 \\ 0 & 0 & 0 & 0 & 0 & \frac{1}{G} \end{bmatrix} \times \begin{bmatrix} \sigma_1 \\ \sigma_2 \\ \sigma_3 \\ \sigma_4 \\ \sigma_5 \\ \sigma_6 \end{bmatrix}$$

When the optical fiber is compressed solely laterally, its stress is

$$\begin{bmatrix} \sigma_1 \\ \sigma_2 \\ \sigma_3 \\ \sigma_4 \\ \sigma_5 \\ \sigma_6 \end{bmatrix} = \begin{bmatrix} -P \\ -P \\ 0 \\ 0 \\ 0 \\ 0 \end{bmatrix}$$

The corresponding strain is: $\begin{bmatrix} \varepsilon_1 \\ \varepsilon_2 \\ \varepsilon_3 \\ \varepsilon_4 \\ \varepsilon_5 \\ \varepsilon_6 \end{bmatrix} = \begin{bmatrix} -\frac{P(1-\mu)}{E} \\ -\frac{P(1-\mu)}{E} \\ \frac{2\mu P}{E} \\ 0 \\ 0 \\ 0 \end{bmatrix}$

Then, the relative change in phase shift can be obtained:

$$\frac{\Delta\phi}{PL} = nk_0\frac{2\mu}{E} + \frac{k_0}{2E}n^3[(1-\mu)P_{11} + (1-3\mu)P_{12}] \tag{3.68}$$

Similarly, the relative phase change $\Delta\phi$ under longitudinal compression and uniform compression can be obtained. The calculation examples and results are

Table 3.1 Influence of external pressure on phase change

	Transverse compression ***P***	Longitudinal compression ***P***	Uniformly compressed ***P***
Stress σ	$\begin{bmatrix} -P \\ -P \\ 0 \\ 0 \\ 0 \\ 0 \end{bmatrix}$	$\begin{bmatrix} 0 \\ 0 \\ -P \\ 0 \\ 0 \\ 0 \end{bmatrix}$	$\begin{bmatrix} -P \\ -P \\ -P \\ 0 \\ 0 \\ 0 \end{bmatrix}$
Strain ε	$\begin{bmatrix} -\frac{P(1-\mu)}{E} \\ -\frac{P(1-\mu)}{E} \\ \frac{2\mu P}{E} \\ 0 \\ 0 \\ 0 \end{bmatrix}$	$\begin{bmatrix} \frac{\mu P}{E} \\ \frac{\mu P}{E} \\ -\frac{P}{E} \\ 0 \\ 0 \\ 0 \end{bmatrix}$	$\begin{bmatrix} -\frac{P(1-2\mu)}{E} \\ -\frac{P(1-2\mu)}{E} \\ -\frac{P(1-2\mu)}{E} \\ 0 \\ 0 \\ 0 \end{bmatrix}$
$\frac{\Delta\phi}{PL}$	$\frac{2k_0 n\mu}{E} + \frac{k_0 n^3}{2E} \times [(1-\mu)P_{11} + (1-3\mu)P_{12}]$	$\frac{-k_0 n}{E} + \frac{k_0 n^3}{2E} \times[-\mu P_{11} + (1-\mu)P_{12}]$	$\frac{-k_0 n(1-2\mu)}{E} + \frac{k_0 n^2}{2E} \times (1-2\mu)(P_{11}+P_{12})$
*	$0.70 + 0.51 = 1.21$	$-2.07 + 0.45 = 1.62$	$-1.37 + 0.96 = -0.41$

* The first term is the value of $\frac{\Delta L}{L}$, and the second term is Δn. $\lambda = 632.8\text{nm}$, and for quartz: n=1.456, $P_{11} = 0.121$, $P_{12} = 0.270$, $E = 7 \times 10^{10} Pa$, $\mu = 0.1$.

listed in Table 3.1. From the calculation results, it can be observed that the change in fiber length contributes more significantly than that of the refractive index, and the two calculation results have opposite signs (except under transverse compression).

To determine the pressure sensitivity of the fiber interferometer, an analysis must be conducted considering the actual multi-layer structure of the fiber: the core, cladding, substrate (typically quartz to minimize loss from the outer coating), primary coating (usually a soft coating to reduce micro-bending loss in the optical fiber), and secondary coating (harder, to preserve the optical fiber's strength). Additionally, the relationship between stress σ and strain ε within the optical fiber must be examined.

$$\begin{bmatrix} \sigma_r^{(i)} \\ \sigma_\theta^{(i)} \\ \sigma_z^{(i)} \end{bmatrix} = \begin{bmatrix} \lambda^{(i)} + 2\mu & \lambda^{(i)} & \lambda^{(i)} \\ \lambda^{(i)} & \lambda^{(i)} + 2\mu & \lambda^{(i)} \\ \lambda^{(i)} & \lambda^{(i)} & \lambda^{(i)} + 2\mu \end{bmatrix} \times \begin{bmatrix} \varepsilon_r^{(i)} \\ \varepsilon_\theta^{(i)} \\ \varepsilon_z^{(i)} \end{bmatrix} \tag{3.69}$$

where, $\sigma_r^{(i)}$, $\sigma_\theta^{(i)}$, $\sigma_z^{(i)}$, and $\varepsilon_r^{(i)}$, $\varepsilon_\theta^{(i)}$, $\varepsilon_z^{(i)}$ represent the components of stress and strain in the ith layer (with the core denoted as i=0, and the cladding as i=1, 2, …) of an optical fiber in polar coordinates. And $\lambda^{(i)}$ is the Lame coefficient, which is related to Youngs modulus $E^{(i)}$ and Poissons ratio $\mu^{(i)}$ as

$$\lambda^{(i)} = \frac{\mu^{(i)} E^{(i)}}{\left(1 + \mu^{(i)}\right)\left(1 - 2\mu^{(i)}\right)}$$

For a cylinder, the strain can be obtained from Lames solution as:

$$\left.\begin{array}{r} \varepsilon_r^{(i)} = U_0^{(i)} + \frac{U_1^{(i)}}{r^2} \\ \varepsilon_\theta^{(i)} = U_0^{(i)} - \frac{U_1^{(i)}}{r^2} \\ \varepsilon_z^{(i)} = W_0^{(i)} \end{array}\right\} \tag{3.70}$$

where $U_0^{(i)}$, $U_1^{(i)}$ and $W_0^{(i)}$ are the constants determined by boundary conditions. Since the stress at the center of the fiber will not become infinite, there should be $U_1^{(0)} = 0$ in the core.

For an optical fiber with m layers, the boundary conditions for determining the three constants of $U_0^{(i)}$, $U_1^{(i)}$ and $W_0^{(i)}$ are as follows

$$\sigma_r^{(i)}\big|_{r=r_i} = \sigma_r^{(i+1)}\big|_{r=r_i}; \quad U_r^{(i)}\big|_{r=r_i} = U_r^{(i+1)}\big|_{r=r_i} \tag{3.71}$$

$$\sigma_r^{(m)}\big|_{r=r_m} = -P; \quad \sum_{i=0}^{m} \sigma_z^{(i)} A_i = -PA_m \tag{3.72}$$

$$\varepsilon_z^{(0)} = \varepsilon_z^{(i)} = \cdots = \varepsilon_z^{(m)} \tag{3.73}$$

$$U_r^{(i)} = \int \varepsilon_r^{(i)} dr \tag{3.74}$$

where, $U_r^{(i)}$, r_i and A_i represent the radial displacement, the radius, and the cross-section area of the i th layer, respectively.

Equation (3.71) demonstrates that the radial stress and displacement are continuous along the interface of each layer; Eq. (3.72) indicates that the applied pressure is considered static; and Eq. (3.73) shows that the axial strains of different layers are equal, with end effects ignored. For the optical fiber, the error from neglecting end effects is less than 1%. Utilizing the aforementioned boundary condition expressions (3.71), (3.72), and (3.69), the values of ε_r and ε_z can be determined, from which the sensitivity value $\frac{\Delta\phi}{(\phi\Delta P)}$ can be calculated using Eq. (3.68).

It is shown that the material of the secondary coating has the greatest influence on the pressure sensitivity of single-mode fiber. The calculation results indicate that the soft cladding of the primary coating has minimal effect on the pressure sensitivity of the interferometer. The outer cladding material of the secondary coating significantly affects the pressure sensitivity. As the outer cladding thickness increases, the pressure sensitivity of the fiber approaches a limit value, which is independent of the cladding material's Young's modulus. When the cladding is thick, the static pressure induces isotropic stress in the fiber, and its magnitude is solely related to the compressibility of the outer cladding (inversely proportional to the bulk modulus). Thus, in the case of a thick outer cladding (approximately 5 mm), the pressure sensitivity of an optical fiber is primarily dictated by the bulk modulus of the cladding and is independent of other elastic moduli. Optical fibers with hard sheaths exhibit less variation in sensitivity with frequency, whereas those with nylon sheaths display the least change. Conversely, fibers with soft synthetic rubber sheaths treated with ultraviolet rays experience the most significant alterations in sensitivity as frequency changes. The coating exhibiting the highest sensitivity is polytetrafluoroethylene (TFE), while the one with the lowest sensitivity is that cured with soft UV light.

It is worth mentioning that the sensitivity of a MZ fiber interferometer in detecting sound waves in the air is significantly higher than that in water. When the surface of the optical fiber is exposed to acoustic wave pressure ΔP, it experiences not only the OPD directly caused by the pressure change but also an additional OPD resulting from the temperature increase of the optical fiber due to the adiabatic process.

$$\frac{\Delta\phi}{\phi} = \frac{1}{\phi}\frac{\delta\phi}{\delta T}\bigg|_p \Delta T + \frac{1}{\phi}\frac{\delta\phi}{\delta P}\bigg|_T \Delta P \tag{3.75}$$

where, $\Delta T = \frac{\delta T}{\delta P}\big|_{\text{surf}}\Delta P$. In addition to the fiber material and shape, the characteristics of the medium surrounding the optical fiber also play a role. For instance, $\frac{\delta T}{\delta P}\big|_{\text{surf}}$ in water and air correspond to 6×10^{-6} K/Pa and 9×10^{-2} K/Pa, respectively. This indicates that the temperature-related term can be completely ignored in underwater acoustic sensing. However, when the bare optical fiber is placed in air, the temperature change term is 2×10^3 times that of the pressure change term, resulting in a measured sensitivity that is one order of magnitude higher than that of underwater acoustic sensing.

3.9 Temperature Induced Phase Modulation of Fiber

The principle of temperature sensing using an MZ interferometer is similar to that of pressure sensing. The only difference is that the cause of the phase change is temperature. For a bare optical fiber with a length of L and a refractive index of n, the relative phase change with temperature is as follows:

$$\frac{\Delta\phi}{\phi\Delta T} = \frac{1}{n}\left(\frac{\delta n}{\delta T}\right) + \frac{1}{\Delta T}\left\{\varepsilon_z - \frac{n^2}{2}\left[(P_{11}+P_{12})\varepsilon_r + p_{11}\varepsilon_z\right]\right\} \tag{3.76}$$

where P_{11} represents the photoelastic coefficient of the core; ε_z denotes the axial strain; ε_r signifies the radial strain. As previously mentioned, optical fibers typically have multi-layer structures, hence the values of ε_z and ε_r are associated with the properties of the outer layer material. Suppose the variation in strain due to temperature alteration of ΔT is

$$\left.\begin{aligned} \varepsilon_r^{(i)} &\to \varepsilon_r^{(i)} - a^{(i)}\Delta T \\ \varepsilon_\theta^{(i)} &\to \varepsilon_\theta^{(i)} - a^{(i)}\Delta T \\ \varepsilon_z^{(i)} &\to \varepsilon_z^{(i)} - a^{(i)}\Delta T \end{aligned}\right\} \tag{3.77}$$

$a^{(i)}$ is the linear thermal expansion coefficient of the material of the ith layer. Substituting Eq. (3.77) into the aforementioned stress-strain relationship (3.69) yields,

$$\begin{aligned} \begin{bmatrix} \sigma_r^{(i)} \\ \sigma_\theta^{(i)} \\ \sigma_z^{(i)} \end{bmatrix} &= \begin{bmatrix} \lambda^{(i)} + 2\mu^{(i)} & \lambda^{(i)} & \lambda^{(i)} \\ \lambda^{(i)} & \lambda^{(i)} + 2\mu^{(i)} & \lambda^{(i)} \\ \lambda^{(i)} & \lambda^{(i)} & \lambda^{(i)} + 2\mu^{(i)} \end{bmatrix} \begin{bmatrix} \varepsilon_r^{(i)} \\ \varepsilon_\theta^{(i)} \\ \varepsilon_z^{(i)} \end{bmatrix} \\ &\quad - \left(3\lambda^{(i)} + 2\mu^{(i)}\right) \begin{bmatrix} a^{(i)}\Delta T \\ a^{(i)}\Delta T \\ a^{(i)}\Delta T \end{bmatrix} \end{aligned} \tag{3.78}$$

The aforementioned Eqs. (3.78) and (3.67) can be employed to determine ε_z and ε_r, subsequently enabling the resolution of $\frac{\Delta\phi}{(\phi\Delta T)}$ through Eq. (3.76).

For example, for a typical SM fiber with four-layer, the boundary conditions are as follows:

$$\begin{aligned} &\sigma_r^{(3)}\big|_{r=d} = 0 \\ &\sigma_z^{(3)}A_3 + \sigma_z^{(2)}A_2 + \sigma_z^{(1)}A_1 + \sigma_z^{0}A_0 = 0 \end{aligned} \tag{3.79}$$

$$\sigma_r^{(3)}\big|_{r=c} = \sigma_r^{(2)}\big|_{r=c},\ \sigma_r^{(2)}\big|_{r=b} = \sigma_r^{(1)}\big|_{r=b},\ \sigma_r^{(1)}\big|_{r=a} = \sigma_r^{(0)}\big|_{r=a} \tag{3.80}$$

$$U_r^{(3)}\big|_{r=c} = U_r^{(2)}\big|_{r=c},\ U_r^{(2)}\big|_{r=b} = U_r^{(1)}\big|_{r=b},\ U_r^{(1)}\big|_{r=a} = U_r^{(0)}\big|_{r=a}, \tag{3.81}$$

$$\varepsilon_z^{(3)} = \varepsilon_z^{(2)} = \varepsilon_z^{(1)} = \varepsilon_z^{(0)} \tag{3.82}$$

Equation (3.68) indicates that radial stress and displacements are continuous; Eq. (3.70) demonstrates that the axial stresses of different layers are equal. The corresponding parameter values $\frac{\Delta\phi}{(\phi\Delta T)}$ can be obtained by utilizing typical parameters of the SM fiber, and the calculation result is as follows.

$$\frac{\Delta\phi}{\phi\Delta T} = 0.71 \times 10^{-5}/^\circ\text{C or } \frac{\Delta\phi}{L\Delta T} = 103\ \text{rad}/(^\circ\text{C}\cdot\text{m})$$

This value is consistent with the actual measurement result.

References

1. Brain C (1983) Optical systems and sensors for measurement and control. J Phys E Sci Instrum 16:978–986
2. Wei J, Shuangchen R et al (2005) New progress in optical fiber sensing. Science Press, Beijing
3. Yuan LB (2002) Multiplexed white-light interferometric fiber-optic sensor matrix with a long-cavity fabry-perot resonator. Appl Opt 41(22):4460–4466
4. Chaoyu Y (1988) Research on high fineness optical fiber ring cavity and its application [D]. Tsinghua University, Beijing
5. Lecot C, Guerin JJ, Lequime M (1993) White light fiber optic sensor network for the thermal monitoring of the state in a nuclear power plant alternator sensors. Proceedings 9th international conference on optical fiber sensors, Florence, Italy, pp 271–274
6. Inaudi D, Elamari A, Pflug L et al (1994) Low-coherence deformation sensors for monitoring of civil-engineering structures. Sensors and Actuators A 44:125–130
7. Butter CD, Hocker GP (1978) Fiber optic strain gauge. Appl Opt 17:2867–2869

Chapter 4
Signal Processing of Fiber Interferometer

4.1 Overview

Fiber optic interferometers, functioning as phase modulation sensors, leverage the phase change of light within optical fibers to characterize variations in measured physical quantities. Owing to the fact that even minute disturbances in physical quantities can induce significant phase changes in light propagating through the fiber, phase modulation fiber optic sensors exhibit the highest sensitivity among all fiber optic sensor types. Concurrently, their high sensitivity makes them susceptible to various external environmental factors, such as temperature, pressure, and vibration.

There are many different ways to classify fiber interferometers. According to the structure of the interference optical path, they can be divided into double-beam, multi-beam, and long-range interferometers; according to the measurement method, they can be divided into direct measurement (such as temperature and strain) and indirect measurement (such as vibration and acoustic waves); according to the type of light source used, they can be divided into laser interferometers and broadband/white light interferometers.

The output signal of the fiber interferometer has periodic and nonlinear characteristics, and complex signal processing methods are required to demodulate the phase signal. Good interference signal processing technology has the following basic characteristics: a linear relationship between the interferometer phase change and the measured physical quantity, uniform sensitivity across the entire measurement range, and the ability to automatically distinguish the direction of phase change when it exceeds 2π. In practical applications, different application scenarios call for different demodulation schemes. This chapter uses the interference optical path structure as a guide, discussing three types of interferometers: double-beam, multi-beam, and long-range. It explores various demodulation technologies suitable for different interference types through relative and absolute measurement methods. The chapter also deeply analyzes the impact of the interferometer light source spectrum on system

Y. Liao et al., *Fiber Optic Interferometer*, Springer-TUP Physics Series,
https://doi.org/10.1007/978-981-95-2558-4_4

performance and introduces two typical photodetector modes: single-ended detection and differential balanced detection. This comprehensive technical framework facilitates a better understanding of fiber interferometer signal processing.

4.2 Signal Processing of Dual-Beam Fiber Interferometer

Fiber Mach–Zehnder (M-Z), Michelson, Sagnac, and Fabry-Pérot (FPI) interferometers all perform double-beam interferometry. Different light sources (laser, broadband, laser-broadband) enable them to perform absolute or relative measurements. which require distinct signal processing methods. Both static absolute and dynamic relative measurements will be discussed separately below.

4.2.1 Laser-Based Interferometric Relative Measurement

Lasers are widely employed as light sources in dual-beam interferometry. Owing to their exceptional monochromaticity and extended coherence length, interferometers utilizing lasers can tolerate significant optical path differences (OPD) between their two arms, thereby enabling ultrahigh sensitivity measurements. A representative example is the fiber-optic M-Z interferometer configuration, as schematically depicted in Fig. 4.1.

The fiber-optic M-Z interferometer comprises two cascaded 3dB directional couplers (labeled Coupler 1 and Coupler 2). Continuous-wave laser light is split into two coherent beams via Coupler 1. These beams propagate through the interferometer's reference and sensing arms, respectively, before undergoing phase-sensitive interference at Coupler 2. The resultant interference pattern is detected by a photodiode, followed by signal conditioning electronics for quantitative analysis.

When the sensing area (i.e., the signal arm of the interferometer) remains undisturbed, the output interference signal approximates a stable DC level with minor low-frequency drift caused by environmental factors such as temperature fluctuations. If a disturbance occurs at a specific location within the sensing area, simultaneous changes in both the geometric length (L) and refractive index (n) of the interferometer arms introduce an additional phase shift to the propagating light waves. This phase modulation directly alters the output optical power of the interferometer, which follows the characteristic interference intensity relationship:

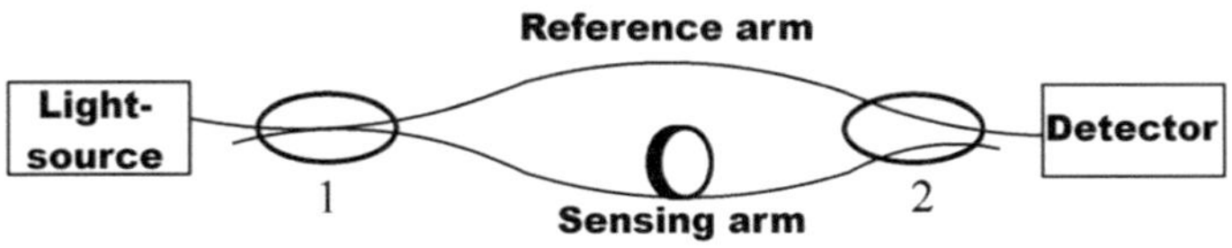

Fig. 4.1 Schematic of fiber-optic M-Z interferometer

$$I_{out} = \frac{1}{2} I_0 e^{-\alpha l} [1 + V \cos(\phi_0 + \Delta\phi)] \tag{4.1}$$

where I_0 is the output power of the light source, a is the loss of the optical fiber, l is the arm length of the interferometer; V is the interference visibility, which is related to the spectral shape of the light source, the OPD of the interferometer and the polarization state of the light wave in the optical fiber. $\phi_0 = \frac{2\pi nL}{\lambda}$ is the initial phase difference of the interferometer; n is the refractive index of the optical fiber, L is the arm length difference of the interferometer, and λ is the laser wavelength.

Environmental perturbations induce variations in three critical parameters of the interferometric system: the geometric length of the interferometer arms (L), the effective refractive index (n), and the laser source wavelength (λ). The resultant phase shift ($\Delta\phi$) in the interferometer can be comprehensively described by the following multiparameter sensitivity relationship:

$$\Delta\phi = \frac{2\pi(\Delta n)L}{\lambda} + \frac{2\pi n(\Delta L)}{\lambda} + \frac{2\pi nL}{\lambda^2}(\Delta\lambda) \tag{4.2}$$

The micrometer-scale wavelength (λ) in the denominator of the phase equation endows laser-based dual-beam interferometers with ultrahigh sensitivity. This configuration is the preferred choice when monitoring relative changes (dynamic quantities) rather than requiring absolute measurements (static quantities).

In order to better illustrate the interferometer signal processing method, we use ϕ to replace $\phi_0 + \Delta\phi$, And simplify Eq. (4.1) as follows:

$$I_{\text{out}} = A + B\cos(\phi) \tag{4.3}$$

Equation (4.3), $A = \frac{1}{2} I_0 e^{-\alpha l}$, $B = \frac{1}{2} I_0 e^{-\alpha l} V$, l is the interferometer arm length, the output optical power of the interferometer exhibits a cosine dependence on the phase, characterized by its non-monotonic periodic behavior. Signal processing in dual-beam interferometry involves extracting the phase parameter ϕ from the measured intensity signal using Eq. (4.3). To address diverse application requirements, this section elaborates on two quintessential demodulation schemes: Orthogonal Detection Method and Phase-Locked Loop (PLL) Method.

The orthogonal detection method employs specialized modulation and demodulation techniques to derive two quadrature components —$\cos(\phi)$ and $\sin(\phi)$— from the right-hand side of Eq. (4.3). These orthogonal signals enable linear reconstruction of the phase parameter ϕ, constituting the fundamental principle of the In-phase/Quadrature (*I/Q*) demodulation method, a widely recognized signal processing paradigm.

The PLL method utilizes negative feedback circuitry to lock the interferometer's operating phase near the quadrature point ($\approx \frac{\pi}{2}$), corresponding to the linear regime of the cosine function in Eq. (4.3). Under small-phase-deviation conditions, $\cos(\phi + \pi/2) = -\sin(\phi) \approx -\phi$, this configuration enables linearized extraction of the phase parameter ϕ. Renowned for its straightforward implementation, the PLL approach is particularly suited for laboratory validation and proof-of-concept demonstrations.

Orthogonal Detection

As described above, orthogonal detection begins by extracting two orthogonal components from Eq. (4.3): $I = \cos(\phi)$ and $Q = \sin(\phi)$. From these, the ratio $\frac{Q}{I} = \tan(\phi)$ is derived, allowing the linear expression of ϕ to be resolved by calculating the arctangent of the phase-dependent tangent value using a lookup table method. Alternatively, the Coordinate Rotation Digital Computer (CORDIC) algorithm can directly compute ϕ from the I and Q components through iterative angle accumulation, bypassing the need for lookup tables entirely.

Three widely adopted methods for extracting I and Q components include: Phase Generated Carrier (PGC) technique, Heterodyne method, and 3×3 coupler approach.

Phase Generated Carrier (PGC) Technique

The operational principle of the Phase Generated Carrier (PGC) technique is illustrated in Fig. 4.2.

The interferometric phase difference can be mathematically expressed as Eq. (4.4):

$$\Delta\phi = \frac{2\pi(\Delta n)L}{\lambda} + \frac{2\pi n(\Delta L)}{\lambda} + \frac{2\pi nL}{\lambda^2} \cdot (\Delta\lambda) \tag{4.4}$$

This phase difference is generated by one of two modulation approaches: Modulating the wavelength of laser source using a single-frequency signal $\Delta\lambda\cos(\omega_c t)$ at frequency ω_c; Modulating the reference arm length using a single-frequency signal $\Delta L\cos(\omega_c t)$ via piezoelectric ceramics (PZT).

These signals interfere at coupler C and, following photoelectric conversion, produce:

$$I_{out} = A + B\cos\left(M\cos(\omega_c \mathrm{t}) + \phi\right) \tag{4.5}$$

where $\cos(\omega_c t)$ represents the applied modulation signal. The term M denotes the modulation index, which scales proportionally with either: The wavelength modulation amplitude $\Delta\lambda$ (for laser modulation), The length modulation amplitude ΔL (for PZT-based modulation).

A schematic of the demodulation process for Eq. (4.5) is provided in Fig. 4.3.

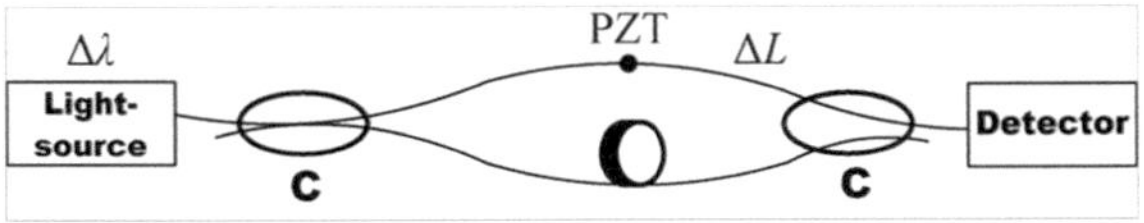

Fig. 4.2 Principle of phase-generated carrier

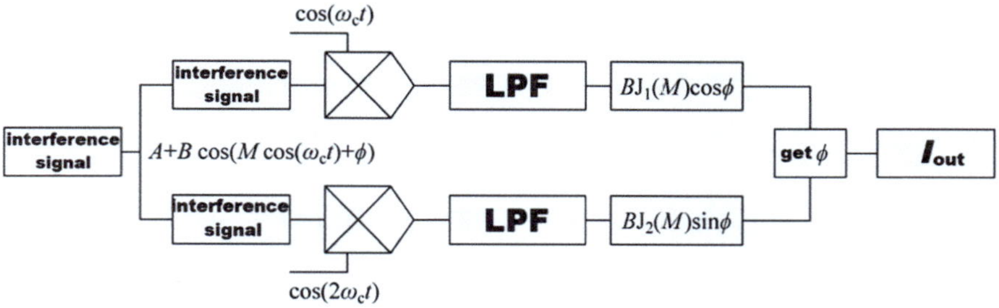

Fig. 4.3 Block diagram of PGC demodulation algorithm (LPF: low pass filter)

The right side of Eq. (4.5) can be expanded into:

$$A + B\left\{\left[J_0(M) + 2\sum_{k=1}^{\infty}(-1)^k J_{2k}(M)\cos(2k\omega_c t)\right]\cos(\phi)\right.$$
$$\left.-2\left[\sum_{k=0}^{\infty}(-1)^k J_{2k+1}(M)\cos((2k+1)\omega_c t)\right]\sin(\phi)\right\} \tag{4.6}$$

Equation (4.6) and the local oscillator signal are independently mixed, then processed through a low-pass filter (LPF) to yield:

$$BJ_1(M)\sin(\phi), BJ_2(M)\cos(\phi) \tag{4.7}$$

By ensuring $M = 2.63$, *i.e.*, $J_1(M) = J_2(M)$, the target phase signal ϕ can be directly extracted from the two orthogonal components of Eq. (4.7).

Heterodyne Method

Unlike the homodyne-based Phase Generated Carrier (PGC) method, the heterodyne scheme utilizes a single laser beam split into two interferometer arms, where each arm is frequency-modulated to carry distinct optical frequencies with a fixed offset (Δf). This principle is illustrated in Fig. 4.4.

As depicted in Fig. 4.4, the output light from the laser source is split into two beams via Coupler 1. One beam propagates through the M-Z signal arm, where an

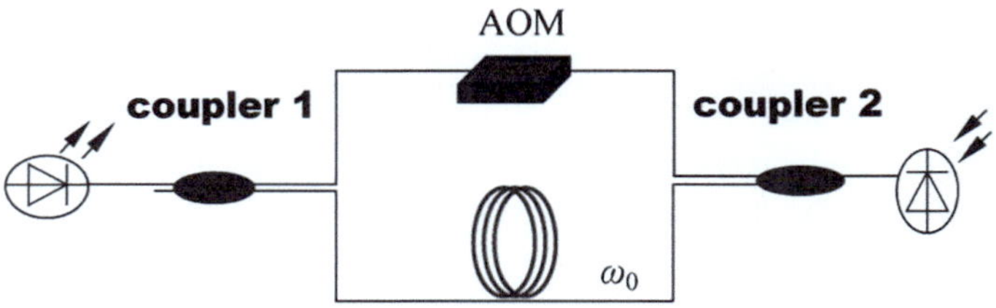

Fig. 4.4 Schematic diagram of heterodyne method

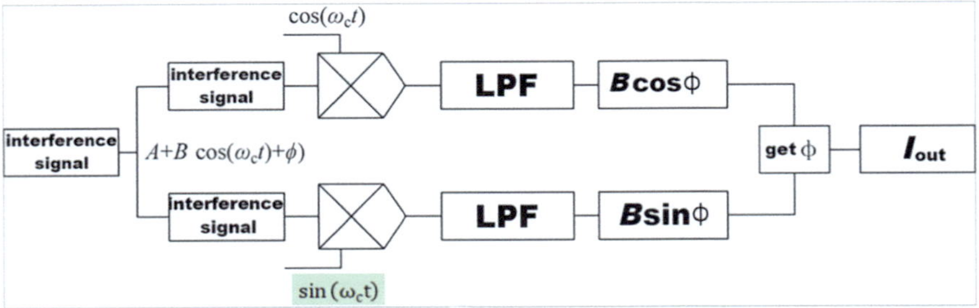

Fig. 4.5 Block diagram of heterodyne demodulation algorithm

electro-optic modulator (EOM) or acousto-optic modulator (AOM) applies a fixed frequency shift (Δf) to the original optical frequency. The second beam serves as the reference arm. Subsequently, the two beams recombine at Coupler 2, and their interferometric output is detected by a photodetector, producing the interference signal described in Eq. (4.8).

$$I_{\text{out}} = A + B\cos(\omega_c t + \phi) \tag{4.8}$$

The heterodyne method generates a frequency-modulated (FM) signal whose carrier frequency originates from the beat frequency of the two interferometer arms. As shown in Eq. (4.8), the bandwidth of this FM signal is significantly narrower than that of the Phase Generated Carrier (PGC) phase-modulated signal described in Eq. (4.6). To extract the two orthogonal components, the signal processing architecture outlined in Fig. 4.5 is employed.

In this process, the local oscillator (LO) signals $\cos(\omega_c t)$ and $\sin(\omega_c t)$ are mixed with I_{out}. The mixed signal is then passed through a LPF to suppress the ω_c component and higher-frequency harmonics, yielding the desired baseband signals.

$$I = \frac{1}{2}B\sin(\phi)\; Q = \frac{1}{2}B\cos(\phi) \tag{4.9}$$

As demonstrated in Eq. (4.9), the heterodyne method offers two key advantages over the Phase Generated Carrier (PGC) technique:

Amplitude Consistency: The two orthogonal components exhibit identical amplitudes, eliminating the Bessel-function-dependent amplitude variations inherent in PGC demodulation. This simplifies subsequent signal processing.

Bandwidth Reduction: The interferometer output signal bandwidth is significantly reduced, thereby relaxing the performance requirements (e.g., sampling rate, dynamic range) for the data acquisition system.

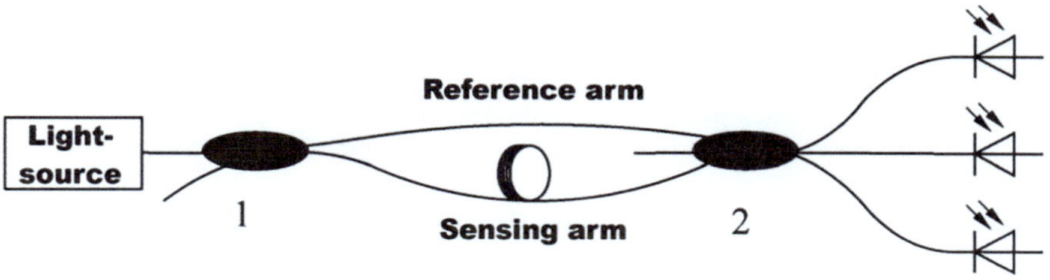

Fig. 4.6 Schematic diagram of 3 × 3 coupler method

3 × 3 Coupler Approach

The 3 × 3 coupler method extracts two orthogonal components directly from the interferometer's output. This technique achieves demodulation without actively modulating the interferometer, distinguishing it from approaches like PGC or heterodyne detection. A schematic of the demodulation process is shown in Fig. 4.6.

Continuous-wave (CW) light from the source propagates through the fiber interferometer, entering via Coupler 1 and subsequently being split by an ideally symmetric 3 × 3 coupler (Coupler 2). The three output ports of Coupler 2 deliver phase-shifted optical signals to three photodetectors. Under the assumption of a perfectly balanced 3 × 3 coupler, the detected signals can be expressed as,

$$\begin{cases} I_1 = I_0\,[M + V\cos(\phi)] \\ I_2 = I_0\left[M + V\cos\left(\phi + \frac{2\pi}{3}\right)\right] \\ I_3 = I_0\left[M + V\cos\left(\phi - \frac{2\pi}{3}\right)\right] \end{cases} \tag{4.10}$$

Here, I_0 represents the laser's output light intensity, and V denotes the interference visibility. To isolate the two orthogonal components (I and Q): The DC term is eliminated by pairwise subtraction of the three output intensities ($I_1 - I_2$, $I_1 - I_3$) in Eq. (4.10). The remaining AC components are amplitude-normalized to derive the final orthogonal signals.

$$Q = \sin(\phi)\, I = \cos(\phi) \tag{4.11}$$

The described amplitude normalization can be implemented through digital computation. While this method is optimized for ideally symmetric 3 × 3 couplers, practical non-ideal couplers require algorithmic compensation (e.g., ellipse fitting or adaptive signal conditioning techniques) to account for fabrication asymmetries. Nevertheless, the core demodulation framework remains valid. For implementation specifics, please refer to relevant literature for a detailed mathematical treatment of non-ideal coupler corrections.

Table 4.1 Comparison of PGC, heterodyne and 3×3 Coupler methods

Method	Advantages	Disadvantages	Applications
PGC	Simple optical path; convenient modulation	Large bandwidth required; demodulation dependent on modulation depth	Applicable for multi-sensor arrays; compatible with digital algorithms; good stability
Heterodyne	Small bandwidth; identical amplitudes of two orthogonal signals	High cost of frequency shifters/ frequency modulation	
3 × 3 Coupler	No need for modulation; larger bandwidth	Requires three detectors; high symmetry 3 channels; complex compensation algorithms	

Comparison of Three Orthogonal Detection Methods

The Phase Generated Carrier (PGC), Heterodyne Detection, and 3 × 3 Coupler Demodulation methods are widely employed for orthogonal detection in fiber optic interferometric systems. Each technique exhibits distinct advantages and limitations in terms of hardware complexity, signal processing requirements, and application suitability. A systematic comparison of these methods is summarized in Table 4.1.

PLL Method [3]

The PLL method stabilizes the interferometer's operating point near quadrature, thereby enforcing a linear relationship between the output signal and the interferometric phase. This principle is mathematically formalized in Eq. (4.12).

$$I_{\text{out}} = A + B\cos\left(\frac{\pi}{2} + \phi\right) \tag{4.12}$$

When the interferometer operates at the quadrature point (Note: the signal amplitude is not inherently restricted to small values), the output signal transitions from the nonlinear cosine response to an approximately linear relationship with the phase perturbation ϕ, provided ϕ remains within a small-signal regime:

$$I_{out} = A - B\phi \tag{4.13}$$

Due to random phase drift between the two interferometer arms, the system deviates from the quadrature point (phase bias of $\pi/2$), resulting in signal fading as the output transitions from linear to nonlinear response. When the total phase difference $\frac{\pi}{2} + \phi$ approaches π (or its integer multiples), the interferometric signal diminishes entirely, rendering detection impossible.

To enable high-sensitivity measurements, the interferometer must be actively stabilized at quadrature using a PLL. This stabilization mechanism defines the core principle of the fiber-optic PLL method, which offers:

Simplified Architecture: Minimal active components and low circuit complexity.

Low Signal Distortion: Linear operation near quadrature minimizes harmonic generation.

Single-Sensor Optimization: Ideal for demodulating individual sensor signals with high fidelity.

To stabilize the interferometer at the quadrature point (phase bias of $\frac{\pi}{2}$), a phase feedback mechanism is implemented through the piezoelectric effect of a piezoelectric transducer (PZT) attached to the interferometer's reference arm. This is achieved by applying a feedback voltage V_f to the PZT, which induces controlled mechanical deformation. The deformation propagates to the reference arm, modulating its optical path length (ΔL) and thereby adjusting the interferometric phase ($\Delta\phi$). The critical challenges in realizing a robust fiber-optic PLL system are.

Precision Voltage Control: Dynamically regulating V_f to maintain quadrature despite environmental disturbances (e.g., temperature fluctuations, mechanical vibrations).

System Stability: Ensuring adaptive compensation for phase drift while preventing PZT from voltage saturation.

A functional block diagram of the PLL architecture, illustrating key components such as the error signal detector, feedback controller, and PZT actuator, is shown in Fig. 4.7.

The working process of a fiber-optic PLL consists of three parts: phase locking, phase tracking, and system stabilization.

Phase Locking: Nonlinear System Analysis, Referring to Fig. 4.7, let the initial phase of the interferometer output signal be ϕ_0, and rewrite Eq. (4.12) as

$$I = A + B\cos[\varphi(t) + \varphi_0] \tag{4.14}$$

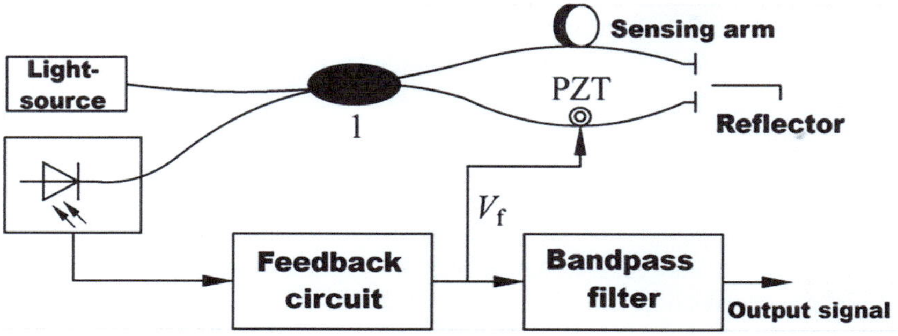

Fig. 4.7 Block diagram of fiber optic phase-locked loop method

$B = kA$, where $k < 1$. This is due to the following factors in the interferometer: polarization state fluctuations, intensity difference between the two arms, finite coherence length of the light source, all contributing to the reduction of interference visibility. After photoelectric conversion, both the DC and AC components exhibit amplitude variations. In fiber-optic PLL systems, the DC component of the interferometer must be eliminated, resulting in $A \approx 0$.

The interferometer signal is processed through a linear feedback system, with its output serving as the control signal for the PZT. Let the impulse response function of the feedback circuit be *h(t)*. According to linear system theory, the signal applied to the PZT is expressed as:

$$v_{\text{FB}}(t) = h(t) * I(t) = h(t) * \{A + B\cos[\phi(t) + \phi_0]\} \tag{4.15}$$

The asterisk (*) denotes the convolution operation. The phase change in the interferometer induced by variations in the PZT voltage is given by:

$$\begin{aligned}\phi_{\text{FB}}(t) &= K_{\text{PZT}} v_{\text{FB}}(t) \\ &= K_{\text{PZT}} h(t) * \{A + B\cos[\phi(t) + \phi_0\end{aligned} \tag{4.16}$$

in the equation, K_{PZT} represents the PZT voltage response. The total phase of the interferometer output signal is given by:

$$\begin{aligned}\phi(t) &= \phi_0 + \phi_{\text{FB}}(t) \\ &= \phi_0 + K_{\text{PZT}} h(t) * \{A + B\cos[\phi(t) + \phi_0]\}\end{aligned} \tag{4.17}$$

To satisfy the interferometer's quadrature condition, the impulse response function $h(t)$ is designed as an integrator, and Eq. (4.17) is reformulated as:

$$\phi(t) = \phi_0 + K_{\text{PZT}} \int_0^t \{A + B\cos[\phi(\tau) + \phi_0]\} d\tau \tag{4.18}$$

This constitutes a nonlinear integral equation for which obtaining an analytical solution is challenging. By differentiating both sides of Eq. (4.18) with respect to time t:

$$\phi'(t) = K_{\text{PZT}} A + K_{\text{PZT}} B\cos[\phi(t) + \phi_0] \tag{4.19}$$

Equation (4.19) establishes the relationship between the interferometer phase and its temporal rate of change. Although it does not explicitly express how the phase difference $\Delta\phi$ evolves over time, it fully characterizes the dynamic behavior of $\Delta\phi$ during the feedback control process. This formulation can be applied to investigate the phase-locking dynamics of the PLL.

The curve is plotted according to Eq. (4.19). Without loss of generality, the initial phase is set to zero, as depicted in Fig. 4.8.

Figure 4.8 depicts the system's state space, each point on the graph represents a distinct system state. Horizontal Axis: Phase difference $\phi(t)$ of the fiber interferometer; Vertical Axis: Time derivative of phase difference $\phi'(t)$.

Directional arrows denote the system's evolutionary trajectory. Above the horizontal axis ($\phi'(t) > 0$): Phase difference increases with time; Below the horizontal axis ($\phi'(t) < 0$): Phase difference decreases with time.

Equilibrium Points (a, b, c...) occur at intersections with the horizontal axis ($\phi'(t) = 0$), indicating: Phase stabilization and dynamic equilibrium.

For equilibrium point b, a positive perturbation induces $\phi'(t) > 0$, driving $\phi(t)$ to increase until reaching point c. Beyond c, $\phi'(t) < 0$ forces $\phi(t)$ to decrease, stabilizing back at c. Conversely, a negative perturbation at b triggers $\phi'(t) < 0$, reducing $\phi(t)$ until arriving at point a. Below a, $\phi'(t) > 0$ restores $\phi(t)$ upward to a. Thus, point b is an unstable equilibrium, while points a and c are stable equilibria.

The intersection pattern between the curve and the horizontal axis is determined by the values of A and B. When $| A |<| B |$ with$A > 0$, the $\phi' - \phi$ curve assumes the configuration shown in Fig. 4.8. Under these conditions, the loop invariably converges to a stable equilibrium point regardless of the initial $\phi(t)$. For instance: If the initial $\phi(t)$ lies between points b and c, the system asymptotically stabilizes at point c. This implies that phase lock is achievable whenever$| A |<| B |$.

During the loop locking process, the interferometer's output phase monotonically converges toward a stable equilibrium point. The phase difference $\phi(t)$ never exceeds 2π during its evolution from the initial value to stabilization.

State Trajectory Dynamics: For any given initial $\phi(t)$, the state point follows the directional arrows in Fig. 4.8 toward equilibrium. The instantaneous velocity $|\phi'(t)|$ decreases as $\phi(t)$ approaches stabilization, exhibiting asymptotic deceleration. Far from equilibrium: rapid transient response; Near equilibrium: exponential relaxation.

Engineering Stabilization Criterion: The system is considered stabilized when $\phi(t) < \delta_{lock}$,, where δ_{lock} is a prescribed phase error threshold.

Equilibrium Calculation: Stable operating points are obtained by setting the left-hand side of Eq. (4.19) to zero:

$$\phi_s = \pi - \arccos \frac{A}{B} + 2m\pi, m = 0, \pm 1, \pm 2, \ldots \tag{4.20}$$

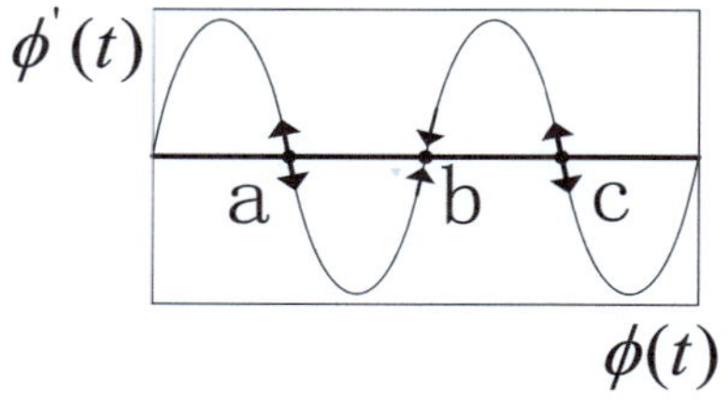

Fig. 4.8 Schematic Diagram of Phase Locking Process in Fiber Optic PLL ($A < B$)

This constitutes the steady-state solution of Eq. (4.19), represents the stabilized phase difference $\Delta\phi$ in the interferometer. It arises from the superposition of the input signal and the PZT feedback signal when the fiber-optic PLL locks. If the condition $|A| \approx 0$ is satisfied, the solution simplifies to:

$$\phi_s \approx \frac{\pi}{2} + 2m\pi, m = 0, \pm 1, \pm 2, \ldots \tag{4.21}$$

Precisely, this defines the quadrature operating point. When $|A| > |B|$ with $A > 0$ (i.e., low interferometer visibility, where AC signal is weaker than DC offset), no stable equilibrium ($\phi'(t) = 0$) exists. The loop cannot lock. $\left|\frac{A}{B}\right|$ ratio governs interference visibility, locking capability, signal distortion and steady-state stability.

Having established the stability mechanism—where equilibrium stems from feedback-phase error balance and is governed by the $|A| < |B|$ criterion—we now pivot to analyzing post-lock dynamics.

phase tracking: Steady-state analysis.

Equation (4.18) represents a nonlinear loop equation. When analyzing key loop characteristics—such as frequency response, tracking behavior, and stability criteria—the system can be linearized under the quadrature operating point (locked state) for simplified analysis.

Figure 4.9 illustrates the linearized model of the fiber-optic PLL, with the following functional blocks:

K_1 (Units: $\frac{\text{V}}{\text{rad}}$): Represents the photoelectric conversion coefficient of the interferometer output signal. Negative sign aligns with Eq. (4.13), counterbalancing the integrator's polarity to ensure negative feedback.

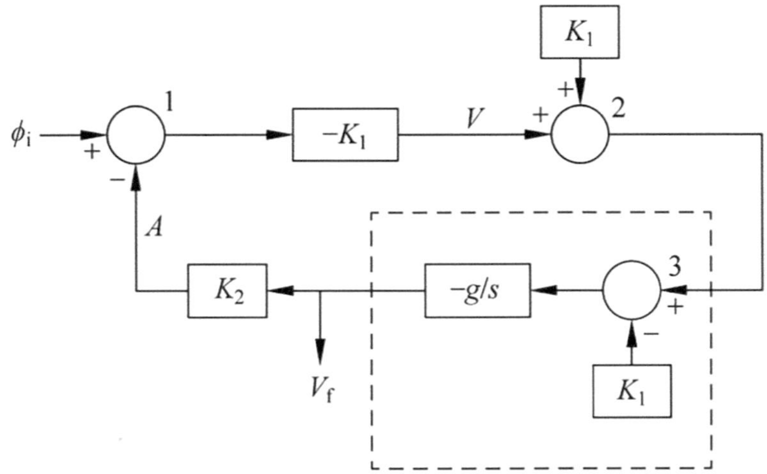

Fig. 4.9 Linear model of fiber optic PLL

K_2 (Units: $\frac{\text{V}}{\text{rad}}$): Denotes the phase modulation coefficient mapping the feedback system's output voltage to the reference arm's phase shift via the PZT.

Adder 2 (Top Right): Represents the DC component in the interferometer output signal, modeled by the photoelectric conversion coefficient K_1.

Subtractor 1 (Top Left): Encodes the OPD between the two interferometer arms.

Subtractor 3 (DC Suppression): The dashed-box subtractor eliminates the DC component in the photoelectric output, implemented via a subtractor circuit or balanced detector (preferred for common-mode noise rejection). Critically, high-pass filters are excluded to avoid low-frequency signal loss essential for phase tracking. This module directly governs parameters A (DC residual) and B (AC visibility), with ideal performance minimizing A (ideally zero) while maximizing B.

Integrator Block ($-\frac{g}{s}$): The Laplace-domain integrator $-\frac{g}{s}$ represents analog integration, where g defines the gain magnitude and the negative sign originates from the operational amplifier's inverting configuration.

Feedback Output Signal V_f: The integrator's output serves dual functions, as the PZT control voltage (adjusting the reference arm phase) and the final detection output.

Figure 4.9 enables analysis of the steady-state phase tracking behavior. Let the measured signal induce an optical path variation in the signal arm, generating a phase shift $\varphi_i(t)$; simultaneously, the PZT voltage adjusts the reference arm path length, producing a compensatory phase shift $\varphi_f(t)$. The closed-loop system is expressed as:

$$\begin{cases} \phi(t) = \phi_i(t) - \phi_f(t) \\ \phi_f(t) = K_2 V_f(t) = -K_2 g \int (-K_1 \phi(t) + K_1 - K_1) dt \end{cases} \tag{4.22}$$

By reorganizing Eq. (4.22) and applying the Laplace transform, we derive.

Equation (4.23) describes the relationship between the input signal ϕ_i and the interferometer output ϕ, termed the loop's error transfer function, which behaves as a first-order high-pass filter with a cutoff angular frequency $K_1 K_2 g$. When the signal's angular frequency ω is significantly lower than $K_1 K_2 g$, the ratio $\left|\frac{\varphi(j\omega)}{\varphi_i(j\omega)}\right|$ approaches zero, enabling the reference arm to effectively compensate for phase variations induced in the signal arm. This mechanism ensures the interferometer operates in a small-signal regime even under large input signal conditions.

$$\frac{\phi(s)}{\phi_{\text{i}}(s)} = \frac{s}{s + K_1 K_2 g} \tag{4.23}$$

The transfer function from φ_i to V_f is given by:

$$\frac{V_f(s)}{\varphi_i(s)} = \frac{K_1 g}{s + K_1 K_2 g} \tag{4.24}$$

Equation (4.24) defines the closed-loop transfer function between the PZT signal V_f and the input phase φ_i, acting as a first-order LPF with a cutoff angular frequency K_1K_2g. For angular frequencies $\omega \ll K_1K_2g$, the ratio $\left|\frac{V_f(j\omega)}{\varphi_i(j\omega)}\right| = \frac{1}{K_2}$, ensuring a linear relationship between V_f and φ_i. By utilizing this voltage as the output signal, phase variations in the signal arm are proportionally extracted, enabling precise interferometric detection even under dynamic operating conditions.

Stability: Destabilization versus Stabilization Mechanisms.

Thermal Drift and Feedback Signal Limit

For fiber-optic interferometers, maintaining small-signal operation under large thermal drift requires proportional scaling of the feedback signal amplitude to accurately track input variations. The feedback network, composed of operational amplifiers and circuits, operates within a fixed voltage range. When thermal drift requires the feedback signal to exceed this range for full compensation, system saturation occurs, preventing effective correction.

Additionally, thermal drift frequencies are typically far lower than signal frequencies. Passing through the integrator, the accumulated drift continuously increases the output, intensifying saturation. This necessitates dedicated reset mechanisms to periodically discharge the integrator.

Thermal drift scales proportionally with interferometer arm lengths. Longer fibers enhance sensitivity but amplify thermal susceptibility. For a given fiber length, expanding the integrator's operational voltage range and employing high-voltage op-amps effectively mitigates thermal drift by accommodating larger offsets without saturation.

Laser Power Fluctuations

Laser power fluctuations, primarily caused by fiber bending-induced losses in the output pigtail, degrade DC cancellation (Fig. 4.9) and disrupt the conditions required for Eq. (4.21), significantly narrowing the system's lock-in range. To mitigate this, meticulous fiber routing is essential to minimize bending-induced intensity variations. As an alternative, incorporating a balanced detector at the output directly eliminates DC components, ensuring the $A \approx 0$ condition is maintained independently of source power drift.

Fiber-optic PLL methods are advantageous for their operational simplicity and clear principles, making them well-suited for laboratory applications. Operating within a linear regime, these systems fully leverage the interferometer's sensitivity to detect weak signals. When used with single or limited sensing probes, they completely eliminate instability arising from source intensity fluctuations.

The PLL techniques discussed here are primarily designed for low-frequency sensor signals (below 1MHz). However, adaptations in digital signal processing extend their applicability to high-frequency scenarios. For example, in narrow-linewidth laser frequency stabilization, integrating PLL with Fast Fourier Transform (FFT) algorithms can reduce initial frequency differences between two lasers from hundreds of MHz to below 1MHz. Detailed implementation guidelines can be found in specialized technical literature.

Having explored laser interferometry for relative measurements, we now turn to interferometers utilizing broadband light sources, which enable absolute measurements.

4.2.2 Fiber-Optic Interferometers Utilize Broad-Spectrum Light Sources for Absolute Measurements

Fiber-optic interferometers employ broadband light sources to measure the absolute OPD between two arms, a principle widely applied in fiber FPI sensors. Broadband light sources include:

Low-coherence sources: Amplified spontaneous emission (ASE) sources, superluminescent diodes (SLEDs), and tungsten-halogen lamps.

Tunable lasers: Swept-wavelength lasers that scan predefined spectral ranges (e.g., the C + L band, 1510 − 1590 nm) to acquire full-frame interference spectra.

Each light source type necessitates a distinct demodulation approach. This chapter introduces two widely adopted methods.

Method 1: Spectral Analysis

This technique utilizes swept-wavelength lasers or finite-coherence-length broadband sources. When the OPD is smaller than the source coherence length, two-beam interference generates fringes in the output spectrum. By analyzing wavelength-dependent interference intensities, the absolute cavity length is determined through the relationship between the wavelength spacing of interference peaks and the physical cavity length.

Method 2: Matched Interferometry

Matched interferometry pairs a fully incoherent white-light source with a reference interferometer to achieve zero OPD condition. The absolute cavity length is derived by reading the pre-calibrated cavity length of the reference interferometer, which resolves phase ambiguity through active optical path matching.

The following sections provide detailed implementations of these two extensively validated techniques.

Spectral Analysis

For a fiber-optic FPI interferometer, the output light intensity is expressed by Eq. (4.25):

$$I = 1 + \gamma(L)\cos\left(\frac{4\pi L}{\lambda} + \phi_0\right) \tag{4.25}$$

Here, $\gamma(L)$ denotes the visibility of interference accounting for the fiber numerical aperture and other attenuation factors, while ϕ_0 represents the initial phase difference between the two interfering beams. The sensor's output signal spectrum manifests as a cosine function of wavelength. Since the interference spectrum is a function of the OPD (which equals twice the cavity length, $2L$), the absolute cavity length of the interferometer can be precisely determined through spectral analysis of the interference pattern. For two wavelengths λ_1 and λ_2 in the light source, the phase difference between them at the interferometer output is given by:

$$\Delta\phi = \phi_1 - \phi_2 = \frac{4\pi L}{\lambda_1} - \frac{4\pi L}{\lambda_2} = \frac{4\pi(\lambda_1 - \lambda_2)}{\lambda_1\lambda_2}L \tag{4.26}$$

From Eq. (4.26), it follows that:

$$L = \frac{\lambda_1\lambda_2}{4\pi(\lambda_1 - \lambda_2)}\Delta\phi \tag{4.27}$$

Once the interferometer's output spectrum is acquired, the sensor's absolute OPD can be determined by selecting two distinct wavelengths and calculating their phase difference. Directly measuring the phase difference between arbitrary wavelengths is challenging; instead, fixed phase relationships, such as the 2π phase difference between adjacent spectral peaks, are leveraged. By detecting the wavelength interval (peak-to-peak spacing) between these peaks, the absolute OPD can be derived.

The interferometer's output spectrum serves as the foundation for spectral analysis, with two primary acquisition methods:

1. Broadband Light Source + Compact Spectrometer: A broadband light source illuminates the system, and a compact spectrometer directly captures spectral information across the target wavelength range (e.g., 1510 – 1590 nm for C + L band applications).
2. Broadband Light Source + Wavelength Scanning + Wavelength Calibration: This method integrates a broadband source with wavelength-scanning functionality, achieved via:

(a) A tunable optical filter paired with a broadband source, or.
(b) A rapidly swept laser scanning a predefined spectral range (e.g., C + L band).
(c) The scanned light (with linearly varying wavelength) is split by a 2 × 2 coupler into two paths: Sensing Path — Light is injected into the FPI interferometer. Calibration Path — Light enters a wavelength calibration system (e.g., an etalon-FBG combination or the Michelson interferometer described in Chapter 3) as a reference. Both paths undergo photoelectric conversion, signal amplification, and analog-to-digital conversion to generate processed interference spectra.

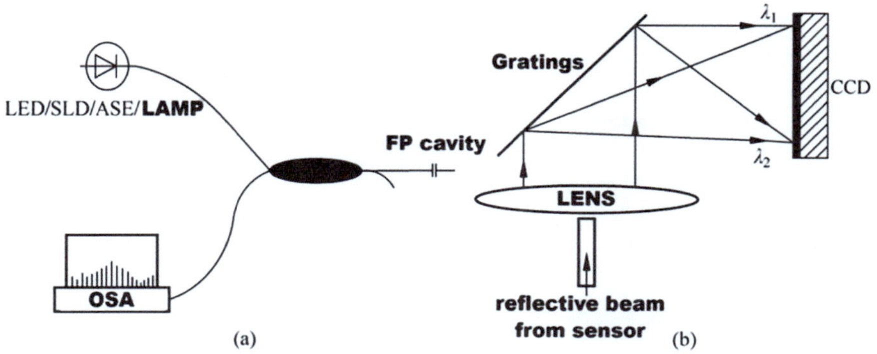

Fig. 4.10 Broad-spectrum light source and miniature spectrometer

The resulting spectrum: with wavelength (horizontal axis) and optical power (vertical axis), enables cavity length calculation (half the interferometer's OPD) using two common approaches:

Peak Spacing Analysis: Similar to time-domain signal processing, but with wavelength instead of time as the axis. The wavelength difference corresponding to the spacing between two adjacent spectral peaks is calculated.

Fourier Transform Method: This frequency-domain approach applies a Fourier transform to the spectrum for analysis.

Both techniques facilitate precise determination of the interferometer's cavity length. First, we detail two spectral acquisition methods, followed by spectral processing techniques.

1. Broadband Source + Compact Spectrometer [5]

This straightforward method is implemented as illustrated in Fig. 4.10a.

Combining a broadband light source with a compact optical spectrum analyzer enables efficient acquisition of interferometer spectra. Traditional spectrometers, however, suffer from drawbacks such as bulkiness, high cost, poor integrability, and complex signal analysis workflows. Modern compact spectrometers address these issues with their small form factor, cost-effectiveness, and ease of integration into custom-developed instruments, thus facilitating system miniaturization. For example, a 1200-line/mm holographic grating disperses incident light onto a linear charge-coupled device (CCD) sensor array. A 12-bit analog-to-digital converter (ADC) then digitizes the spectral intensity data, which is transmitted to a computer for processing. The internal structure of such a compact spectrometer is illustrated in Fig. 4.10b.

In this configuration, broadband light (e.g., from a tungsten-halogen lamp) is injected into the interferometer. Reflected light is collimated by a lens, diffracted by the grating, and projected onto the CCD array, where each pixel records the intensity of a specific wavelength. Sufficiently high CCD sampling rates even enable dynamic

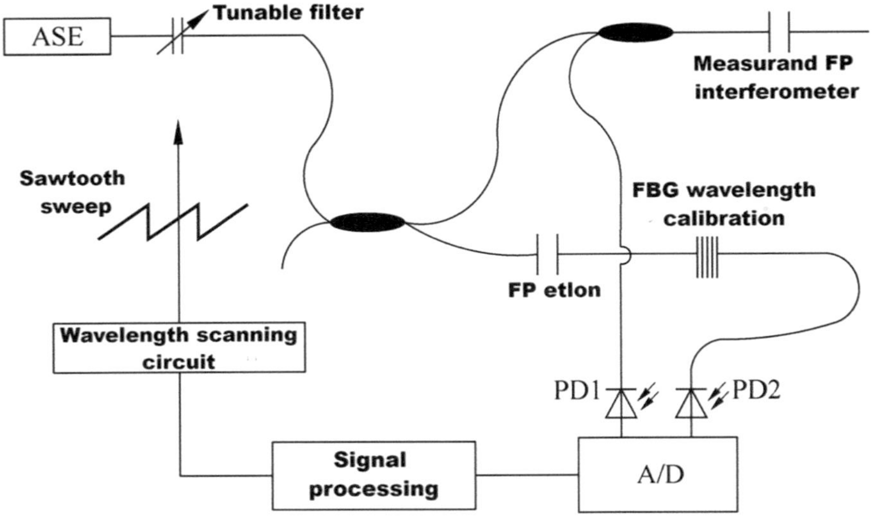

Fig. 4.11 Block diagram of the tunable light source method

signal measurement. A critical consideration is the uneven spectral distribution of broadband light sources. To mitigate this, the acquired reflection spectrum is normalized by dividing it by the pre-characterized source spectrum during the calibration process.

2. Broadband Light Source + Wavelength Scanning + Wavelength Calibration [2]

A schematic diagram of this interference spectrum acquisition method is shown in Fig. 4.11. A tunable light source—comprising either a broadband source integrated with a tunable optical filter or a wavelength-swept laser—emits wavelength-scanned light, which is split into two optical paths via a fiber coupler:

Path 1 (Sensing Channel):Light enters the interferometer, and the reflected signal is detected by photodetector PD1, serving as the raw sensing signal containing interference information.

Path 2 (Calibration Channel): Light propagates through a FPI etalon followed by a fiber Bragg grating (FBG) filter. The transmitted light, which carries wavelength-calibrated reference information, is detected by PD2 to generate a wavelength-marker signal.

The signal processing system drives a sawtooth wave generator circuit, which outputs a linearly increasing voltage to tune the wavelength of the light source. Concurrently, the data acquisition system synchronously samples the output signals from both PD1 and PD2, feeding them into the central processing unit for joint analysis.

The raw output signal from PD1, representing the interferometer's unprocessed spectral response, is depicted in Fig. 4.12a.

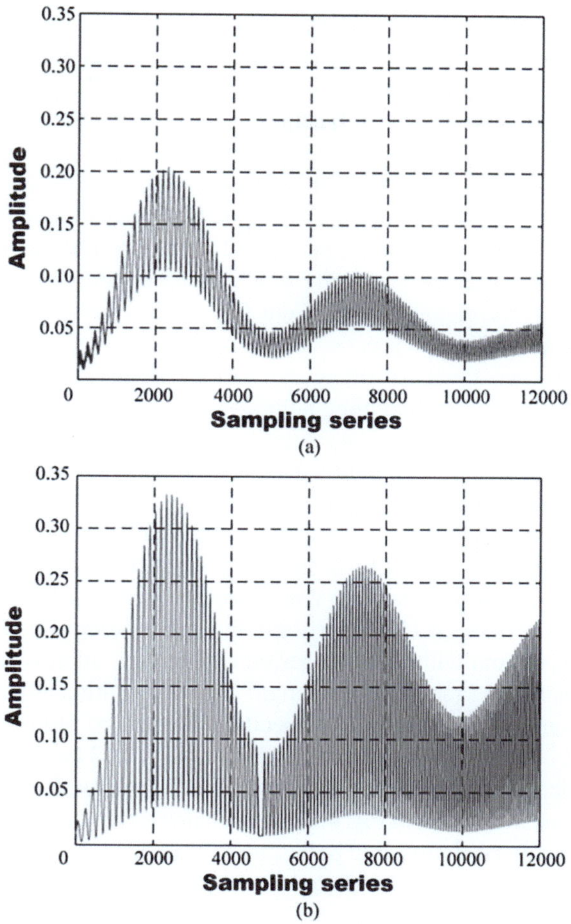

Fig. 4.12 Spectrum obtained by wavelength scanning method **a** interferometer reflected signal **b** Etalon reflected signal

In Fig. 4.12, the horizontal axis represents sampling points at uniform time intervals. As illustrated, the spectral signal exhibits a sparse distribution in the shorter-wavelength region (first half) and a dense distribution in the longer-wavelength region (second half), with significantly non-uniform periodicity. This poses a challenge for accurately determining the wavelength difference corresponding to a 2π phase shift.

The fundamental cause of this issue stems from the hysteresis effect of the wavelength-scanning device: the relationship between the scanned wavelength and the applied sawtooth drive voltage deviates from ideal linearity, while data acquisition occurs at constant time intervals. As a result, the acquired spectral signal displays uneven sampling in the wavelength domain, manifesting as periodicity distortion.

To acquire spectral signals with equal wavelength spacing, wavelength-scanning nonlinearity must be calibrated using a fiber FPI etalon as a wavelength reference, paired with a broadband FBG centered at wavelength λ_{FBG}. The FBG suppresses the etalon's spectral line at λ_{FBG}, ensuring that each remaining interference peak of the

FPI etalon corresponds to a precisely known wavelength. The combined transmission spectrum of the FPI etalon and FBG is detected by PD2, serving as a wavelength-calibration reference. As depicted in Fig. 4.12b, both PD2 (calibration channel) and PD1 (sensing channel) acquire signals at constant time intervals.

To obtain a spectral signal with uniform wavelength intervals, the horizontal axis of Fig. 4.12a must be converted from time-based sampling points to actual wavelengths. Both Figs. 4.12a and b initially share the same time-based sampling axis, but the spectral peaks in Fig. 4.12b inherently represent a true wavelength difference corresponding to a 2π phase shift. By leveraging this relationship, the true wavelength values from Fig. 4.12b can be mapped to the time-based axis of Fig. 4.12a. This process establishes a direct wavelength-to-time correlation, enabling the horizontal axis of Fig. 4.12a to be redefined in terms of absolute wavelength rather than arbitrary sampling points.

Let the sensing data series from Fig. 4.12a be denoted as $D(t)$, where t represents time, and the calibration data series from Fig. 4.12b as $R(t)$. Given the known cavity length of the FPI etalon and the characteristic wavelength of the FBG, the spectral wavelength λ_t corresponding to time t can be accurately determined. By establishing a mapping from t to λ_t, the wavelength corresponding to $D(t)$ at each time instance is obtained, generating a new data series indexed by wavelength. As observed, both plots in Fig. 4.12 exhibit non-uniform periodicity. To achieve a wavelength-expressed spectrum with uniform periodicity, interpolation is applied to regions with sparser sampling (shorter wavelength intervals), ultimately rendering the entire spectrum uniformly periodic. The specific method is as follows:

Let the total wavelength range in Fig. 4.12 be denoted as λ, divided into N wavelength points with an interval of $\Delta\lambda$, where $N = \lambda/\Delta\lambda$. For the corresponding spectrum $D(t)$ in Fig. 4.12a, original wavelength points retain their measured values, while newly inserted points are zero-padded. In regions where the spectrum appears dense (shorter wavelength intervals) in Fig. 4.12a, more zero-padding is applied, whereas sparser regions (longer wavelength intervals) require fewer zeros. The entire spectrum is then processed with FIR low-pass filtering to eliminate high-frequency artifacts introduced by zero-padding, yielding a smoothed spectrum. After this interpolation algorithm, the spectrum (illustrated in Fig. 4.13) achieves uniform periodicity. Figure 4.13a shows the spectrum prior to low-pass filtering, and Fig. 4.13b presents the filtered spectrum with consistent periodicity.

As described earlier, we define the spectral wavelength interval as $\Delta\lambda$ and set $\Delta\phi = 2\pi$.

$$L = \frac{\lambda_1\lambda_2}{4\pi(\lambda_2 - \lambda_1)}\Delta\phi \approx \frac{\lambda^2}{2\Delta\lambda_{\mathrm{MM}}} \tag{4.28}$$

Here, $\Delta\lambda_{\mathrm{MM}}$ denotes the wavelength spacing between two adjacent spectral peaks, with λ representing the central wavelength at which this spacing is measured. L signifies the unknown cavity length to be determined. By taking the derivative of both sides of Eq. (4.28), the cavity length resolution can be derived as follows:

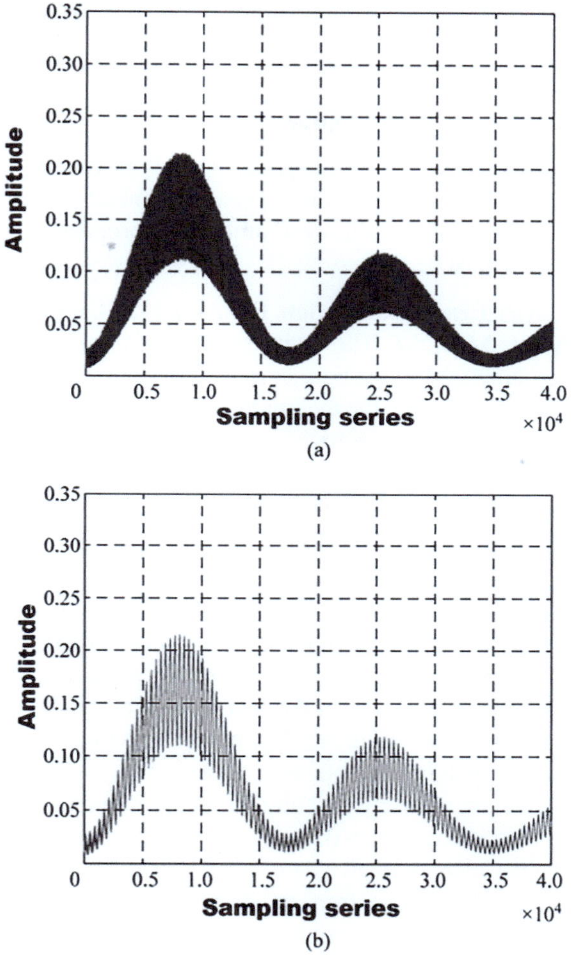

Fig. 4.13 Interferometric spectrum interpolation and filtering **a** post-interpolation unfiltered spectrum **b** post-interpolation low-pass filtered spectrum

$$dL = \frac{1}{2}\left(\frac{\lambda}{\Delta\lambda_{MM}}\right)^2 \Delta(\Delta\lambda_{MM}) = \frac{1}{2}\left(\frac{\lambda}{\Delta\lambda_{MM}}\right)^2 \Delta\lambda \tag{4.29}$$

Here, $\Delta\lambda$ represents the wavelength interval set during interpolation.

As shown in Eq. (4.26), the output spectrum of a dual-beam interferometer is a cosine function of the cavity length. The fringe peak positions exhibit broad and flat profiles, complicating accurate localization when measuring the fringe spacing $\Delta\lambda_{MM}$. However, the Fourier transform of a cosine function yields a delta function (impulse function) in the frequency domain, characterized by an extremely narrow spectral bandwidth. By performing a Fourier transform on the interferometer spectrum and identifying the frequency point with the maximum amplitude, the period of the interference spectrum—i.e., the peak-to-peak wavelength spacing—can be calculated using this frequency point, as expressed in Eq. (4.30).

$$f = \frac{1}{\Delta\lambda_{MM}} = \frac{2L}{\lambda^2} \tag{4.30}$$

Based on Eq. (4.28), where f denotes the frequency component corresponding to the delta function obtained after Fourier transform, $\Delta\lambda_{MM}$ is the wavelength spacing between adjacent spectral peaks, and λ is the central wavelength at the corresponding point, the cavity length L can be determined by calculating f.

Before performing the Fourier transform on the interference spectrum, two key issues must be addressed:

Cavity Length Resolution: From Eq. (4.30), it is evident that f is proportional to L, hence $\Delta f \propto \Delta L$. The cavity length resolution ΔL is proportional to the frequency resolution Δf of the Fourier transform. Whether using a scanning method or a spectrometer directly, the measured spectrum is sampled at discrete wavelengths. When performing a Fourier transform on such data, it becomes a discrete Fourier transform (DFT), typically implemented via the FFT algorithm in practice. The frequency resolution of the DFT is defined as:

$$\Delta f = \frac{f_S}{N} \tag{4.31}$$

where: f_s: Sampling frequency, N: Number of sampling points,

For Scanned Light Source Interference Spectra: The sampling rate is proportional to the acquisition rate of the data acquisition system and inversely proportional to the wavelength scanning rate. The number of sampling points scales with the acquisition time, which in sawtooth wave scanning is proportional to the light source's spectral width and inversely proportional to the scanning rate.

For Spectrometer-Based Systems: The sampling rate is inversely proportional to the wavelength interval between adjacent CCD pixels. As shown in Fig. 4.10b, the farther the CCD is from the diffraction grating, the higher the sampling rate. The number of sampling points is proportional to the number of CCD pixels.

Spectral Windowing: The spectral data obtained via the aforementioned methods are finite-length arrays. Direct Fourier transform application to such data induces spectral leakage due to truncation effects. For finite-length data, a windowing process is typically required prior to Fourier transform: the data array is multiplied by a discrete window function to mitigate abrupt amplitude changes at the array edges. This ensures smooth transitions to zero at both endpoints, eliminating artificial high-frequency spectral components caused by sharp discontinuities. Different application scenarios demand different window functions—omitting windowing (equivalent to a rectangular window) results in significant amplitude errors in Fourier transform outputs.

Importance of Window Selection in Interferometric Spectroscopy: In interferometric signal processing, selecting an appropriate window function is critical for precise spectral analysis. To accurately extract peak information of the primary spectral feature, enhancing amplitude resolution is essential. The Blackman-Harris window is particularly advantageous here, offering high amplitude resolution and low sidelobe levels. When applied to the original spectral signal, it suppresses sidelobes, broadens the main peak, and improves the frequency resolution of the primary feature, thereby reducing spectral peak misidentification errors. Experimental results show that cavity length errors are reduced by two orders of magnitude after implementing the Blackman-Harris windowing process.

Determining Cavity Length Thus far, we have discussed windowing the interference spectrum and the process of resolving cavity length prior to performing the Fourier transform. The windowed spectrum is subsequently processed using the established Fast Fourier Transform (FFT) algorithm, which computes the discrete Fourier transform (DFT) across various computing platforms. This generates a cavity length spectrum with cavity length as the horizontal axis, as described in Eq. (4.30).

Extracting the interferometer's cavity length from this spectrum can be achieved most simply by identifying the spectrum's maximum value. This approach is intuitive and computationally efficient; however, relying on a single extremum makes it highly susceptible to random noise. For cavity lengths in the hundreds of micrometers range, this method achieves nanometer-level precision.

Theoretical improvements to measurement accuracy involve averaging multiple spectra to reduce random noise at spectral peaks. While straightforward to implement, this approach significantly degrades real-time performance.

Waveform Similarity Fitting addresses both noise sensitivity and real-time requirements without depending on spectral maxima. The measured spectrum is compared against a pre-stored set of reference spectra, each corresponding to a specific cavity length, and the actual cavity length is determined by identifying the most similar reference spectrum [5].

Similarity fitting can be implemented via cross-correlation or least-squares minimization between the measured and reference spectra. Both methods exhibit comparable computational complexity and accuracy but reduce errors by an order of magnitude relative to direct extremum detection, achieving sub-nanometer precision for cavity lengths in the hundreds of micrometers. Cross-correlation outperforms least-squares in noise immunity, while least-squares yields smaller errors under high signal-to-noise ratios. This discrepancy arises because cross-correlation better preserves periodic signal characteristics—noise impacts amplitude more than periodicity, making cross-correlation inherently more robust.

Within similarity fitting approaches, the accuracy of pre-stored reference spectra is critical. In practice, this approach typically constructs the FPI interferometer's reflection spectrum based on a theoretical model, as shown in Eq. (4.32) [5].

$$I_R = I_0 \begin{bmatrix} R_1 + (1 - R_1)^2 R_2 \eta^2 (w_0, w) \\ +2\sqrt{R_1 R_2}(1 - R_1)\eta(w_0, w)\cos\left(\frac{4\pi G}{\lambda} + \pi\right) \end{bmatrix} \tag{4.32}$$

Here, R_1 and R_2 are the end-face reflectivities at the incident and reflective ends, respectively, and G is the FPI cavity length. The coupling coefficient between the two beams is given by.

$\eta(w_0, w) = \frac{2ww_0}{w^2+{w_0}^2}$, where $w_0 = a\left(0.65 + \frac{1.619}{V^{1.5}} + \frac{2.879}{V^6}\right)$ is the mode field radius of the fundamental fiber mode, and a is the fiber core radius. The normalized fiber frequency is $V = \frac{2\pi aNA}{\lambda}$. The Gaussian mode field radius in free-space propagation is $w(l) = w_0\sqrt{1 + \frac{l^2}{{l_R}^2}}$, where $l = 2G$ and $l_R = \frac{\pi {W_0}^2}{\lambda}$ is the Rayleigh length.

For each FPI cavity, reflectivities R_1 and R_2 are monitored during fabrication and stored in memory. Using Eq. (4.32), G is fixed to calculate the interference spectrum for the corresponding cavity length; G is then incremented stepwise to compute the spectrum for the next cavity length until all spectra for the target cavity length range are generated and cached in memory. The core principle of this approach is to distribute the effect of multiple averaging across each frequency point of the measured spectrum, thereby mitigating random noise impacts on individual frequency components. Once acquired, pre-stored waveforms are fixed in memory, enabling real-time operations like least-squares fitting or cross-correlation without latency from repeated averaging, which significantly enhances real-time performance.

The comprehensive methodology for absolute OPD, measurement in interferometers via spectral analysis has now been fully delineated. While a system capable of absolute measurement can theoretically perform relative measurements, the compromised real-time performance of spectral analysis in absolute measurement raises considerations about its suitability for dynamic relative signal measurements—evaluations must be conducted on a case-by-case basis.

Matched Interferometry

This method utilizes a matched interferometer, such as a Fizeau interferometer, as shown in Fig. 4.14.

Similar to the FPI interferometer, the Fizeau interferometer comprises two closely positioned optical planes between which light reflects to generate interference. Its key distinction from FPI interference lies in the wedge-shaped structure of the two planes. When the angle between the planes is θ, the gap $l(x)$ at any position x within the wedge is described by:

$$l(x) = x \cdot \tan\theta \tag{4.33}$$

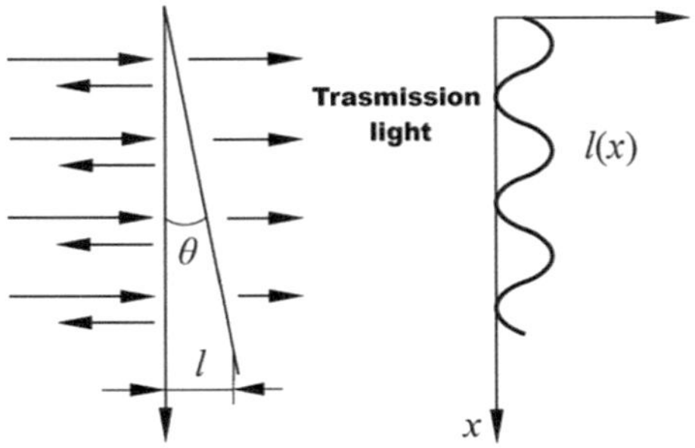

Fig. 4.14 Fizeau interferometer [6]

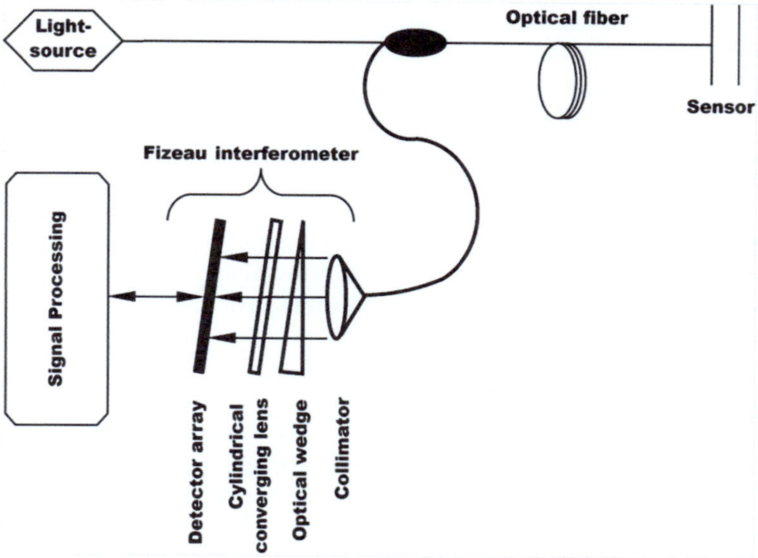

Fig. 4.15 FPI interferometer demodulation system using Fizeau interferometer

At different positions x, this gap $l(x)$ introduces varying OPD for the incident light. The Fizeau interferometer employs a matched interferometer configuration, with its principle illustrated in Fig. 4.15.

The Fizeau interferometer functions as a matched interferometer, forming an optical cross-correlator structure with the test interferometric sensor. A low-coherence broadband light source (wavelength-scale coherence length), such as a halogen-tungsten lamp, is injected into the test interferometer (with cavity length L) via a 2×2 coupler. Reflected light diverges at the output of the fiber, where a collimating lens converts the divergent beam into parallel light to satisfy the Fizeau interferometer's input requirements. Given that each CCD pixel is much smaller than the optical wedge width at its corresponding position, a cylindrical focusing lens concentrates optical energy along the wedge's width direction into a single-pixel area to ensure sufficient intensity. The light, focused by the cylindrical lens into a linear spot parallel to the wedge's length axis, is incident on the Fizeau interferometer. At the position where the Fizeau wedge gap equals L, the total OPD between the two interferometers becomes zero, causing cross-correlation to peak and light intensity to reach a maximum. Projecting the Fizeau output onto the CCD array generates a white-light interference pattern: at positions deviating from zero OPD, cross-correlation between the test interferometer and Fizeau device vanishes, causing intensity to drop abruptly to zero. By calibrating the Fizeau wedge gap and identifying the CCD peak-intensity position, absolute measurement of the test interferometer's OPD is achieved.

When external disturbances alter the interferometer's OPD, the Fizeau interferometer's maximum-transmittance position shifts, and the CCD line array detects the corresponding movement of the highest-intensity pixel. This enables dynamic absolute measurement of OPD.

Compared to the spectral scanning method described earlier, this system's primary advantage is the absence of moving parts, theoretically ensuring excellent long-term stability. However, the optical wedge in the Fizeau interferometer represents a critical and challenging component: based on current photodetector line array specifications, its wedge angle must be approximately 0.1° with machining accuracy of 1 arcminute. Additionally, alignment precision for the collimating lens, cylindrical lens, and CCD array must meet extremely stringent standards.

Beyond optical component fabrication and assembly, the CCD array is another critical element. The system's measurement rate depends on the CCD's line readout speed, measurement accuracy on the precision of locating the maximum-intensity position, and dynamic measurement range on the total pixel count of the CCD line array. Commercial CCDs currently achieve line readout rates up to tens of kHz, pixel sensitivity exceeding $10\frac{V}{\mu J}$, and single-sensor resolutions as high as 4096 pixels, making them well-suited for dynamic absolute strain measurement [15].

Broadband Frequency-Domain Interferometric Demodulation for Subwavelength-Thickness Thin Films

1. Theoretical Basis of Single-Peak Wavelength Shift in White-Light Frequency-Domain Interferometry

For nanoscale-thickness thin films, the OPD is extremely small, resulting in an interference spectrum with only one peak. Therefore, a demodulation scheme based on the wavelength shift of this single peak must be developed. The theoretical foundation of white-light interferometry primarily stems from the partial coherence theory of light. The interference spectrum of two coherent beams passing through a Michelson interferometer can be expressed as:

$$S(\lambda) = \frac{1}{2}S_0(\lambda)\left\{1 + Re\left[\mu_{12}(\lambda)\cos\left(\frac{2\pi}{\lambda}\Delta L\right)\right]\right\} \tag{4.34}$$

where $S_0(\lambda)$ denotes the source spectrum and ΔL represents the OPD between the two interferometer arms. $Re[\mu_{12}(\lambda)]$—the real part of the complex spectral coherence function for the frequency-domain interference fringes—is a constant. At zero phase, $\mu_{12}(\lambda)$ equals 1, and Eq. (4.34) can be rewritten as:

$$S(\lambda) = \frac{1}{2}S_0(\lambda)\left[1 + \cos\left(\frac{2\pi}{\lambda}\Delta L\right)\right] = \frac{1}{2}S_0(\lambda)[1 + \cos(\Delta\phi)] \tag{4.35}$$

where $\Delta\phi$ denotes the phase difference between the two interferometer arms. Assuming the source spectrum $S_0(\lambda)$ follows a Gaussian distribution with a central wavelength of 1550 nm and a full width at half maximum (FWHM) $\Delta\lambda = 50nm$, the coherence length of the source is:

$$L_c = \frac{\lambda_0^2}{\Delta\lambda} \tag{4.36}$$

The calculated coherence length is 48.05 μm, and the source spectrum can be expressed as:

$$S_0(\lambda) = \exp\left(-\frac{(\lambda - \lambda_0)^2}{2\Delta\lambda^2}\right) \tag{4.37}$$

Thus, Eq. (4.35) can be rewritten as:

$$S(\lambda) = \frac{1}{2}\exp\left(-\frac{(\lambda - \lambda_0)^2}{2\Delta\lambda^2}\right)[1 + \cos(\Delta\phi)] \tag{4.38}$$

Equation (4.38) constitutes the theoretical basis of white-light frequency-domain interferometry.

In frequency-domain interferometry, interference fringes remain detectable even when the OPD exceeds the source's coherence length. This phenomenon originates from the fact that a white-light source spectrum can be modeled as a superposition of numerous monochromatic components, each with infinite coherence length. When a diffraction grating (e.g., in a spectrometer) detects the interference spectrum, its dispersion effect splits broadband white light into narrowband spectral components, thereby modifying the effective coherence length of each channel. With typical diffraction gratings, the effective coherence length per channel can reach millimeter scale—far exceeding the white-light source's intrinsic micrometer-scale coherence length. Consequently, even if the OPD between interfering beams surpasses the source's coherence length, interference fringes remain observable, as illustrated in Fig. 4.16.

In frequency-domain white-light interferometry simulations, when the OPD between the two arms of the Michelson interferometer is zero, the output interference spectrum matches the source spectrum exactly, with its peak wavelength at λ_0. Figure 4.17 illustrates how the interference spectrum evolves as the OPD increases.

As the OPD rises from $\Delta L = 0$ nm (solid line in Fig. 4.17a), the interference spectral peak undergoes a redshift (shift toward longer wavelengths), as evident in the curves for $\Delta L = 400$ nm and 600 nm. Simultaneously, the peak intensity gradually decreases until it disappears (blue curve marked with asterisks). Further increasing the OPD causes the interference spectrum to reappear: the peak intensity increases again while the wavelength blueshifts (triangular and circular markers). This cyclical

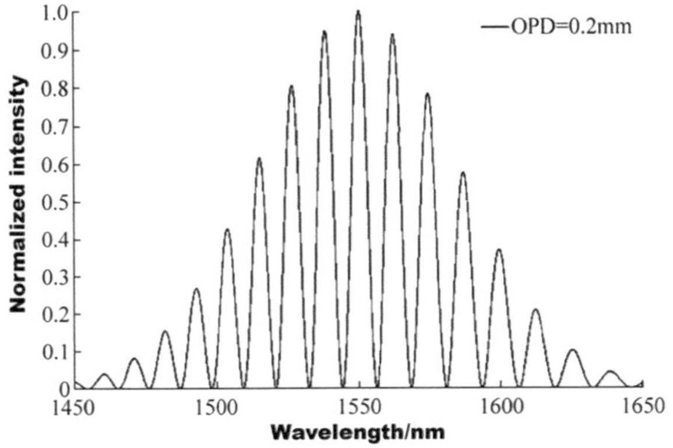

Fig. 4.16 White light interference, spectrum OPD greater than coherence length

intensity modulation and alternating wavelength shift (redshift to blueshift) persist until two distinct interference peaks emerge (Fig. 4.17b), with the higher-intensity peak defined as the central wavelength.

In Fig. 4.17b, as the OPD increases, the dominant peak's intensity decreases while the secondary peak's intensity increases, with both peaks gradually redshifted. When the two peaks equalize in intensity (indicating a phase difference of π), further OPD increases trigger an abrupt transition from redshift to blueshift—termed the spectral switch. Beyond this point, continued OPD increases drive the peak wavelength to blueshift until the two peaks merge into a single peak that fully overlaps with the original source spectrum.

Figure 4.18 depicts the normalized interference peak intensity as a function of OPD. As OPD increases, the peak intensity decreases initially and then increases, exhibiting periodic variation. Within each period, a specific OPD exists where the interference peak intensity reaches its minimum—the valley position.

To facilitate the understanding of the normalized spectral shift (NSS), Eq. (4.39) is defined as:

$$NSS = \frac{\delta\lambda}{\lambda_0} = \frac{(\lambda - \lambda_i)}{\lambda_0} \tag{4.39}$$

where λ_0 is the central wavelength of the light source, and λ_i denotes the peak wavelength of the interference spectrum. When $NSS = 0$, the interference spectrum coincides with the source spectrum. As illustrated in Fig. 4.19, a negative *NSS* signifies a redshift of the peak wavelength, whereas a positive value corresponds to a blueshift. The abrupt transition from redshift to blueshift in the interference spectral peak wavelength is termed the spectral switch, occurring at the OPD corresponding to the minimum spectral peak intensity—i.e., the valley position depicted in Fig. 4.19.

Figures 4.18 and 4.19 show that when the OPD varies within a small range, the peak wavelength of the interference spectrum exhibits a linear relationship with the OPD. This linearity enables demodulation of the OPD, facilitating measurements of physical quantities associated with

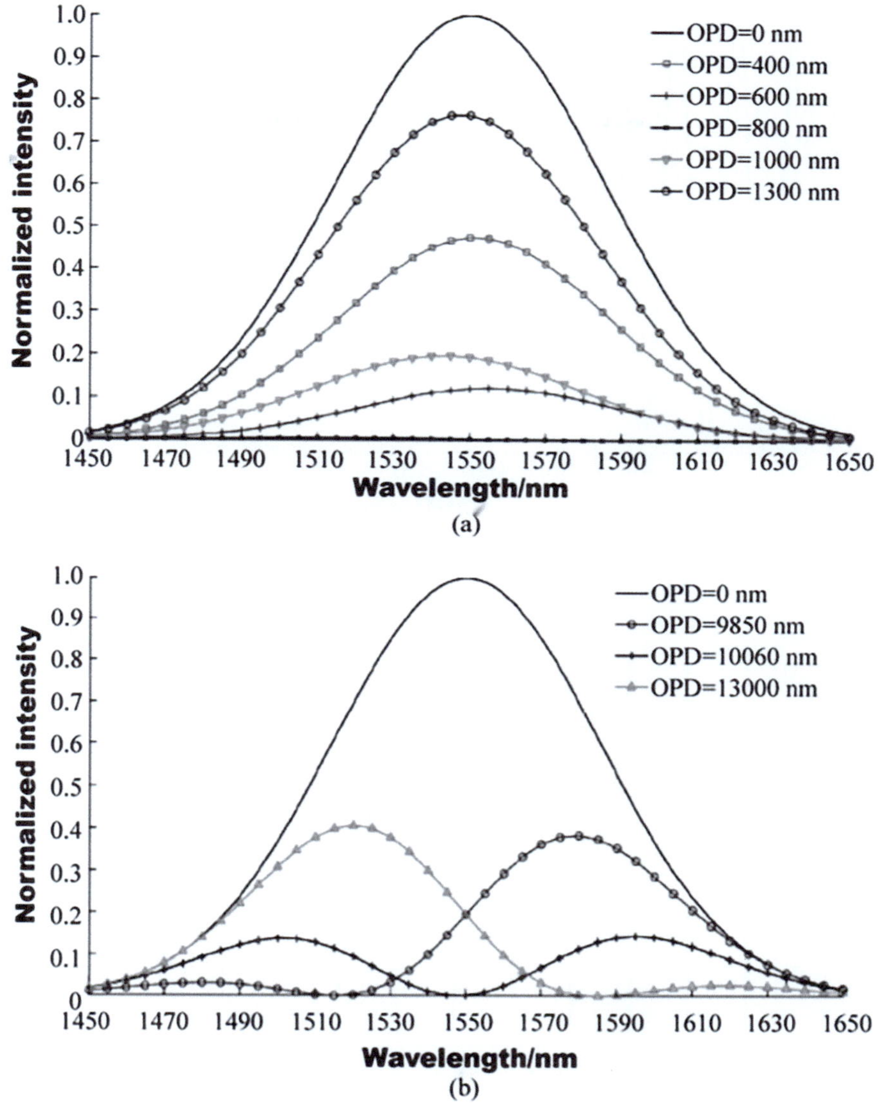

Fig. 4.17 Variation of interference spectrum with OPD **a** only one interference peak **b** two interference peaks appear

OPD changes—such as thin-film thickness and temperature. Experimental results for germanium films demonstrate that with a spectrometer resolution of 4 pm, the thickness measurement resolution reaches 0.4 nm.

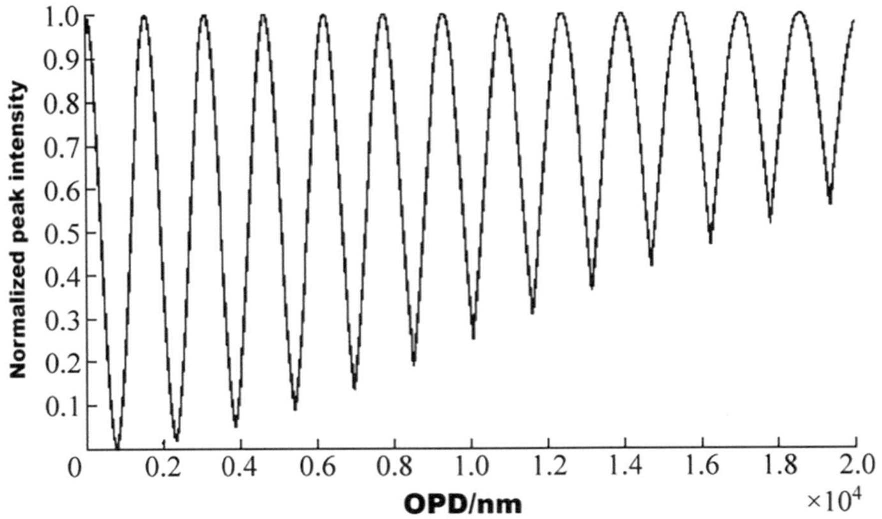

Fig. 4.18 Curve of normalized peak intensity versus OPD

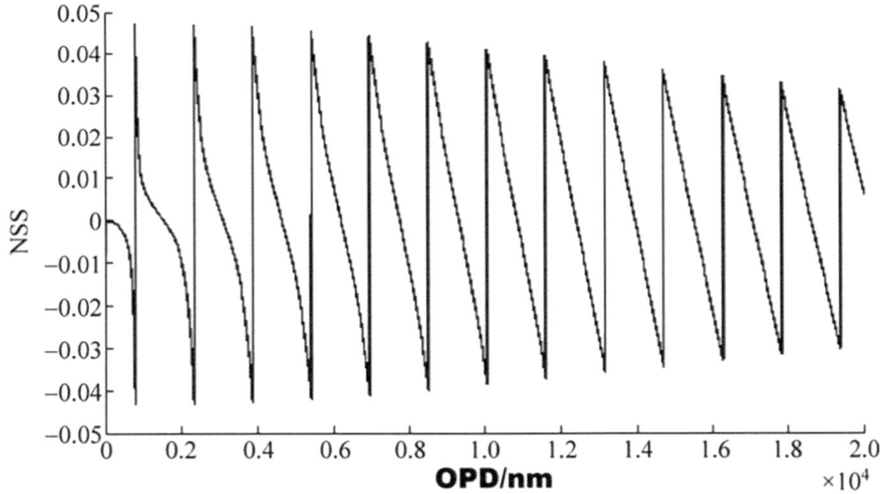

Fig. 4.19 Curve of NSS versus OPD

4.3 Multi-beam Fiber Interferometer Signal Processing

Fiber optic interferometers excel at detecting extremely weak signals, primarily because the optical wavelength is typically on the submicron scale: the interferometer's output phase is inversely proportional to wavelength and directly proportional to OPD. Consequently, even minute OPD variations—equivalent to one

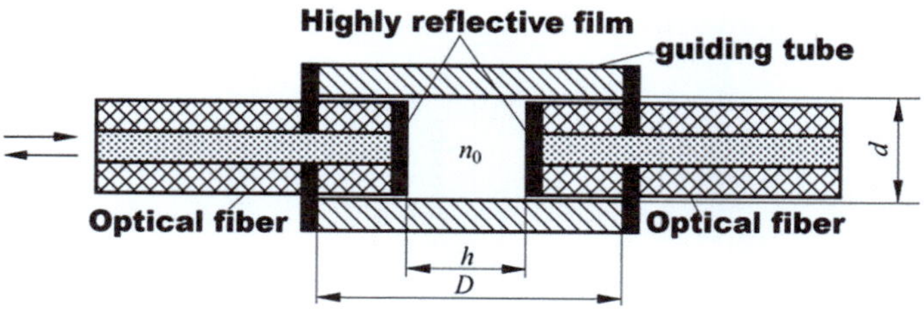

Fig. 4.20 Multi-beam FPI interferometer

wavelength—induce significant phase fluctuations (2π). In practice, sensitivity is enhanced by increasing interferometer arm length to amplify signal interaction with the interferometric structure.

For dual-beam configurations (e.g., M-Z, Michelson, or Sagnac interferometers), longer arms are achieved by directly lengthening fiber paths in both arms. However, in dual-beam FPI interferometers, increasing OPD drastically increases signal light attenuation, severely degrading interference signal visibility or even eliminating interference altogether.

To realize compact, high-sensitivity interferometric sensors, multi-beam interference principles are employed. By increasing the reflectivity of the two interferometric end faces, light undergoes multiple reflections before interfering, effectively extending the effective arm length. Under identical dimensions, multi-beam interferometers achieve higher sensitivity than dual-beam counterparts. Figure 4.20 illustrates the schematic of a multi-beam FPI interferometer.

Optical coatings are commonly employed to enhance the reflectivity of fiber end faces, with their interference characteristics described by Eq. (4.40); see Eqs. (1.24) and (1.25) in Chapter 1, Sect. 1.2.3.

$$I_T = \frac{1}{1 + F\sin^2\left(\frac{\delta}{2}\right)}; I_R = \frac{F\sin^2\left(\frac{\delta}{2}\right)}{1 + F\sin^2\left(\frac{\delta}{2}\right)} \tag{4.40}$$

Here, I_T and I_R denote the transmitted and reflected intensities of the interferometer, respectively. The finesse is defined as $F = 4R/(1-R)^2$, and $\delta = \frac{4\pi}{\lambda}n_0 h\cos(\theta)$ represents the phase difference induced by the interferometer's OPD, where R is the reflectivity of the two end faces. As R approaches 1, the finesse F increases rapidly. Figure 4.21 illustrates the curves of I_T and I_R versus δ under different finesse conditions.

Similar to dual-beam interferometers, multi-beam interferometers are capable of performing both absolute and relative measurements.

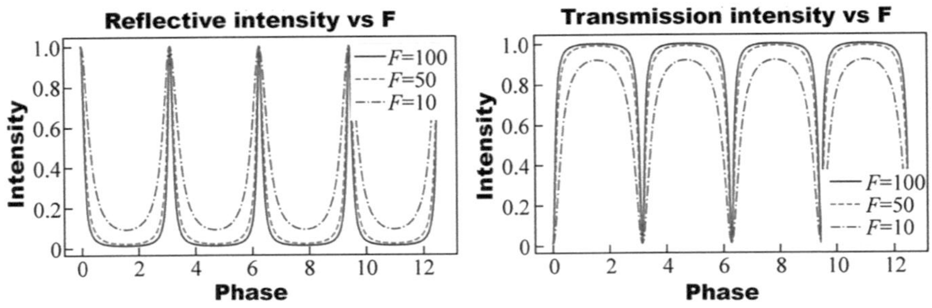

Fig. 4.21 Transmission and reflection spectra of multi-beam FPI interferometer

4.3.1 Absolute Measurement

As depicted in Fig. 4.21, increasing the finesse F leads to more beams participating in interference, yielding sharper transmitted spectral peaks and flatter spectral baselines near zero—markedly distinct from the sinusoidal spectral variations of dual-beam interference.

In the signal processing of dual-beam FPI interferometers (discussed earlier), the broad, sinusoidal spectral peaks are too shallow for direct detection, necessitating Fourier transforms to convert spectra from the wavelength domain to the cavity-length domain. In this transformed domain, narrow cavity-length spectral lines facilitate cavity length determination. For high-finesse multi-beam interferometers, however, the wavelength-domain spectral peaks are inherently narrow and well-defined, enabling direct measurement of wavelength intervals between peaks to achieve absolute cavity length measurements without Fourier transforms. The spectral acquisition methods (e.g., spectrometer-based as in Fig. 4.10 or wavelength-scanning as in Fig. 4.11) remain identical to those for dual-beam interferometers, as detailed in the preceding section.

4.3.2 Relative Measurement

Under identical dimensions, multi-beam interferometers exhibit superior sensitivity to weak signals compared to dual-beam counterparts—a feature particularly pronounced in relative measurements of faint signals. As illustrated in Fig. 4.21, for high-finesse multi-beam interferometers, minute fluctuations in cavity length near sharply peaked spectra induce drastic variations in interference intensity. By detecting these intensity changes, multi-beam interferometers enable highly sensitive measurements of extremely weak signals. Figure 4.22 depicts an experimental setup for high-precision measurements using a multi-beam interferometer.

The ASE light source in the figure has a 35 nm linewidth (1529–1565 nm) and 20 mW output power. A multi-beam interferometer constructed from SMF-28

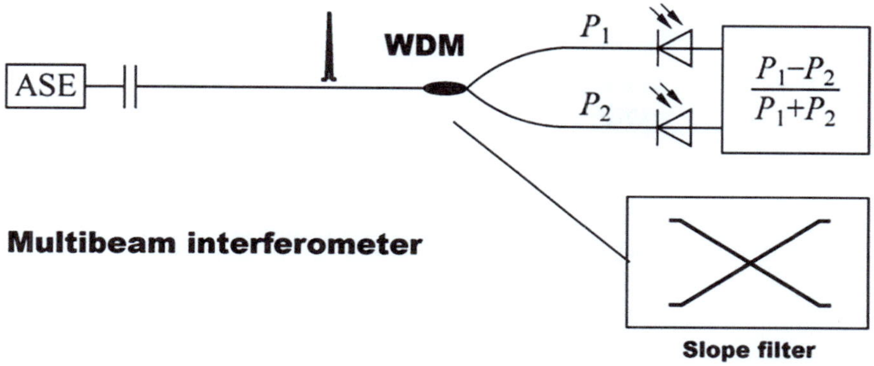

Fig. 4.22 Multi-beam interference demodulation system using WDC

single-mode fiber generates a forward-output resonance peak. The cavity length is designed to be sufficiently short to ensure only one resonance peak exists within the source's spectral range, i.e., the free spectral range (FSR) exceeds the system's spectral linewidth, monitored using a spectrometer during fiber splicing. Minute strain or temperature variations induce resonance peak wavelength shifts. For an air-gap cavity length of 20 μm, the FSR is 52 nm, larger than the source linewidth (40 nm). After coating fiber end faces with multilayer dielectric films, the reflectivity reaches $R = 97\%$. The interferometer output is fed into a wavelength division coupler (WDC) with two channels operating at 1520 nm and 1580 nm. Within the source's 1525–1565 nm range, the WDC's coupling ratio varies linearly with wavelength at a nearly constant rate of 0.23 dB/nm: the 1520 nm channel shows a linear decrease in coupled power with increasing wavelength, while the 1580 nm channel exhibits the opposite trend. Thus, the WDC functions effectively as a slope filter. The transmission peak's central wavelength is adjusted during splicing using low-power fusion current, with the optimal central wavelength corresponding to the intersection of the WDC's two channel coupling ratio–wavelength curves, as shown in Fig. 4.23.

The optical intensities at the two output ports of the WDC exhibit an inverse relationship with wavelength. After being connected to photodetectors PD1 and PD2, respectively, they are processed according to Eq. (4.41):

$$P = \frac{P_1 - P_2}{P_1 + P_2} \tag{4.41}$$

This normalized ratio calculation minimizes intensity fluctuations. When the central wavelength is selected near the intersection point in the figure, the system achieves its highest sensitivity.

This scheme offers simple circuitry and low cost. System sensitivity depends on the slope filter's performance and circuit detection accuracy, while the dynamic range is determined by the interferometer's free spectral range (FSR) and the source's

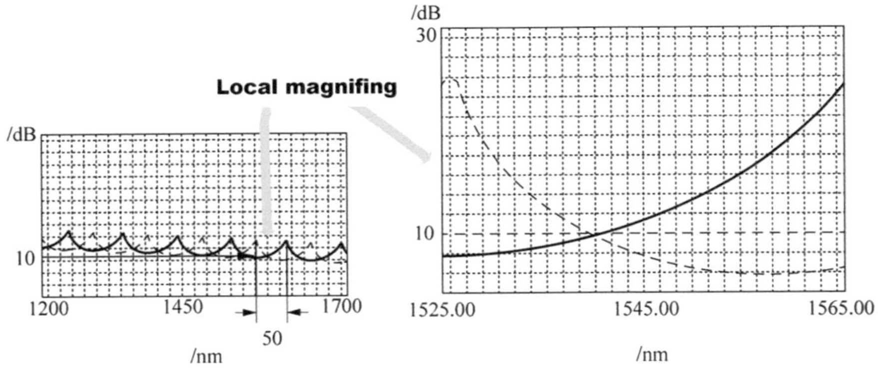

Fig. 4.23 Curve of two-channel coupling ratio versus wavelength for WDC

spectral width. A wider FSR and broader source spectrum enable a larger dynamic range.

This configuration provides high precision and can measure high-frequency dynamic signals. It is typically used for dynamic measurements of high-frequency, low-amplitude signals: since the slope filter is a passive component, the system's operating frequency is limited solely by the circuitry.

4.4 Long-Range Fiber Interferometer Signal Processing

Long-range fiber interferometer signal processing fundamentally relies on dual-beam fiber interferometry, with its distinct feature lying in the interferometer's extremely long arm lengths—extending to several kilometers or even tens of kilometers. Such interferometers are primarily employed in distributed fiber sensing, enabling a single continuous fiber to serve as thousands of sensors. The entire sensing system operates passively, resists corrosion and electromagnetic interference, and can both detect and transmit environmental signals, representing the apex of fiber-optic sensing applications.

The two most prominent configurations for long-range fiber interferometers are: High-sensitivity sensor architectures based on the M-Z scheme. Continuous multi-point sensor architectures based on distributed schemes.

Signal processing methods for M-Z-based systems were detailed in Chapter 3. Below, we introduce two signal processing techniques for distributed schemes:

Phase-sensitive Optical Time Domain Reflectometry ($\phi-$OTDR), Relies on OPD detection to resolve phase variations induced by environmental perturbations along the fiber.

Optical Frequency Domain Reflectometry (OFDR), Utilizes optical frequency difference detection to achieve high spatial resolution (sub-millimeter) for distributed sensing.

4.4.1 ϕ-OTDR

ϕ-OTDR, or Phase-sensitive Optical Time-Domain Reflectometry, integrates fiber interferometry and optical time-domain reflectometry (OTDR) technologies. It is applied in distributed acoustic sensing and is commonly known as Distributed Acoustic Sensor (DAS) technology.

Basic Scheme

ϕ-OTDR Exploits the interference of backward Rayleigh scattering signals generated as narrow-linewidth laser pulses propagate through an optical fiber to detect physical quantity changes along the fiber. By combining the collected interference signals with optical time-domain reflectometry (OTDR), it enables the acquisition of all acoustic signals along the fiber. Figure 4.24 illustrates a typical schematic of ϕ-OTDR.

The key distinction from the interferometers discussed in previous chapters lies in the structure of Fig. 4.24: here, the light source operates in pulsed mode rather than continuous wave. A high-coherence, narrow-linewidth laser pulse with pulse width ΔT occupies a fiber length $\Delta l = c_f \Delta T$ at any instant, where c_f denotes the speed of light in the fiber. When this optical pulse is injected into the test fiber via the circulator, it serves as a probe of length Δl, propagating along the fiber to detect position-specific information that returns to receiver C through backward Rayleigh scattering. At receiver C, an unbalanced M-Z compensating interferometer configures the system into an interferometric structure with an initial OPD of zero.

This pulsed operation resembles conventional optical time-domain reflectometry (OTDR) but differs in its use of a high-coherence laser source. Consequently, ϕ-OTDR's detection mechanism diverges from conventional OTDR's intensity-based approach: from emission to reception, the optical pulse in ϕ-OTDR effectively forms a dual-beam interferometer that relies on phase detection—hence the "ϕ" in its name. Figure 4.25 illustrates the interference mechanism.

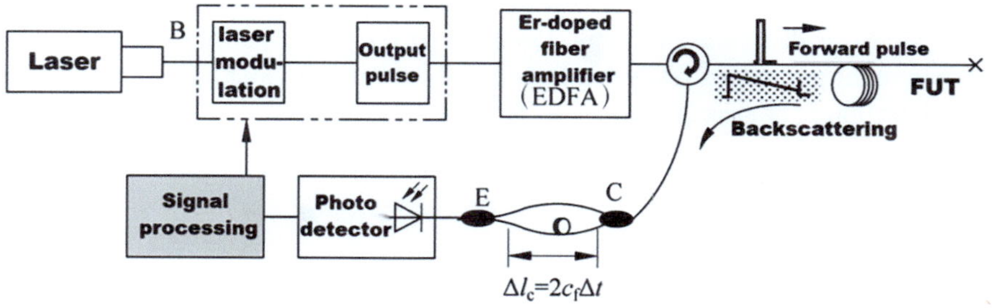

Fig. 4.24 Typical structure of Φ-OTDR

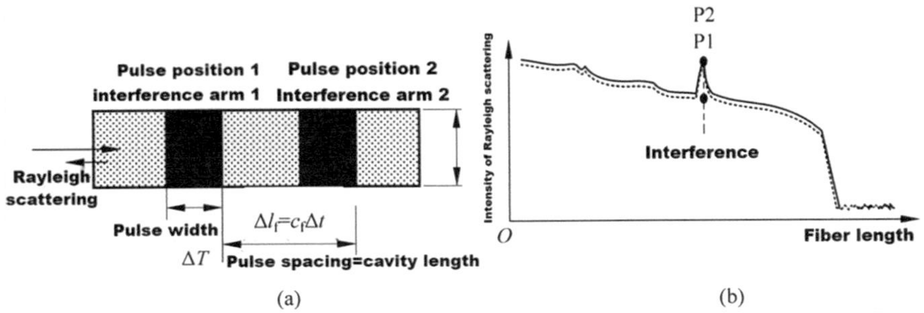

Fig. 4.25 Interference principle of Φ-OTDR

Consider Fig. 4.25a: A laser pulse of width ΔT propagates through the fiber at speed c_f , generating continuous backward Rayleigh scattering to form a backscattered Rayleigh curve. Pulse positions 1 and 2 correspond to two distinct fiber locations separated temporally by Δt and spatially by $\mathbf{\Delta} l_f = c_f \mathbf{\Delta t}$. Scattered light P1 from position 1 travels through the long arm of the M-Z compensating interferometer at point C (Fig. 4.24), while P_2 from position 2 traverses the short arm. These pulses interfere at point E, where their Rayleigh scattering curves temporally overlap (Fig. 4.25b), yielding zero OPD. Essentially, pulses emitted simultaneously return synchronously to the receiver. Acting as dual interferometric arms, perturbations between positions 1 and 2 (i.e., within the interferometric cavity) propagate backward for real-time acquisition, enabling spatially resolved disturbance monitoring. As noted, the compensating interferometer's OPD defines the sensitive fiber length for zero-OPD operation, while the pulse width governs the interferometric arms' energy.

This simplified model divides the pulse into P1 and P2. In reality, the pulse comprises infinitely narrow segments whose collective interference determines the net effect. Rayleigh scattering introduces positional randomness to the interfering light; P1 and P2 here represent averaged leading/trailing segment positions for conceptual clarity.

Section 4.2.1 introduced quadrature detection and PLLs for dual-beam interferometers. However, distributed long-range systems partition the fiber into numerous independent interferometers, rendering PLL-based feedback impractical. Thus, quadrature detection is empirically adopted. The three methods in 4.2.2 apply equally to distributed architectures and are analyzed below.

Phase-Generated Carrier (PGC) Method [7]

The PGC method entails introducing an OPD modulation mechanism—such as a PZT piezoelectric ceramic—into the compensating interferometer in Fig. 4.24, as indicated in the left dashed box of Fig. 4.26a.

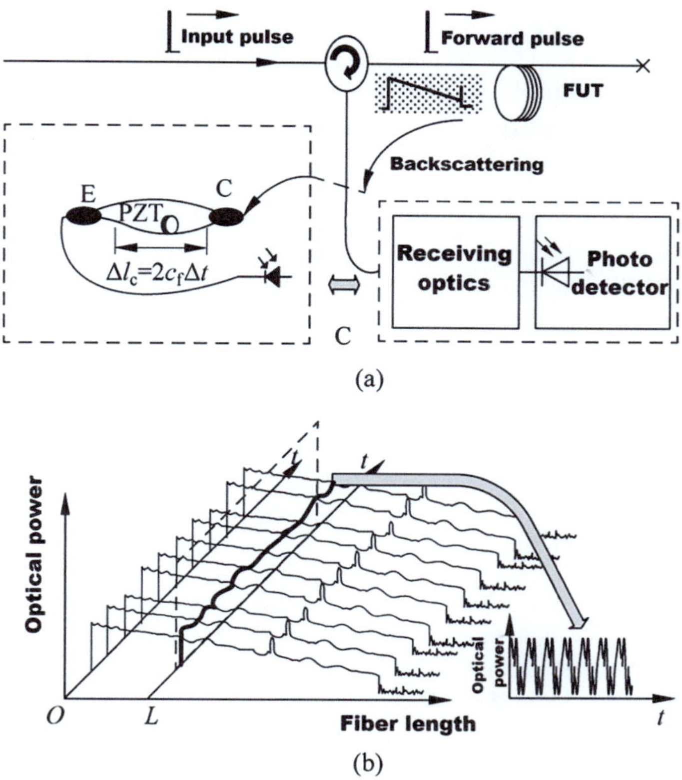

Fig. 4.26 Principle of Φ-OTDR based on PGC

One arm of the interferometer is wound around a PZT, to which a sinusoidal carrier modulation signal is applied, generating the interference signal in Eq. (4.42):

$$I = A + B\cos(M\cos(\omega_c t) + \phi) \tag{4.42}$$

where $\cos(\omega_c t)$ the modulation signal. The modulation depth M is set to 2.63 so that $J_1(M) = J_2(M)$, with M proportional to the amplitude of the modulation voltage applied to the PZT.

The backward Rayleigh-scattered pulses from the fiber are divided into leading P1 and trailing P2 segments, with P1 arriving Δt ahead of P2. These pulses pass through a 2×2 coupler into the compensating interferometer. Upon exiting the interferometer, the P2 segment from the short arm and the P1 segment from the long arm temporally overlap, forming a balanced interferometer where the equivalent fiber interferometer and the receiver-side matched interferometer exhibit zero OPD. This configuration mitigates phase noise caused by the laser's finite coherence length. Applying the PZT to either interferometer arm does not affect the system's operational principle.

The interferometer's output signal undergoes photoelectric conversion, and the data acquisition system uses the optical pulse as a synchronization trigger to collect and store the entire OTDR curve for subsequent processing.

After injecting a single optical pulse into the fiber and completing n-point OTDR signal acquisition, the system emits the next pulse. The interval between consecutive pulses—determined by the total fiber length L_{total}— is calculated as $T_S = 2\frac{L_{total}}{c_f} = n\Delta T$, which equals the pulse's round-trip propagation time in the fiber. This interval defines the Nyquist period for time-domain signals at each fiber point, thereby limiting the vibration signal bandwidth. Following each pulse emission, n-point Rayleigh scattering signals are acquired and repeated m times, forming an $m \times n$ matrix where each row represents an OTDR scattering curve. In Sect. 4.4.1, this matrix will be processed to extract OPD variations induced by external physical changes at corresponding fiber locations. Figure 4.26b schematically illustrates this process.

Heterodyne Method [8]

In contrast to the PGC method, the heterodyne approach modulates the laser frequency in one or both arms of the interferometer to generate a beat frequency ω_c in the interference signal, serving as the carrier for quadrature detection. As shown in Fig. 4.27, the experimental system operates as follows:

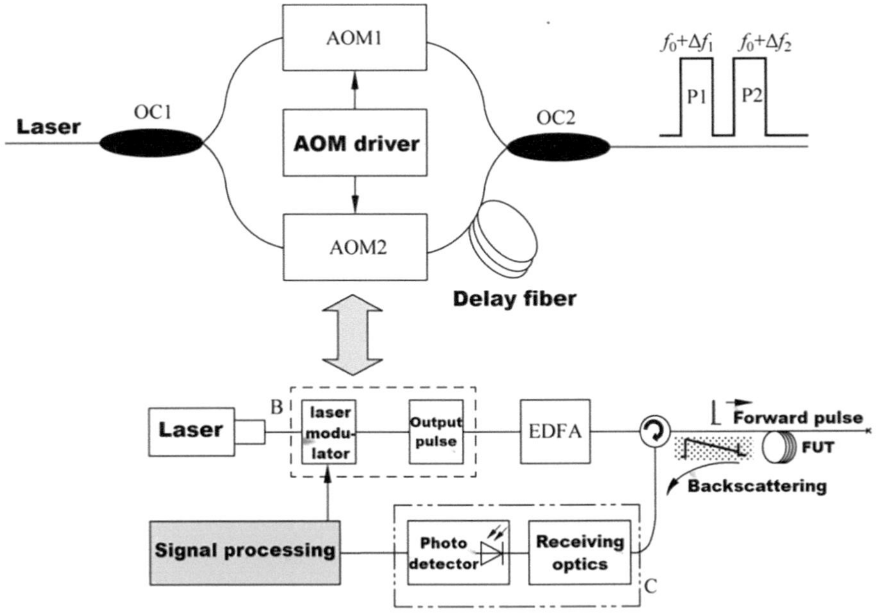

Fig. 4.27 Interference principle of Φ-OTDR based on heterodyne method

A narrow-linewidth light source emits continuous laser light at a central frequency f_0. A 2 × 2 coupler (OC1) splits the light into two paths:

- The first path passes through an acousto-optic modulator (AOM1), generating a pulsed signal P1 with a central frequency $f_1 = f_0 + \Delta f_1$.
- The second path passes through another acousto-optic modulator (AOM2), producing a pulsed signal P2 with a central frequency $f_2 = f_0 + \Delta f_2$.

After passing through a fiber delay line, the second path recombines with the first via coupler OC2. The combined pulses P1 and P2 propagate into the sensing fiber, with backward Rayleigh scattering generating interference signals at the receiver. Due to the frequency difference between P1 and P2, the interference signal exhibits a beat frequency ω_c, described by Eq. (4.43):

$$I = A + B\cos(\omega_c t + \phi) \tag{4.43}$$

where $\omega_c = 2\pi(f_1 - f_2)$, and ϕ represents phase variations induced by external disturbances.

In this heterodyne scheme, the receiver omits the compensating interferometer, as the delay is introduced at the pulse output stage. Both pulses originate simultaneously from the light source and return to the receiver synchronously, with one pulse traversing an additional distance equivalent to half the delay line length in the output path. This ensures a zero-OPD interferometric configuration, where the spatial separation between P1 and P2 in the fiber defines the sensor's sensitive region.

In Fig. 4.27, the beat frequency arises from the difference in modulation frequencies of the two AOMs. An alternative heterodyne implementation uses a continuous-wave reference light path split directly from the source, interfering with the Rayleigh-scattered signal light. Here, the beat frequency corresponds to the modulation frequency of a single AOM. This approach enhances interference signal amplitude due to the reference light's high intensity and non-attenuating nature, though the laser coherence length requirement increases with sensing distance as the OPD between reference and signal paths scales with fiber length.

Data acquisition and storage follow the same protocol as the PGC method, employing the signal processing techniques outlined in Sect. 4.2.2.

3 x 3 Coupler Method [1, 4, 9]

The 3 × 3 coupler method dispenses with the need for carrier modulation in the interference signal, imposes the least stringent requirements on signal processing speed, and reduces computational complexity relative to the PGC and heterodyne methods under identical sampling rates. However, due to the lack of phase or frequency modulation-induced spectral shifting, it is more prone to $\frac{1}{f}$ circuit noise. Figure 4.28 illustrates the system structure utilizing a M-Z interferometer as the matched interferometer.

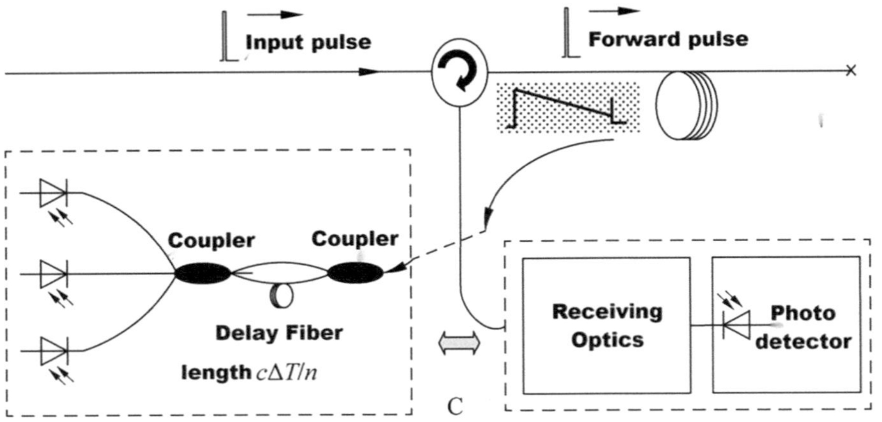

Fig. 4.28 Φ-OTDR interference principle based on a 3 × 3 Coupler

The 3 × 3 coupler method shares similarities with the PGC method, except that the receiver-side matched interferometer (at point C) does not require a PZT for carrier generation. Interference signals from the symmetric 3 × 3 coupler are directed to three separate photodetectors. Following data acquisition, each detector's output is stored as an independent row, yielding a $3m \times n$ matrix (in contrast to the $m \times n$ matrix in the PGC method).

The preceding sections have sequentially discussed three quadrature detection methods. Additional approaches, such as direct 90°phase shifting to generate orthogonal components [10], exist for users to select based on practical needs.

Compared to conventional fiber interferometers, long-range interferometers form an interferometer array, effectively functioning as a sensor array in practice. Extracting signals from such arrays necessitates specialized signal processing methods.

Array Signal Processing

As previously discussed, the pulsed long-range fiber interferometer operates as a sensor array. For each emitted laser pulse, the receiver continuously acquires backward Rayleigh scattering, conducting a single acquisition for each sensor array unit to generate a vector of length N, where $N = \frac{T_s}{\Delta T}$. Here, T_s represents the time interval between consecutive pulses, and $\frac{1}{\Delta T}$ denotes the system's data acquisition rate. After the laser emits M pulses, the system captures M vectors of length N, forming an $M \times N$ matrix (or a $3M \times N$) matrix for the 3 × 3 coupler configuration), as shown in Eq. (4.44):

$$I = \begin{Bmatrix} I_{11} & \dots & I_{1N} \\ \dots & \dots & \dots \\ I_{M1} & \dots & I_{MN} \end{Bmatrix} \tag{4.44}$$

The element I_{ij} in matrix $\boldsymbol{I}$ represents the interference signal resulting from Rayleigh scattering at the j-th point (where the fiber has a total of $N = \frac{T_s}{\Delta T}$ points) during the i-th pulse injection.

The time interval between I_{ij} and I_{i+1j} is T_s, corresponding to a data sampling frequency of $f_s = \frac{1}{T_s}$ for each point in the optical fiber. Per the Nyquist sampling theorem, the maximum frequency of the interference signal is $f_s = \frac{1}{2T_s}$. In the PGC and heterodyne methods, a carrier signal with frequency ω_c (generally at least three orders of magnitude higher than f_s is introduced. The j-th column vector $\left[I_{1j} \dots I_{ij} \dots I_{Mj}\right]'$ in Eq. 4.45 is obtained by decimating a continuous signal sequence with frequency ω_c (for heterodyne; for PGC, the frequency is even higher) at a ratio of *N:1*. This is equivalent to sampling a signal of frequency ω_c at a rate of f_s, which inevitably causes spectral aliasing. Therefore, in the configurations shown in Fig. 4.3 or Fig. 4.5, each row of the matrix in Eq. 4.44 must first be mixed with a local oscillator signal (for PGC, the local oscillator signal is $\cos(\omega_c t)$, $\cos(2\omega_c t)$; for heterodyne, the local oscillator signal is $\cos(\omega_c t)$, $\sin(\omega_c t)$), followed by low-pass filtering. This transforms the signals in Eq. 4.42 or Eq. 4.43 into the form $I = \cos(\phi)$, $Q = \sin(\phi)$, such that each original row becomes two rows, and the matrix in Eq. 4.44 is separated into matrices $\boldsymbol{I_I}$ and $\boldsymbol{I_Q}$. Finally, by processing the corresponding columns of the $\boldsymbol{I_I}$ and $\mathbf{I_Q}$ matrices, the vector for each column is derived, representing the desired disturbance signal.

The selection of M determines the frequency resolution of the system. According to signal processing theory, the frequency resolution of the data acquisition system is defined as $f_\delta = \frac{f_s}{M}$. With a larger M, the minimum resolvable frequency interval of the sensor array is reduced, leading to finer spectral measurements. However, under the same data conversion rate and fiber length, the required storage space is also increased proportionally with larger M.

The pulse width, which typically exceeds or equals the sampling interval of the data acquisition system, determines the spatial resolution of the sensor array. When the pulse width is set equal to the sampling interval ΔT, the spatial resolution ΔL is given by $\Delta L = c_f \times \frac{\Delta T}{2}$, where c_f is the speed of light in the fiber (typically 2×10^8 m/s). The factor of 2 arises from the doubled optical path due to round-trip propagation in the fiber. A smaller ΔT results in higher spatial resolution—for example, a 10 ns pulse width corresponds to a spatial resolution of 1 m. However, ΔT cannot be reduced indefinitely. Since the optical pulse is equivalent to a narrow pulse modulated by constant-frequency Lightwave, its frequency-domain representation is described by a sinc function convolved with the approximately monochromatic laser spectrum, thereby broadening the laser's spectral width. The magnitude of this broadening is inversely proportional to the pulse width; thus, reducing the pulse width degrades the laser's frequency characteristics and increases the interferometer's phase noise. Additionally, smaller ΔT demands higher data acquisition capabilities and increases the complexity of the acquisition system.

The dynamic range of the sensor array's detected signals is determined by the effective number of bits of the AD converter. As is recognized, the intensity of laser light propagating in an optical fiber is gradually attenuated along the transmission direction. When the fiber is sufficiently long, a significant intensity difference is exhibited between light returning from the near and far ends of the fiber. For a fixed effective number of bits in the AD converter, if the receiver sensitivity is set too high, large signals may be saturated; whereas if the sensitivity is set too low, small signals cannot be detected. Therefore, the effective number of bits of the AD converter is considered one of the critical factors that are governed by the system's dynamic range.

The system's dynamic range is also determined by the full-scale range and noise floor of the photoelectric conversion circuit. A larger full-scale range and lower noise floor are associated with a greater dynamic range, which allows for longer measurable fiber lengths. To expand the full-scale range, gain control is introduced at the front-end of the photoelectric conversion. Typically, the power dynamic range can be extended by 20 dB through single-stage gain control. In the technical scheme illustrated in Fig. 4.27, identical attenuation characteristics in the fiber are experienced by the two beams forming the interferometer. For a fiber loss of 0.2 dB/km, the loss of the interference signal is equivalent to 0.8 dB/km (owing to the round-trip propagation of both beams). A 20 dB extension is translated into over 20 km of additional measurement range. In the Fig. 4.27 scheme, the interferometric coherence noise is not scaled with fiber length. Thus, through increasing the electrical signal gain, the system's dynamic range can be enhanced without the introduction of additional phase noise.

In long-range fiber optic interferometric applications, the noise floor is primarily originated from the ASE noise of optical amplifiers, the broadband and thermal noise of transimpedance amplifiers (TIA), and the aliasing noise in distributed operation. ASE noise, as a common—mode noise at the photoelectric receiver, is typically suppressed by using balanced detection in conjunction with optical filters (the design of balanced detectors is detailed in Sect. 4.6.2). TIA operating in high—frequency modes are made susceptible to electromagnetic interference and thermal effects, where the system's minimum detectable light intensity is determined by their broadband and thermal noise. The aliasing noise in distributed operation is caused by the conflict between spatial resolution and total fiber length: higher spatial resolution requires narrower optical pulses, and the frequency bandwidth of the low—pass—filtered *I/Q* components is broadened by these narrower pulses. For a 40 km fiber with 10 m spatial resolution, the *I/Q* components exhibit a minimum frequency of 10 MHz, yet the final sampling rate is reduced to 2.5 kHz. This drastic down sampling from 10 MHz to 2.5 kHz inevitably leads to spectral aliasing, and an elevated noise floor for the phase calculated from the aliased *I/Q* signals is resulted in.

The system's dynamic range is also influenced by the injected optical power. The Rayleigh backscattering signal is weakened by insufficient power, while nonlinear effects are induced by excessive power, and careful parameter optimization is required. Substantial data volumes are generated by distributed fiber sensing: for a 10 km fiber with 100 Msps sampling and 16-bit resolution per point, 10,000 data points are contained in a single Rayleigh scattering trace. A buffer capacity of approximately 160 Mbit is necessitated by acquiring 1,024 pulses. Real-time processing is achieved through a ping—pong buffering scheme, where memory is partitioned into two sections—one for continuous data writing and the other for parallel computation. The required storage capacity is doubled (e.g., 640 Mbit in this case). When a 3×3 coupler is employed, the storage demand is increased threefold to ~2 Gbit.

To meet real-time processing requirements for high-speed (enhanced spatial resolution) and high-capacity (improved frequency resolution and dynamic range) applications, two approaches are commonly adopted:

FPGA-based embedded systems: These have FPGA integrated with large computational buffers for high-speed processing. While being cost-effective, compact, and energy-efficient, higher development complexity is posed by this solution.

GPU-accelerated computing: GPU are paired with computers to enable flexible programming, but higher costs, larger size, and increased power consumption are entailed. Additionally, compatibility challenges are introduced by interfacing GPU with photoelectric modules.

4.4.2 OFDR

OFDR (Optical Frequency Domain Reflectometry), as the name implies, is a technique designed to process reflected signals in the frequency domain through the combination of frequency-modulated continuous wave (FMCW) technology and dual-beam fiber interferometry. Backward Rayleigh scattering signals in the fiber are analyzed in the frequency domain, and the frequency shifts of Rayleigh scattering at each position are demodulated.

An OFDR system employing tunable wavelength interferometry achieves high sensitivity and spatial resolution for distributed temperature and strain measurements. On standard fibers spanning tens of meters, millimeter-level spatial resolution is attained, with strain and temperature measurement accuracies reaching 1 $\mu\varepsilon$ and 0.1 °C, respectively.

The basic system configuration is introduced in this subsection, followed by a discussion of specific signal processing methods [11, 12].

Basic Configuration

The system structure of OFDR is shown in Fig. 4.29a, and its working principle is described in Fig. 4.29b.

The optical path of OFDR is constructed using a Michelson interferometer configuration. A linearly frequency-swept narrowband continuous-wave light is split by a

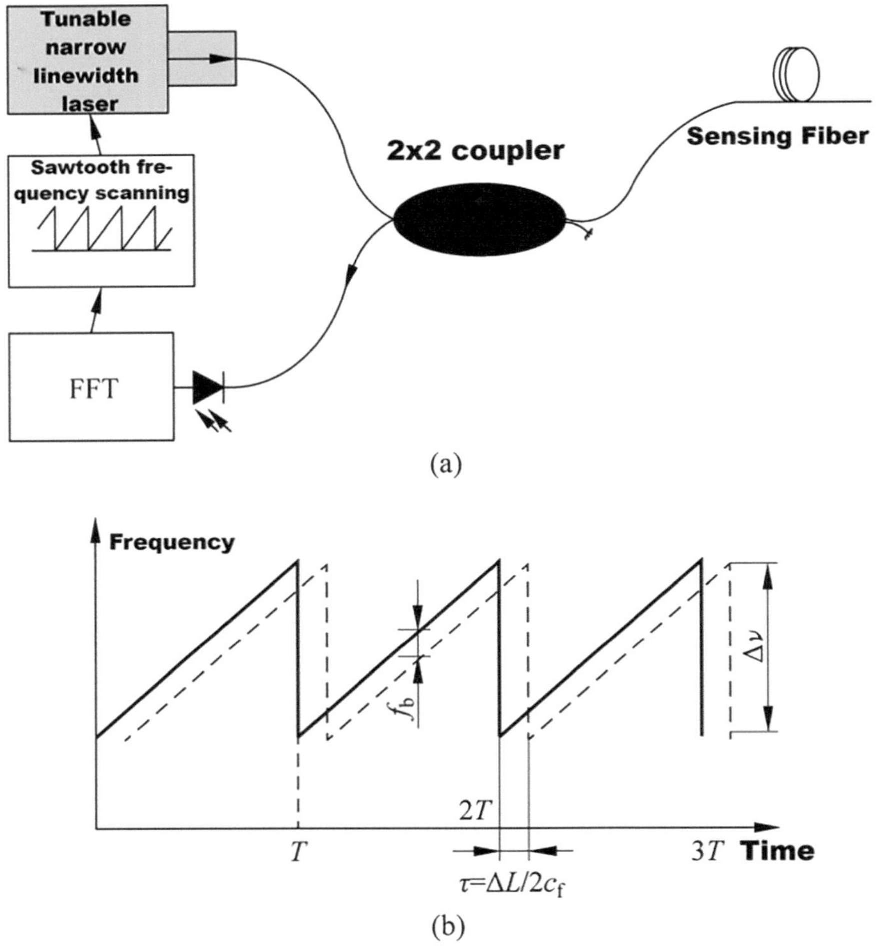

Fig. 4.29 Principle block diagram of OFDR

2 × 2 coupler into the reference arm and sensing arm of the interferometer. The Rayleigh scattering signals and Fresnel end-face reflections from the sensing arm (i.e., the test fiber) are interfered and mixed with the Fresnel end-face reflection from the reference arm at the 2 × 2 coupler. The resulting beat-frequency signal after photoelectric conversion is characterized as a linear frequency-modulated signal. The frequency of the beat signal is proportional to the position of the scattering point (or the final reflective end-face) in the sensing arm, while its amplitude is indicative of the scattering intensity at that point. The spectral shift of the Rayleigh scattering at each point corresponds to axial elongation or contraction of the fiber, which is linearly related to external physical quantities (e.g., temperature, strain). By analyzing the spectrum of the interferometric signal, scattering distribution information across the

entire sensing array is obtained. Figure 4.29b illustrates the relationships between the sweep period T, frequency sweep range Δf, beat frequency f_b, time delay difference τ, and OPD ΔL between the sensing and reference arms.

This method differs from the previously described ϕ-OTDR, in the optical phase expression $i(2\pi ft + kr)$, ϕ-OTDR relies on the ikr term related to optical path length, whereas OFDR utilizes the $i2\pi ft$ term associated with optical frequency. By generating frequency-domain probes through the beat-frequency signals of the two arms, the position of each point in the interferometer's sensing arm is indicated. The principle of the OFDR method is expressed in Eq. (4.45).

$$\begin{aligned} f &= f_0 + \gamma t \\ E(t) &= E_0 \exp\left[i\left(2\pi f_0 + \pi\gamma t^2 + \varphi_t\right)\right] \\ \tau &= 2\left(\frac{\Delta L}{c_f}\right) \\ f_b &= \gamma t = \frac{2\gamma\Delta L}{c_f} \end{aligned} \tag{4.45}$$

f_0 is defined as the initial frequency of the laser sweep. The frequency sweep rate of the laser is denoted by γ. The electric field vector of the laser at time t is represented by $E(t)$. The instantaneous phase of the light source at time t is φ_t. The OPD between a specific point in the sensing arm and the reference arm is corresponded to by ΔL. The speed of light in the fiber is denoted as c_f. The time delay difference between the scattered signal from the measured point in the sensing arm and the reflected signal from the reference arm's end—face is denoted as τ. The beat frequency between the scattered signal at the measured point in the sensing arm and the reflected signal from the reference arm's end—face is denoted as f_b.

The signal processing method of OFDR is characterized by frequency-domain analysis of the interferometer's output signal, centered on Fourier transform. After the light source is linearly frequency-modulated using a sawtooth wave and injected into the interferometer, the output signal is contained with beat frequency components ranging from DC (corresponding to the nearest end of the sensing arm) to the maximum beat frequency (corresponding to the farthest end of the sensing arm). To extract the signal at each point in the sensing arm, the following steps are typically performed:

Data Acquisition: Two sets of time-domain data are separately acquired—one reference signal a (from the undisturbed fiber) and one measured signal b.

FFT Transformation: Both signals a and b are transformed via Fast Fourier Transform (FFT) to obtain their frequency-domain representations A and B, by which the relationship between the signal spectrum and fiber positions is established.

Data Window Selection: At the same position P in A and B, data windows of length N centered at P are selected.

Inverse FFT: The selected N-length data windows in A and B are inversely transformed via IFFT to generate new time-domain data a' and b'.

Cross-Correlation Analysis: a' and b', which contain the spectral shift information of Rayleigh scattering at position P, are subjected to cross-correlation to determine the frequency shift of b relative to a. Since the magnitude of the frequency shift is proportional to external stress or temperature, the measured external parameters can be derived through this relationship.

The following discussion focuses on key concepts in OFDR signal processing—measurement range, spatial resolution, measurement accuracy, and data acquisition rate—which are interrelated.

When designing an OFDR system, the measurement range (i.e., the length of the sensing arm, denoted as L) must first be determined. For simplicity, the reference arm length is assumed to be zero, making L half of the Michelson interferometer's OPD. Once L is determined, the system's maximum beat frequency $f_B(max)$ can be derived from Eq. (4.45), as shown in Eq. (4.46):

$$f_B(max) = \frac{2L}{c_f}\gamma \tag{4.46}$$

γ denotes the laser frequency sweep rate in Hz/s, where c_f is the speed of light in the fiber. The system's maximum beat frequency dictates the necessary AD conversion rate. Per the Nyquist sampling theorem, the system's sampling rate f_s is given by Eq. (4.47):

$$f_S = 2f_B(max) = \frac{4L}{c_f}\gamma \tag{4.47}$$

Next, the spatial resolution is discussed. Based on the sampling frequency defined in Eq. (4.47), the transformation from N time-domain sampling points to N frequency-domain points results in a frequency resolution of:

$$\Delta f = \frac{f_S}{N} = \frac{f_S}{T/\Delta T} = \frac{1}{T} \tag{4.48}$$

where T is the duration corresponding to N sampling points, and $\Delta T = \frac{1}{f_s}$ is the AD sampling interval. As adjacent spectral lines in the frequency spectrum correspond to spatially adjacent points in the fiber, the frequency resolution directly determines the OFDR spatial resolution. Their relationship is expressed in Eq. (4.49):

$$\frac{\Delta L}{\Delta f} = \frac{L}{f_B(Max)} \Rightarrow \Delta L = \frac{L\Delta f}{f_B(Max)} = \frac{L\frac{1}{T}}{\gamma\frac{2L}{c_f}} = \frac{c_f}{2\gamma T} = \frac{c_f}{2f} \tag{4.49}$$

In the last equation on the right side, c_f is the speed of light in the fiber; $f = \gamma T$ represents the total frequency bandwidth swept by the light source during the data sampling duration. Equation (4.49) describes the theoretical highest spatial resolution of OFDR, which can reach submillimeter-level precision.

Finally, let us discuss measurement accuracy. The determining factors include light source intensity, phase noise of the light source, frequency modulation linearity, and circuit noise in the data acquisition system.

If the light source intensity is too weak, both the Rayleigh scattering signal and interference signal will diminish. Since OFDR systems measure the intensity rather than the phase of interference signals, insufficient light intensity significantly degrades system accuracy. Conversely, excessive light intensity introduces nonlinear effects in optical fibers, generating additional frequency components in Rayleigh scattering signals and creating new errors. Therefore, light intensity must be carefully balanced.

During linear frequency modulation of the light source, non-ideal characteristics in modulation systems cause deviations from linear frequency tuning. This nonlinearity disrupts the linear relationship between interference signal frequency and distance. To compensate, specialized optical path structures have been proposed. The OFDR system exhibits an extremely large OPD between its two arms—as the reference arm is typically short, this difference essentially equals the full length of the signal arm, often reaching tens or hundreds of meters. This imposes stringent requirements on the coherence length of the light source [13].

Typical noise sources in data acquisition circuits include: Thermal noise in front-end photoelectric conversion circuits, power frequency noise from AC power supplies, switching power supply noise for digital chips, clock jitter in analog-to-digital converters, AD thermal noise, ground bounce noise in digital circuits. These noise components superimpose on interference signals, degrading measurement accuracy, mitigation strategies involve:

Circuit design stage: adding power decoupling capacitors to suppress ground bounce and power supply noise, optimizing transimpedance resistance in photoelectric circuits (noise $\propto \sqrt{R}$, signal $\propto R$) to reduce thermal noise, employing high-stability temperature-compensated clocks to minimize phase noise (supplemented with clock stabilization IC when necessary).

PCB layout: Implementing electromagnetic compatibility considerations, installing heat sinks for AD chips (practical experience shows significant thermal noise reduction).

4.5 Influence of Light Source Spectrum

As the energy source for fiber-optic interferometers, the light source plays a critical role in interference systems and often constitutes the primary cost component of the system. The quality of the light source largely determines the overall system performance. It is therefore essential to analyze the light source characteristics that affect interferometer performance metrics. This section will first examine laser sources, followed by broad-spectrum sources.

4.5.1 Laser Light Source

This subsection provides a preliminary discussion on the relationship between laser coherence and interferometer noise from a signal processing perspective.

Laser light, with its superior monochromaticity, is the most common light source for interferometers. Ideal monochromatic light would constitute an infinite sinusoidal wave, generating interference fringes with infinitely high orders and permitting unlimited OPD in interferometers – an idealized scenario diverging from practical light sources.

As introduced in Chap. 1, light coherence theory explains that practical light sources exhibit spectral broadening due to: Natural broadening: Caused by finite atomic transition times, represented by a sinc function spectral profile; Doppler broadening: Resulting from thermal motion-induced Doppler shifts, characterized by Gaussian distribution; Collisional broadening: Arising from atomic collisions interrupting radiation, described by Lorentzian distribution.

In practice, natural broadening is rarely observable, with actual spectra typically exhibiting mixed Gaussian–Lorentzian profiles.

The concepts of coherence time and coherence length were established to quantify light source monochromaticity. Coherence time (Δt) represents the average duration of finite wave trains emitted by the source, while coherence length ($\Delta L = c\Delta t$) denotes the statistical average propagation distance during this interval.

Interferometric phase detection depends critically on coherence length: When OPD remain below ΔL, interfering wave trains maintain fixed phase relationships, producing high-visibility interference signals. When OPD exceeds ΔL, random phase variations between uncorrelated wave trains lead to incoherent superposition, reducing interference visibility to zero.

Historically, the coherence length-visibility relationship has been extensively studied. Michelson experimentally demonstrated through visibility curves that: Single-peak spectra exhibit approximately exponential visibility decay with increasing OPD; Multi-peak spectra show exponentially decaying sinusoidal visibility modulation.

Modern fiber interferometer design emphasizes the coherence length-phase noise relationship. The coherence length (ΔL) is determined by wave train duration (Δt). Laser manufacturers typically specify linewidth parameters (Δf or $\Delta\lambda$), with their relationships expressed in Eq. (4.50):

$$\Delta L = c\Delta t \approx \frac{c}{\Delta f} = \frac{\overline{\lambda}_0^2}{\Delta\lambda} \tag{4.50}$$

where ΔL denotes coherence length (unit: m), c the speed of light (m/s), Δt coherence time (s), Δf source frequency bandwidth (Hz), and $\overline{\lambda}_0$ the mean wavelength (typically in nm).

For an interferometer with OPD nl, assuming the light source spectrum follows a Gaussian distribution centered at wavelength λ_0 with variance σ^2, as expressed in

Eq. (4.51):

$$f(\lambda) = \frac{1}{\sqrt{2\pi}\sigma} \exp\left[\frac{-(\lambda - \lambda_0)^2}{2\sigma^2}\right] = \frac{1}{\sqrt{2\pi}\sigma} \exp\left[-\frac{(\Delta\lambda)^2}{2\sigma^2}\right] \tag{4.51}$$

The phase difference corresponding to an OPD nl is given by $\frac{2\pi}{\lambda_0}nl$. The phase noise at the broadened spectral line $\Delta\lambda'$ is expressed by Eq. (4.52):

$$\Delta\phi = \frac{2\pi nl}{\lambda_0^2}\Delta\lambda' \tag{4.52}$$

The corresponding RMS phase noise is given by Equation:

$$\Delta\phi = \frac{2\pi nl}{\lambda_0^2}\sigma = \frac{2\pi nl}{c}\frac{f_{\text{FWHM}}}{2.355} \tag{4.53}$$

Here, f_{FWHM} represents the full width at half maximum (FWHM) typically specified for lasers. For Gaussian spectra, $f_{FWHM} \approx 2.355\sigma$. In practical applications, $f_{FWHM} \approx \sigma$ can be adopted for order-of-magnitude estimation of phase noise.

The Gaussian spectrum assumption in the above discussion serves analytical convenience. Actual spectra exhibit hybrid Gaussian–Lorentzian characteristics, where Lorentzian profiles decay more slowly than Gaussian profiles and lack a finite variance. In practice, FWHM is used as a variance estimator and substituted into Eq. (4.53) for phase noise magnitude estimation.

Taking the NKT Photonics BASIK X15 industrial fiber laser (Denmark) as an example [14]: Linewidth on the order of Hz, coherence length $c/\Delta f \approx 10^5$ km.

For a fiber interferometer with 10 km OPD, phase noise remains at the mrad level. Reducing the path difference to 10 m lowers phase noise to the μrad level.

In fiber-optic interferometers discussed in this book, point sensors exhibit minimal OPD between arms, resulting in correspondingly low phase noise. Distributed interferometers like Φ-OTDR, despite their extended fiber lengths, employ zero path difference configurations through proper optical design, rendering them insensitive to path-difference-induced phase noise. This effectively nullifies the ikr term in the optical phase expression $i(2\pi ft + kr)$. However, distributed interferometers operate with pulsed light sources rather than continuous-wave illumination. When narrow-linewidth sources are pulse-modulated, their frequency-domain spectra become convolved with a sinc function, broadening the original line spectrum into a sinc pattern. Beat frequencies generated between these broadened spectral components introduce low-frequency noise, corresponding to non-zero contributions from the $i2\pi ft$ term in the phase expression. As the interacting frequency components are closely spaced, their beat signals manifest as low-frequency noise that degrades sensor performance in low-frequency regimes. Additional phase noise originates from temporal jitter in the optical pulses. In severe cases, this noise can render signal demodulation impossible.

OFDR represents the most phase-noise-sensitive interferometric configuration, as its effective path difference equals the full signal arm length. For a laser with Hz-level linewidth and 100 m signal arm length, phase noise remains around 10 μrad. In contrast, semiconductor lasers with MHz-level linewidths generate phase noise exceeding 1 rad even with 10 m arms, causing interferograms to exhibit prohibitively low SNR and become virtually undetectable.

4.5.2 Broadband Light Sources

This subsection discusses broadband light sources primarily used in absolute measurement fiber FPI interferometers. In practice, key factors to evaluate include the source's power, spectral profile, and bandwidth. The required optical power depends on the optical path and detector sensitivity, while the operating wavelength range is determined by the detector's working window. Different bandwidths are selected based on specific performance requirements. For methods involving scanning or tunable filtering, linearity of the scanning process must also be considered.

Common broadband sources include light-emitting diode (LED), superluminescent diodes (SLD/SLED), ASE sources, and halogen-tungsten lamps. Table 4.2 compares mainstream commercial broadband sources.

As can be seen from Eqs. (4.25) and (4.26) in Sect. 4.2.2.1, the OPD L and the spectral width $\Delta\lambda$ constituting the phase fluctuation term in the interferometer's output signal exhibit a multiplicative relationship. This is analogous to the relationship between frequency and time in the temporal function $\sin(\omega t)$. When performing Fourier transform on the spectrum to convert from the wavelength domain to the cavity length domain, the spectral resolution and cavity length resolution satisfy the uncertainty principle as shown in Eq. (4.54):

$$\Delta L \cdot \Delta\lambda \geq C \tag{4.54}$$

Table 4.2 Characteristics of broadband light sources

Type	Optical power (mW)	Spectral range (nm)	3 dB bandwidth (nm)	Fiber coupling
LED	1–50	400–1700	20–100	Standard (SMA905/FC)
SLD/SLED	1–15	650–1700	50–300	Integrated (FC/APC)
ASE	10–200	1250–1650 (C/L-band)	30–100	High-efficiency (FC/APC)
Halogen-Tungsten	100–5000	360–2500	500–2000	Requires external fiber

As the constant C, the broader the spectral width $\Delta\lambda$ of the light source, the greater the number of spectral peaks (i.e., the number of spectral periods). After transforming to the cavity length domain, a smaller ΔL corresponds to finer cavity length resolution. Imagine the spectrum being truncated by a rectangular window: a longer rectangular window reduces truncation effects, leading to less spectral leakage in the transformed result, concentrated energy in the main characteristic peak, suppressed sidelobes, and finer cavity length resolution.

The spectrum of a broadband light source typically exhibits a bell-shaped structure, which differs significantly from the ideal flat spectrum. Two methods are commonly used to mitigate the impact of spectral shape discrepancies on the interferometer's output. The first involves using multiple light sources with different wavelength ranges simultaneously; careful calibration can achieve a flattened spectrum. The second method normalizes the interferometer's spectrum by simultaneously measuring the light source's spectrum. The first method is costlier, so the latter is often preferred in practice.

Finally, regarding the optical power of the light source: in a dual-beam FPI interferometer with uncoated end faces, the reflectivity at the incident end is 4%. The reflectivity at the reflective end decays with increasing distance due to diffraction effects. Typically, coating is applied to enhance reflectivity, bringing the total interferometric optical power to approximately 8%. When using commercial-grade Positive-Intrinsic-Negative (PIN) diode to detect the interferometer's spectrum, the sensitivity is on the order of nW. Assuming negligible optical transmission loss, an mW-level light source is sufficient. In multi-beam interferometers, the required optical power is even lower than in dual-beam configurations due to stronger participating interference signals.

For CCD-based detection schemes, the situation differs slightly. The responsivity R of each CCD pixel is typically specified in units of $\frac{\mathrm{V}}{\left(\frac{\mu J}{cm^2}\right)}$ or $\frac{\mathrm{V}}{\mathrm{lx \cdot s}}$. To calculate the required optical power P, parameters such as the CCD's dark voltage V_d, saturation voltage V_S, pixel dimensions $A = L \cdot W$, total number of linear array pixels N, and exposure time T per pixel must be considered, as shown in Eq. (4.55).

$$\frac{P \cdot T}{A} \cdot R > V_d \tag{4.55}$$

The physical significance of Eq. (4.55) is that during the exposure time, the voltage generated by input light illuminating each pixel must exceed the dark voltage—defining the minimum input optical power threshold—while remaining below the saturation voltage, which serves as the maximum power limit. Equation (4.55) offers an order-of-magnitude estimation for the light source's optical power requirements.

Taking a high-quality linear array image sensor from Dalsa Inc. (United States) as an example [16], each linear array comprises $N = 4096$ pixels with a size of $A = 10\mu m \times 10\mu m$, responsivity $R = 13.8\frac{\mathrm{V}}{\frac{\mathrm{mJ}}{cm^2}}$, typical dark voltage 1mV, saturation voltage 900mV, and exposure time 40ns. Substituting these parameters into Eq. (4.55) yields the following expressions in Eq. (4.56):

$$\frac{P(W)}{10^2(\mu m^2)} \times 40 \times 10^{-9}(s) \times 13.8\left(\frac{V}{\frac{\mu J}{cm^2}}\right) > 10^{-3}(V)$$
$$\frac{P(W)}{10^2(\mu m^2)} \times 40 \times 10^{-9}(s) \times 13.8\left(\frac{V}{\frac{\mu J}{cm^2}}\right) < 0.9(V) \tag{4.56}$$

Here, P denotes optical power in watts, T represents time in seconds, A is pixel area in μm^2, V signifies voltage in volts, and R stands for responsivity in $\frac{V}{\frac{\mu J}{cm^2}}$.

From Eq. (4.56), the example indicates that each pixel requires a minimum of 2nW of optical power, totaling 8μW for all 4096 pixels to surpass the CCD's dark voltage threshold. The maximum allowable power per pixel is 2 μW (or 8mW for the entire array) to avoid signal saturation.

As illustrated, high-quality CCD—similar to PIN diodes—exhibits nW-level sensitivity per pixel. In practical implementations, after accounting for post-fabrication optical path losses in the interferometer, the required light source power can be determined using the principles outlined here, ensuring robust system design.

4.6 Photoelectric Conversion in Fiber-Optic Sensors

Photoelectric conversion is a fundamental stage in fiber-optic sensing applications, situated at the front end of signal processing. The design quality of the photoelectric conversion circuit critically impacts signal bandwidth, signal-to-noise ratio (SNR), and distortion. Depending on the application scenario, photoelectric conversion typically employs either single-end detection or differential balanced detection (balanced photodetector, BPD). The primary devices used are PIN photodiodes and avalanche photodiodes (APD). Subsection 4.6.1 discusses single-end detection structures, first addressing PIN-based configurations and then APD-based configurations. Subsection 4.6.2 covers differential balanced detection structures.

4.6.1 Single-End Detection Structures

Single-end detection utilizes either PIN photodiodes or APD. For PIN-based designs, three typical configurations are illustrated in Fig. 4.30.

Figure 30a shows a passive structure where the PIN photodiode acts as a constant current source. Its output photocurrent I_{PD} is proportional to the input light intensity and flows through the load resistor to generate an output voltage $V_{out} = R_{load} \cdot I_{PD}$. By measuring$V_{out}$, the light intensity variation can be observed. This structure requires no external power supply and consists of only two components. For observing low-frequency dynamic light intensity changes, an oscilloscope probe can directly clamp the PIN's anode and cathode pins, utilizing the oscilloscope's input resistance

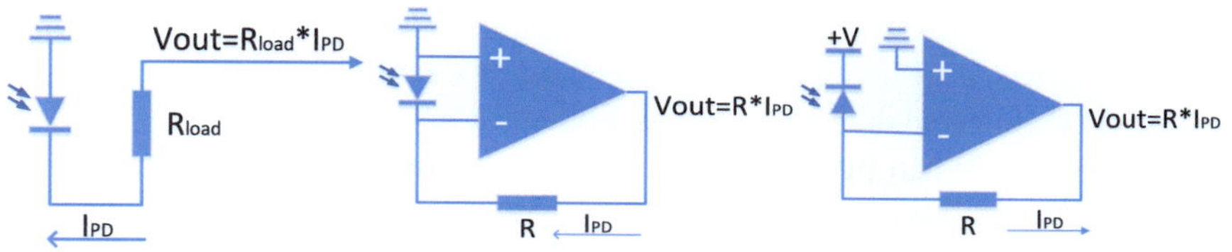

(a) Passive (b) Photovoltaic Mode (c) Photoconductive Mode

Fig. 4.30 Three application structures of PIN photodiode

(approximately 1 MΩ) to complete the photoelectric conversion. This setup is convenient for laboratory use. However, at high frequencies, the PIN's internal capacitance reduces its internal resistance, breaking the constant current assumption under large external resistances and causing nonlinearity between V_{out} and I_{PD}.

Figure 30b illustrates the photovoltaic mode, where the PIN operates similarly to a solar cell. By connecting an operational amplifier (op-amp) and a resistor R to form a transimpedance amplifier (TIA), the input impedance seen from the PIN's cathode is given by:

$$R_{in} = \frac{R}{1 + A_O} \tag{4.57}$$

where A_O is the op-amp's open-loop gain (typically 10^5 or higher). Thus, R_{in} approximates zero, much smaller than the PIN's internal resistance, ensuring constant current condition even at high frequencies and large R. The circuit bandwidth is limited by the op-amp's gain-bandwidth product (GBW). As frequency increases, A_O decreases, causing R_{in} to rise. At frequencies equal to GBW, A_O=1, leading to $R_{in} = 0.5R$, which invalidates the constant current assumption for large R. Additionally, increasing R raises R_{in}, further degrading performance.

Figure 30c introduces a reverse bias voltage at the PIN's cathode, reducing its junction capacitance and enhancing its high-frequency internal resistance. This extends the PIN's constant current range and improves bandwidth, making it the preferred structure for high-frequency applications.

APD-Based Structures (Fig. 4.31).

Fig. 4.31 Photodetector conversion structure using APD diode

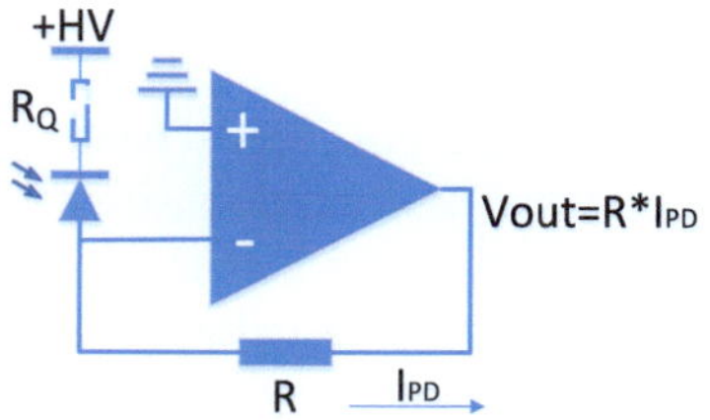

APD circuits differ from PIN circuits by requiring a high-voltage reverse bias (HV). For Si-based APD (visible light detection), HV ranges from 100 to 200 V; for InGaAs APD (infrared detection), HV is around 50 V.

Linear Mode: When biased slightly below the avalanche voltage, (e.g., $HV \sim HV - 3$), APD exhibits a temperature-dependent gain increase (by an order of magnitude) proportional to the bias voltage. Therefore, temperature control circuits are often used to stabilize gain.

Geiger Mode: For single-photon counting, APD is biased above the avalanche voltage. A quenching resistor (R_Q) is inserted between the cathode and HV to reduce the APD voltage below the avalanche threshold after photon detection, resetting it for subsequent events.

TIA analysis for APD adheres to analogous principles as that for PIN, where APD is operated in linear mode by most fiber interferometer applications.

The previous sections introduced various photoelectric conversion circuit configurations and their operational bandwidths. In fiber-optic sensor applications, in addition to bandwidth, the circuit's weak-light detection capability, noise performance, and stability are equally critical.

Weak-light detection capability depends on two factors.

First, photoelectric conversion efficiency. APD, leveraging the avalanche effect, achieve conversion efficiencies an order of magnitude higher than PIN diodes. Thus, APD is preferred for extremely weak light inputs. However, APD noise increases with bias voltage, making PIN more suitable for stronger light conditions.

Second, input bias current (I_{Bias}) of the TIA's Op-Amp: In an ideal op-amp, I_{Bias}=0, ensuring all photocurrent flows through the transimpedance to generate voltage. In reality, commercial op-amps exhibit I_{Bias} as low as fA levels, far below typical photodiode dark currents (nA range). Thus, the minimum detectable current is typically limited by the photodiode's dark current.

Third, challenges with high-impedance nodes: Low I_{Bias} implies high input impedance, which enhances weak-light sensitivity but increases susceptibility to noise currents. These include leakage currents from environmental factors (e.g., humidity, soldering residues), dense pin layouts, long high-impedance traces, and electrostatic/electromagnetic interference. For example, static charge from a nearby person can induce ~ 100 fA of interference. Mitigation strategies include: Minimizing high-impedance trace lengths (from PIN/APD outputs to the op-amp's inverting input). Routing slots near high-impedance paths in space-constrained designs. Avoiding power planes beneath high-impedance traces. Shielding the inverting input with a guard ring (equipotential to the non-inverting input) to suppress stray capacitance. Placing ceramic capacitors (leakage currents up to pA levels) downstream of the transimpedance.

Key Trade-offs: Signal bandwidth, weak-light sensitivity, noise, and stability are interrelated. Enhanced sensitivity increases noise pickup, wider bandwidth broadens noise spectra, and higher complexity degrades stability. To balance these while meeting bandwidth and sensitivity requirements: Apply filters at the TIA output. Use op-amps in larger packages to isolate high-impedance nodes and reduce stray capacitance. Connect a pF-level capacitor in parallel across the transimpedance (weighing leakage trade-offs) to suppress oscillation. Implement proper decoupling capacitors at op-amp power pins. Address capacitive loads (e.g., coaxial cables) to prevent high-bandwidth oscillations.

With single-end detection methods covered, the next section will discuss differential balanced detection.

4.6.2 Differential Balanced Detection

Equations (3.5) and (3.6) in Chap. 3 reveal that two differential optical signals are generatedat the output of the fiber interferometer,expressedin Eq. (4.58) as:

$$\begin{cases} I_{out} = A + B\cos(P) \\ I_{outb} = A - B\cos(P) \end{cases} \tag{4.58}$$

The term A in Eq. (4.58) consists of two components: the DC component from the interferometer's output and optical path noise. In distributed applications, optical amplification methods such as Erbium-doped Optical Fiber Amplifier (EDFA) are often employed. While amplifying the optical signal, EDFA also generates significant ASE noise, which manifests as white noise and is present in both I_{out} and I_{outb} of Eq. (4.58). When the optical gain is high, the ASE noise can exceed the signal by 2–3 orders of magnitude, making effective detection challenging with single-end PIN or APD configurations. The noise component in A is nearly identical in both the upper and lower parts of Eq. (4.58), acting as common-mode noise. To address this, a differential detection method known as balanced detection is proposed, as illustrated in Fig. 4.32.

Figure 4.32 illustrates the balanced detection method. Here, the common-mode components A (including the DC component and ASE noise) from photodiodes PIN1 and PIN2 are attenuated through subtraction, while the differential AC components $B\cos(P)$ are enhanced. The photoelectric conversion is achieved via the transimpedance R. When the characteristics of the two photodiodes are matched, the output voltage Vout eliminates the common-mode components and doubles the signal component. This common-mode rejection is performed directly at the photoelectric conversion stage, preventing the TIA from saturating due to excessive common-mode light intensity and thereby preserving the amplification of useful signals. Balanced detection significantly improves the receiver's dynamic range and signal-to-noise ratio (SNR).

Balanced detectors are primarily used for differential detection of interferometer output signals. The design considerations discussed for single-end detection—signal bandwidth, weak-light sensitivity, noise, and stability—also apply to balanced detectors. Due to the differential nature of balanced detection, PIN pairs with closely matched parameters are typically selected. In practice, for long-distance (tens of

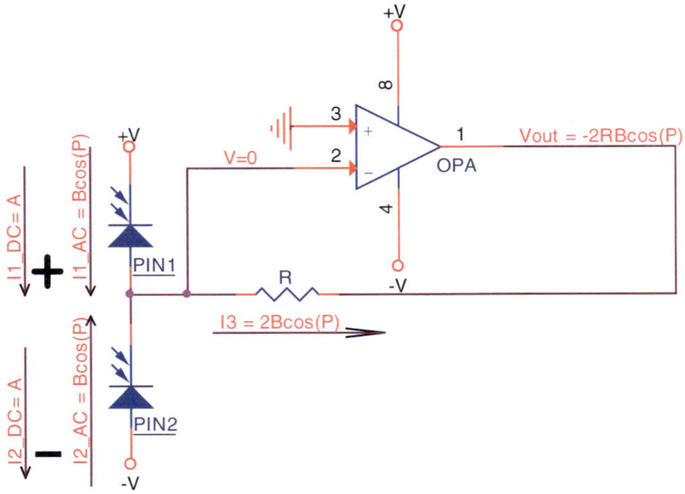

Fig. 4.32 Schematic diagram of balanced detector principle

kilometers) Φ-OTDR applications, balanced detectors using PIN diodes demonstrate sufficient sensitivity while achieving substantial noise suppression.

References

1. Mei Z, Lü HF, Wen XY et al (2021) Improved ellipse fitting algorithm and vibration sensing phase demodulation. Acta Optica Sinica 41(24):2412001
2. Zhang HY (2011) Time-division multiplexing technology for fiber-optic ydrophones based on matched interferometer structures. Tsinghua University, Beijing
3. Kuang W (2005) Research on key issues of fiber-optic interferometric sensors. Tsinghua University, Beijing
4. Yu L (2015) Design and application of high-precision fiber Fabry-Perot cavity sensors. Tsinghua University, Beijing
5. Jiang Y, Tang CJ (2009) Principles and applications of fiber Fabry-Perot interferometers. National Defense Industry Press, Beijing
6. Jin W, Ruan SC (2005) New advances in fiber-optic sensing technology. Science Press, Beijing
7. Bao X, Chen L (2012) Recent progress in distributed fiber optic sensors. Sensors 12(7):8601–8639
8. Fang GS, Xu TW, Feng SW et al (2015) Phase-sensitive optical time domain reflectometer based on phase-generated carrier algorithm. J Lightwave Technol 33(13):2811–2816
9. He XG, Xie SR, Liu F et al (2017) Multi-event waveform-retrieved distributed optical fiber acoustic sensor using dual-pulse heterodyne phase-sensitive OTDR. Opt Lett 42(3):442–445
10. Masoudi A, Newson TP (2017) High spatial resolution distributed optical fiber dynamic strain sensor with enhanced frequency and strain resolution. Opt Lett 42(2):290–293
11. Wang YN, Zhang L, Wang S et al (2016) Coherent Φ-OTDR based on I/Q demodulation and homodyne detection. Opt Express 24(2):853–858
12. Song J (2014) Optical frequency domain reflectometry: sensing range extension and enhanced temperature sensitivity. University of Ottawa, Ottawa Canada

13. Fu CL, Peng ZW, Li PF (2023) Research advances in OFDR-based distributed fiber-optic temperature/strain/shape sensing. Laser Optoelectron Prog 60(11):1106007
14. Qin J, Zhou Q, Xie WL et al (2015) Coherence enhancement of a chirped DFB laser for frequency-modulated continuous-wave reflectometry using a composite feedback loop. Opt Lett 40(19):4500–4503
15. NKT Photonics (2019) KOHERAS BASIK X15.ultra narrow band fiber laser, product datasheet [M]. BASIK Datasheet LG21 Feb19. Pdf. NKT Photonics. https://www.photonicsolutions.co.uk/upfiles/koheras
16. Teledyne DALSA Inc. (2008) Teledyne DALSAIT-P1–4096E-R image sensors, product datasheet[M]. Teledyne DALSA Inc. https://www.doa88.com/p-1788089946189.html

Chapter 5
Applications of Fiber-Optic Interferometers

Physical quantities like pressure and temperature, when acting upon an optical fiber, can directly induce phase changes in the light wave within the fiber. This phenomenon enables phase modulation in fiber vibration/acoustic sensors, pressure sensors, temperature sensors, or fiber rotation sensors (fiber gyros). Among these, the fiber optic gyro and the fiber optic hydrophone are the most representative and have been successfully commercialized in the evolution of fiber optic sensor technology.

Nevertheless, certain physical quantities that do not directly affect the fiber can still induce changes in stress and/or temperature within the fiber through the interaction with sensitive materials. This, in turn, causes a phase change in the transmitted light wave within the fiber. For instance, a fiber optic magnetic field sensor can be created by splicing or coating a film of magnetostrictive material onto the fiber. Similarly, a fiber current sensor can be produced by attaching an electrostrictive material to the fiber, and a fiber accelerometer can be formed by affixing a mass block to the fiber. Additionally, a specialized coating on the optical fiber can serve as a fiber-optic chemical or biosensor for specific chemical reactions or biological processes. For instance, depositing a 10-μm-thick layer of palladium on a SMF can create a fiber-optic hydrogen sensor, and so forth.

The working mechanism and analytical models of various fiber interferometers have been detailed in the previous sections, and this chapter focuses on problems and available solutions in various applications.

5.1 Fiber Optic Underwater Acoustic Sensor

A fiber optic underwater acoustic sensor, commonly referred to as a hydrophone, is an essential type of fiber optic pressure sensor. The stress effect of underwater sound waves on the fiber alters its intensity, phase, or polarization state, from which acoustic parameters can be deduced inversely.

Y. Liao et al., *Fiber Optic Interferometer*, Springer-TUP Physics Series,
https://doi.org/10.1007/978-981-95-2558-4_5

Fiber optic hydrophone can not only overcome the shortcomings of piezoelectric ceramic/PVDF hydrophone in constituting a towing array, but also possess unique advantages: (1) a broad applicable frequency bandwidth (several Hz to tens of kHz). The upper operating frequency of piezoelectric hydrophone is limited by the resonance frequency of sensitive components and wave size, and the poor low-frequency sensitivity of piezoelectric components restricts the lowest detection frequency (20 Hz); (2) It possesses high acoustic voltage sensitivity, being 2 to 3 orders of magnitude greater than the best piezoelectric ceramic hydrophones. It is influenced by fluid static pressure and operating frequency and can function under high hydrostatic pressure of 300 atmospheres (1 atm = 101 kpa). (3) It remains unaffected by strong electromagnetic fields, is compact and lightweight, and offers flexibility in shape design. (4) It combines the functions of sensing and information transmission into one, facilitating the construction of sensing networks. In summary, the fiber optic hydrophone has become the most ideal acoustic detector in contemporary marine technology.

Since its inception, dozens of fiber-optic hydrophones have been reported, which can be categorized into three types based on the photo-acoustic modulation mechanism: intensity, polarization, and phase modulation (interferometric type). Although amplitude-modulated fiber hydrophones have engineering applications, they are unsuitable for towing arrays due to excessive noise, poor stability, and low sensitivity. The interference type is the most widely used hydrophone.

Fiber optic interference hydrophones can be classified into two types: dual-beam interference and multi-beam interference. Further subdivisions include Michelson, MZ, FP, and Sagnac [1, 2]. Among these, the Michelson interference type (as shown in Fig. 5.1) has been the most successfully applied. The subsequent discussion focuses primarily on the research and applications of the Michelson interference hydrophone [3–7].

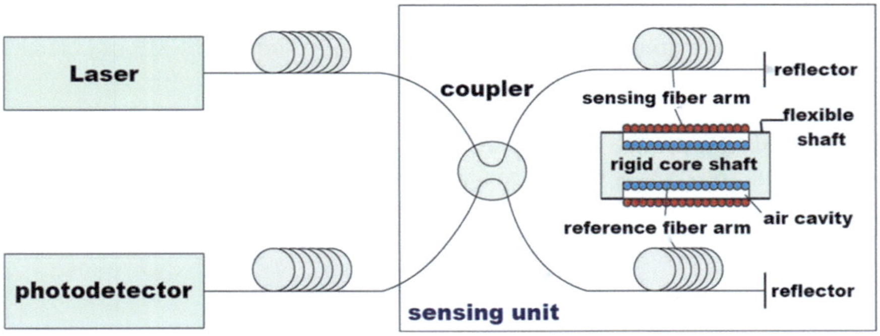

Fig. 5.1 Schematic diagram of the Michelson interference fiber hydrophone system

5.1.1 Design and Preparation of Fiber Optic Hydrophone Sensing Unit

The sensing arm and reference arm of the fiber hydrophone sensing unit should exhibit good symmetry to ensure the unit has a robust noise resistance capability. There are two commonly used structures: the multi-layer coaxial structure (Fig. 5.2a) and the serial coaxial structure (Fig. 5.2b). The design of a fiber hydrophone sensing unit mainly involves the following four key issues.

1. Sensitization and desensitization of optical fibers

To enhance the phase shift sensitivity of the optical fiber hydrophone (which corresponds to the minimum detectable sound field pressure), it is necessary to increase the sensitivity of the sensing arm fiber and decrease the sensitivity of the reference arm fiber. For example, the sensing arm selects an optical fiber that is sensitive to sound pressure, and the fiber's coating material is conducive to improving phase shift sensitivity, such as using Teflon as the coating layer. Conversely, the reference arm fiber is coated with a metal film, for example, an aluminum film. Additionally, the design of the two-arm fiber lengths in the interferometer must meet sensitivity requirements and be as short as possible to minimize volume and noise.

2. Selection of the supporting cylinder for the sensing unit, and the lining and external protection coating for the interference arms

To ensure the pressure resistance and tensile strength of the sensing unit, the main structure is typically designed with a support cylinder as a bracket, constructed from hard materials such as aluminum alloy or hard rubber wood. Outside the support cylinder, two lining cylinders are wrapped in turn with different fibers to serve as the sensing and reference arms.

The requirements for lining material are: good adhesion between the material and the optical fiber, and compatibility with the support cylinder to meet the sensitivity and signal processing needs. The reference arm can select piezoelectric ceramic materials to satisfy the phase compensation or modulation/demodulation requirements of the PZT. To enhance the phase shift sensitivity of the sensing arm, acoustic-sensitive

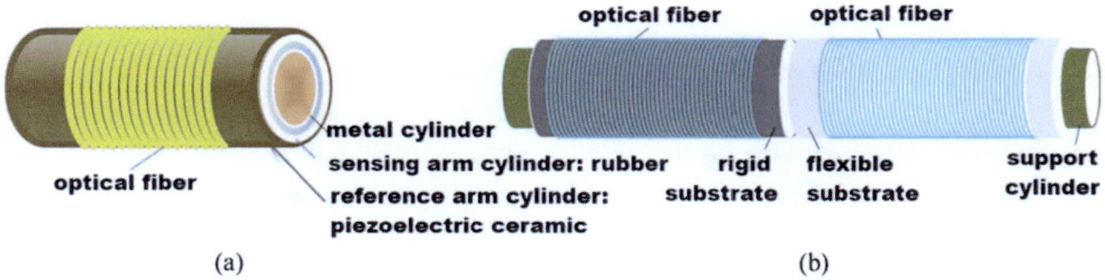

Fig. 5.2 Structure of two commonly used optical fiber hydrophone units

linings such as rubber, nylon, and polymer are utilized. Although theoretical calculations suggest that PTFE could improve phase shift sensitivity, it is too smooth to serve as a lining for the signal arm.

For external protection materials: the sensing arm fiber can directly contact with an acoustically lossless liquid (such as certain oils) or an acoustic rubber that matches the acoustic impedance. The outer protective layer of the reference arm fiber employs a sound insulation material, or metal such as aluminum or nickel, and it is more effective to retain a thin air layer between the outer protective layer and the reference arm optical fiber.

3. Winding of the interference arm optical fiber

To minimize transmission light loss, the optical fiber should be wound in such a way that it is arranged closely and parallelly in single or double layers. The reference arm fiber should be pre-stressed and wound onto a piezoelectric ceramic for compensation or phase modulation. With the advancement of fiber optic gyro and hydrophone technology, optical fiber winding machines have matured and are commercially available, suitable for the automatic winding of polarization/depolarization optical fiber rings.

4. Separation of the fiber sensing unit from the signal processing unit

Similar to the traditional tow cable detector, the fiber sensing unit and the signal processing unit should be separated to ease installation and maintenance, akin to the wet end and dry end. The sensing unit is housed within the tower, whereas the light source and signal processing unit are integrated into the digital package. Only the essential transmission fiber connects the packaged sensing unit to the exterior.

5.1.2 *Phase Shift Sensitivity of Composite Structured Fiber Hydrophones*

Due to the differing working principles, the sensitivity index of a fiber optic hydrophone differs from that of a piezoelectric hydrophone. The key parameters that describe the sensitivity of fiber optic hydrophones include.

Phase-Shift Sensitivity Versus Normalized Sensitivity

The fiber interference hydrophone utilizes phase shift as a parameter to describe its sensitivity characteristics, known as the fiber hydrophone phase shift sensitivity M_p. This can be expressed as: $M_p = \frac{\Delta\phi}{\Delta P}$, with the unit being rad/Pa, meaning the phase change of the fiber hydrophone under a unit pressure change. This directly reflects the sensitivity of the fiber itself and the composite structure of the probe. To eliminate the effects of the light source wavelength and fiber length, the defined parameter, normalized phase shift sensitivity M_0 is commonly used and expressed as follows:

$$M_0 = \frac{\Delta\phi}{\phi\Delta P} = \frac{M_p}{\phi} = \frac{M_p}{2\pi nL/\lambda} \tag{5.1}$$

where, λ represents the wavelength of the light source, n is the refractive index of fiber core, L is the sensitive fiber length in unit Pa, $\Delta\phi$ represents the phase change of the fiber per unit length under unit pressure, and M_p is a measure of the sound-light conversion efficiency of the sensor.

Minimum Detectable Sound Field Pressure Threshold P_{min}

The minimum detectable sound field pressure threshold defined as the minimum sound pressure detected by a fiber optic hydrophone. Set φ_{min} the phase detection accuracy of the system with unit in rad, the minimum sound pressure P_{min} can be expressed as: $P_{min} = \frac{\phi_{min}}{M_p} = \frac{\lambda\phi_{min}}{2\pi nLM_0}$. The common minimum detectable sound pressure level, P_{min}, is represented in dB·μPa. 0 dB·μPa corresponds to 1 μPa, and the conversion formula is $P_{min}(dB \cdot \mu Pa) = 20\lg\frac{P_{min}}{P_0}$, (where, $P_0 = 1$ μPa $= 10^{-6}$ Pa).

Sound Pressure Sensitivity

Acoustic reception voltage sensitivity is commonly referred to as sound pressure sensitivity for short. The sound pressure sensitivity of the traditional hydrophone M_p is defined as the output voltage value M_p (V/μPa) of the hydrophone system generated by a change in unit sound pressure, measured in dB. The corresponding reference is: 0 dB for an output voltage of 1 V generated by 1 μPa. Similarly, to establish the same parameter for fiber-optic hydrophones, thus $M_V = M_pGRI_0Q$. Where G represents the sensitivity of the photodetector (A/W), R denotes the load resistance (Ω) of the photodetector, I_0 is the average output power of the fiber interferometer, and Q signifies the contrast of the interference stripe. Considering the factors of circuit conversion and amplification, sound pressure sensitivity cannot directly reflect the performance of the interferometer. However, it does adopt the sensitivity concept and measurement approach of the traditional piezoelectric hydrophone. Consequently, the pressure sensitivity can be directly compared with that of the traditional hydrophone.

The unit of M_V is also dB, the conversion formula is $M_V(\text{dB}) = 20\,lg\frac{M_V}{M_{V_0}}$. where $M_{V_0} = 1V/\mu Pa$. The sound pressure sensitivity of common piezoelectric hydrophones is about −200 ~−180dB.

To enhance the sensitivity of a hydrophone, it is necessary to coat the sensing fiber or embed special materials to increase its sensitivity. The fiber sensing unit with enhanced sensitivity is known as the composite structure hydrophone unit. The acoustic sensing optical fiber features a coating layer that is sensitive to sound pressure, while the inner lining and outer protective layers are also made of acoustic materials such as PTFE and polyurethane. The subsequent example will demonstrate how to calculate the phase shift sensitivity using such a composite structure optical fiber sensing unit.

Example 5.1 Calculate the phase shift sensitivity of a composite structure fiber hydrophone.

According to the concept and definition of phase shift sensitivity and normalized sensitivity of a fiber hydrophone, as expressed in Eq. (5.1), the phase change of a composite structure fiber hydrophone under sound pressure is calculated.

According to the photoelastic effect, the phase shift sensitivity of an uncoated SM fiber can be expressed as:

$$\frac{\Delta\varphi}{\Delta P} = \frac{L\beta}{E}(1-2\mu)\left[\frac{n^2}{2}(2p_{12}+p_{11})-1\right] \tag{5.2}$$

Specifically, $\Delta\varphi$ represents the phase change of the interferometer output corresponding to the sound pressure variation of ΔP over a fiber length of L; $\beta = \frac{2\pi n}{\lambda}$ denotes the propagation constant of light, E is the Young's modulus, and μ is the Poisson ratio of the fiber, while n is the refractive index of the fiber core, and p_{11} and p_{12} are the photo-elastic coefficients of the fiber. For SM fused silica fiber, E, μ, p_{11} and p_{12} are constant, substituting $n = 1.456$, $E = 7.3 \times 10^{10}$ N/m^2, $p_{11} = 0.121$, $p_{12} = 0.27$, $\mu = 0.17$ into Eq. (5.2), we obtain $\frac{\Delta\varphi}{\Delta P} = 2.48 \times 10^{-11}\frac{L}{\lambda}$. The core diameter of the actual SMF ranges from 4 to 9 μm, the cladding diameter is 125 μm, and the coating diameter is 250 μm, which means the coating layer thickness is 62.5 μm. Figure 5.3 illustrates the simplified fiber cross-section in theoretical calculation.

In Fig. 5.3, when an external pressure P acts on the composite structure of fiber, it is necessary to consider the radial and axial stresses of zones 1 and 2, as well as their resulting displacements. By applying certain boundary constraints and substituting r_1, r_2, E_1, E_2, n_1, and n_2 into the stress and strain equations, a set of five linear equations and their numerical solutions can be obtained.

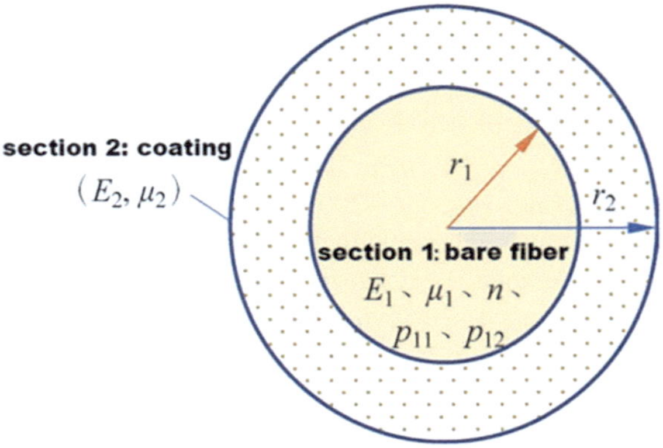

Fig. 5.3 Fiber cross-section diagram

In the cylindrical coordinate system, the radial stress of a circular symmetric cross-section can be expressed as:

$$\sigma_r = \frac{A}{r^2} + B(1 + 2lgr) + 2C$$

$$\sigma_\theta = -\frac{A}{r^2} + B(3 + 2lgr) + 2C$$

where, in zone 1: $\sigma_r = \sigma_\theta = 2D$, $\sigma_z = G$; in zone 2: $\sigma_r = \frac{A}{r^2} + 2C$, $\sigma_\theta = -\frac{A}{r^2} + 2C$, $\sigma_z = F$. A, C, D are constants, while $B = 0$.

The five boundary constrains applied for the composite structure shown in Fig. 5.3 are: the radial stress at $r = r_2$ equals to the external force P; the radial stress zone 1 and 2 are equal at $r = r_1$, and the radial displacement are matched; the axial strains of zone 1 and 2 are equal far from the ends. The axial pressure applied on the composite structure equals to the integral of the axial pressure σ_z through zone 1 and 2:

$$\sigma_{r_2}(r_2) = -P; \quad \sigma_{r_1}(r_1) = \sigma_{r_2}(r_1); \quad u_{r_1}(r_1) = u_{r_2}(r_1); \; \varepsilon_{z1} = \varepsilon_{z2};$$
$$-\pi r_2^2 P = \pi r_1^2 \sigma_{z1} + \pi\left(r_2^2 - r_1^2\right)\sigma_{z2}$$

where, axial displacement $u = \int \varepsilon_r dr$, and $\varepsilon_i = \frac{1}{E}\left[\sigma_i - \mu\left(\sigma_j + \sigma_k\right)\right].i, j, k = r, \theta, z.$

A group of equations obtained is expressed by a matrix as:

$$\begin{bmatrix} 1 & 2 & 0 & 0 & 0 \\ R^2 & 2 & -2 & 0 & 0 \\ -E(1+\mu_2)R^2 & 2(1-\mu_2)E & -2(1-\mu_1) & -E\mu_2 & \mu_1 \\ 0 & 0 & 0 & R^2 - 1 & 1 \\ 0 & 4E\mu_2 & -4\mu_1 & -E & 1 \end{bmatrix} \begin{bmatrix} X \\ Y \\ Z \\ V \\ W \end{bmatrix} = \begin{bmatrix} -1 \\ 0 \\ 0 \\ -R^2 \\ 0 \end{bmatrix} \tag{5.3}$$

where,

$$X = \frac{Ar_2^2}{P}, Y = \frac{C}{P}, Z = \frac{D}{P}, V = \frac{F}{P}$$

$$W = \frac{G}{P}, \; R = \frac{r_2}{r_1}, E = \frac{E_1}{E_2}$$

The values of X, Y, Z, V and W can be calculated based on the thickness and material parameters of the coating layer. The phase change $\Delta\varphi$ produced by a fiber length of L subjected to a pressure of P is,

$$\Delta\varphi = \beta\Delta L + L\Delta\beta \cong \beta\varepsilon_z L + Lk_0\Delta n = \beta\varepsilon_z L - \frac{1}{2}Lk_0 n^3 \Delta\left(\frac{1}{n^2}\right) \tag{5.4}$$

where, $\Delta\left(\frac{1}{n^2}\right)$ is the change of the fiber's refractive index. Since there is no tangential strain in the optical fiber, then

$$\Delta\left(\frac{1}{n^2}\right)_i = \sum_{j=1}^{3} p_{ij}\varepsilon_j \tag{5.5}$$

Assume the optical fiber material is uniform and isotropic, its photoelastic coefficient tensor can be written as

$$p_{ij} = \begin{bmatrix} p_{11} & p_{12} & p_{12} \\ p_{12} & p_{11} & p_{12} \\ p_{12} & p_{12} & p_{11} \end{bmatrix} \tag{5.6}$$

When x or y polarization propagates along the z-axis, the change in the refractive index of the fiber is as follows:

$$\Delta\left(\frac{1}{n^2}\right)_{x,y} = (p_{11} + p_{12})\varepsilon_r + p_{12}\varepsilon_z \tag{5.7}$$

An expression for the phase-shift sensitivity can be obtained:

$$\begin{aligned}\frac{\Delta\varphi}{\Delta P}\frac{E_1}{\beta L} = W&\left\{(1 + \frac{n^2}{2}\left[(p_{11} + p_{12})\mu_1 - p_{12}\right]\right\} \\ &+ Z\left\{-4\mu_1 - n^2\left[(p_{11} + p_{12})(1 - \mu_1) - 2\mu_1 p_{12}\right]\right\}\end{aligned} \tag{5.8}$$

For typical silica fiber composite structures, available:

$$\left.\frac{\Delta\varphi}{\Delta P}\frac{E_1}{\beta L}\right|_{\substack{\text{fuse}\\\text{silica}}} = 0.784W - 1.173Z \tag{5.9}$$

For fused silica fiber without coating, the stress is isotropic and equals to pressure, i.e., $G = -P$, $2D = -P$; $W = -1$, $Z = -1/2$. Substituting to Eq. (5.9), we get:

$$\left.\frac{\Delta\varphi}{\Delta P}\frac{E_1}{\beta L}\right|_{\substack{\text{uncoated}\\\text{fiber}}} = -0.198; \quad \left.\frac{\Delta\phi}{\Delta P}\frac{E_1}{\beta L}\right|_{\text{si\_resin}} = -0.20 \tag{5.10}$$

Equation (5.10) aligns perfectly with the calculation results of Eq. (5.2), indicating that the case without a coating layer is merely a special instance of Eq. (5.10). With a coating layer, the values of W and Z can be calculated through substitution of r_2, E_2, v_2, and subsequently inserted into Eq. (5.9) to determine the phase shift sensitivity of the composite structure.

The standard communication fiber features a soft coating material, specifically silicone resin, with a Young's modulus of 3.5×10^6 N/m^2 and a Poisson ratio of 0.499. The cladding diameter of the single-mode (SM) fiber is 125 μm, and the outer diameter of the coating layer is 250 μm, resulting in a ratio (R) of 2. The calculations are as follows:

$$\left.\frac{\Delta\phi}{\Delta P}\frac{E_1}{\beta L}\right|_{\text{Si\_ resin}} = -0.20 \tag{5.11}$$

It is evident that the phase shift sensitivity of silicone resin coated fiber is nearly equivalent to that of bare silica fiber, suggesting that the coating of standard communication fiber does not impact its phase shift sensitivity. If the coating material is altered to PTFE (Young's modulus: 3.06×10^9 N/m^2, then $E = 23.86$, Poisson ratio $\mu_2 = 0.317$), the calculated W and Z are W = −2.0372 and Z = −0.6128, respectively. Then

$$\left.\frac{\Delta\varphi}{\Delta P}\frac{E_1}{\beta L}\right|_{\text{PTFE}} = -0.8784 \tag{5.12}$$

Relationship of Phase Shift Sensitivity to Coating Thickness and Material Parameters

Compared with Eq. (5.10), the PTFE-coated fiber exhibits approximately four times the sensitivity of the uncoated silica fiber. Comparative calculations for various materials indicate that with a thin coating, PTFE achieves greater phase shift sensitivity. Furthermore, according to Eqs. (5.3) and (5.9), the variation curve of the phase shift sensitivity with respect to E, μ_2 and R are calculated with numerical method and shown in Figs. 5.4 and 5.5. Figure 5.4 illustrates the curve of phase shift sensitivity versus E and R when $\mu_2 = 0.317$. It is evident that for thin coatings (for example, $R = 2$), the phase shift sensitivity becomes saturated at low values of E, and as the coating thickness increases, the phase shift sensitivity varies approximately linearly with E.

For a specific coating material, there is a limit to the coating thickness, meaning that beyond a certain point (for example, R = 30 for silica fiber), the phase shift sensitivity will not significantly increase with additional thickness.

As illustrated in Fig. 5.5, when E is held constant ($E = 20$), the phase shift sensitivity varies significantly with μ_2. The smaller the value of μ_2, the greater the sensitivity. However, there is a limit to the coating thickness ($R = 30$). Beyond this thickness, the sensitivity will no longer increase with additional coating. With this feature, the reference and guide fibers can be desensitized. Suitable coating materials include glass, aluminum, nickel, and others, and the control of coating thickness must be considered. The surface area of many rubbers is approximately 0.5 m^2. As shown in Fig. 5.5, rubber is not an effective coating material for enhancing the phase

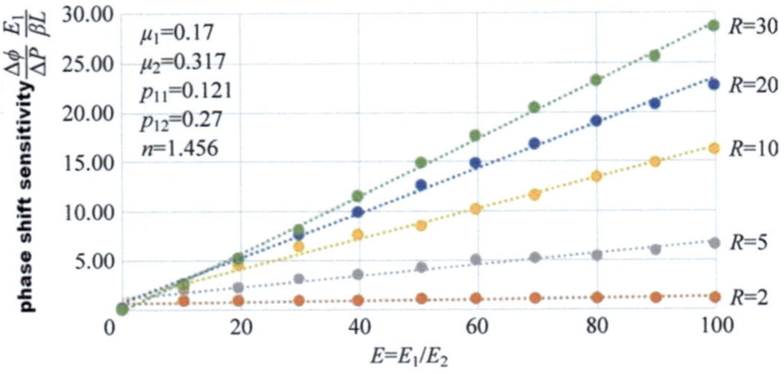

Fig. 5.4 Phase shift sensitivity versus E for different coating thicknesses

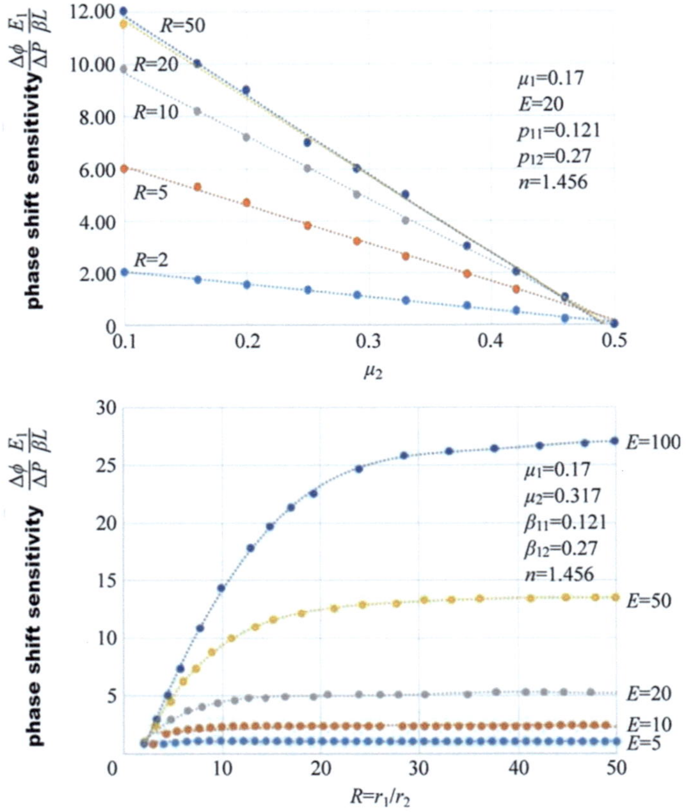

Fig. 5.5 Phase shift sensitivity with μ_2 and R for different coating thicknesses

shift sensitivity of fiber composite structures. In fact, the phase shift sensitivity may initially decrease and then increase with the coating thickness for certain rubbers with specific Young's modulus values.

5.1.3 Lightsource and Noise in the Fiber Underwater Acoustic Sensor

Requirements for Light Sources

High frequency stability: The phase noise introduced by the light source's inherent frequency instability significantly increases and rises linearly with the interferometer's OPD. Semiconductor lasers have relatively low frequency stability. Approximately, the phase noise with an Optical Path Difference (OPD) of 1 m is about three orders of magnitude greater than that of zero OPD. This implies that to ensure a phase detection accuracy of 10^{-6} radians, the OPD between the two arms of a double-beam interferometer, such as a Mach–Zehnder (MZ), must be less than 1 mm. Acoustic sensors with fiber arm lengths ranging from 10 to tens of meters typically require a complex regulation structure, multiple steps, and precision length control instruments, such as a low coherence reflectometer.

Small intensity fluctuation: otherwise, the amplitude modulation introduced by the PGC demodulation technology itself will be greatly amplified, making it impossible to distinguish between the sound field signal and the interference signal when amplitude noise is severe.

Narrow line width and sufficient coherence length. For a fiber hydrophone system utilizing PGC demodulation, the stability and frequency modulation characteristics of the light source present a pair of contradictory factors that must be balanced. Techniques such as OPD modulation or PZT feedback circuit phase drift compensation are not favorable for realizing multiplexing.

Noise Control

The noise control of the fiber optic hydrophone system is crucial when detecting ultra-low amplitude signals. Potential sources of environmental noise include:

2. Random phase drift, often at low frequency, is caused by factors such as ambient temperature. The selection of temperature-insensitive coating materials can enhance the system's resistance to environmental interference.
3. Polarization noise. In the context of dual beam interference, polarization noise arises when a delay occurs between the two interference arms, a phenomenon caused by birefringence. Techniques to suppress polarization noise encompass the use of Faraday rotating mirrors and time-division compensation, among others.

4. The irregular movement of water flow, such as turbulence, also introduces system noise.

Signal Processing of the Fiber-Optic Hydrophone

Signal processing of a fiber optic hydrophone involves converting the hydrophone's optical output into an electrical signal that is proportional to the amplitude of the sound field. Concerning the factors that produce random phase shifts in the working environment, the processing of small phase signals to combat fading is the key issue. Currently, five basic detection systems have been developed for hydrophones:

(1) Passive zero-difference detection (HOM) technology;
(2) DC phase tracking zero difference detection (PTDC) technology;
(3) AC phase tracking zero-difference detection (PTAC) technology;
(4) Heterodyne detection (HET) technology;
(5) Synthetic heterodyne detection (CHET) technology.

They have advantages and disadvantages, and need to be chosen in experiments.

Multiplexing and Consistency of Multi-hydrophones

The practical application of a towing cable necessitates the formation of a multi-detector array, which in turn requires the multiplexing of the light source, probe, and detection circuit. Currently, there are numerous multiplexing schemes available, yet selecting the appropriate one for practical use is not straightforward. To ensure the consistency of multiple detectors, it is necessary not only to maintain process consistency during the manufacture of individual probes but also to properly control the gain of the optical detector and the signal processing circuit for each probe.

5.2 Fiber-Optic Vector/Acceleration Sensor

Interferometric fiber optic hydrophones can be classified into scalar and vector sensors. The scalar optical fiber hydrophone measures sound pressure, a scalar parameter that reflects the water density around the sensor. It is the most common type of hydrophone; thus, the term "optical fiber hydrophone" typically refers to the scalar fiber acoustic pressure sensor. The other type, known as an optical fiber accelerometer or vector hydrophone, measures the gradient of sound pressure or particle acceleration, endowing the sensor with vector characteristics. The physical models of fiber optic hydrophones and acceleration sensors are similar to those of dynamic models, except that the incentive element of the system differs. Consequently, the definitions of phase shift sensitivity and normalized sensitivity for fiber optic accelerometers are altered in comparison to interferometric fiber hydrophones (refer to Table 5.1).

Table 5.1 Interferometric fiber acceleration sensor: performance parameters

Parameter	Definition	Unit
Phase shift sensitivity	Sensitivity of the OPD of the sensing unit (fiber interferometer) to the perceived acceleration (g gravitational acceleration)	rad/g
Resonant frequency/ bandwidth	Reflects the frequency characteristics and directly affects the operating frequency range of the sensor	Hz
Minimum detectable acceleration	Noise limits of the sensor system. The detected acceleration values when the amplitude of the signal equals to system noise, is a function of frequency	g/ $\sqrt{\text{Hz}}$
Maximum detectable acceleration	Demodulation limit of the sensor output interference signal, i.e., the maximum acceleration measured without distortion after demodulation, and is a function of frequency	g/Hz
Dynamic range	The acceleration range detected by the system is also a function of frequency	dB/ Hz
Cross-crossing sensitivity/ cross-scrambling ratio	As a vector sensor, the cross-talk ratio reflects the directionality of the acceleration sensor	

The phase-modulated acceleration sensor offers high detection accuracy and a large dynamic range, making it particularly suitable for high-performance accelerometers. Like other interference sensors, phase accelerometers encompass the primary types of fiber interferometers.

The typical structure of the MZ acceleration sensor is depicted in Fig. 5.6. Within the sensing system, the two arms of the MZ interferometer function as the reference arm and the sensing arm, respectively. The compliant cylinder, wound with optical fiber, and the inertial mass collectively constitute the "mass-spring" sensing unit. The acceleration signal ultimately induces a phase shift in the optical fiber (the sensing arm). The output light from the interferometer passes through the second coupler to reach the photodetector, and the measured acceleration value is obtained following the demodulation of the phase signal. However, designing and miniaturizing the MZ sensor is challenging, as is the isolation of the reference arm within the interferometer.

The Michelson interference sensor effectively addresses the aforementioned shortcomings and is the most commonly used phase-modulated acceleration sensor. The optical fiber arms of the fiber-optic acceleration sensor interferometer are coated with reflective films and wound around a compliant cylinder to create a push–pull structure (refer to Fig. 5.6). When the acceleration signal acts on the sensor probe, the phase changes in the two interference arms are of equal amplitude but opposite in phase; whereas the phase changes in both arms due to variations in ambient temperature are identical, resulting in a differential signal. The acceleration can be obtained by demodulating the differential signal. This push–pull structure possesses a robust noise suppression capability, and the subsequent analysis will concentrate on the acceleration sensor of this variety.

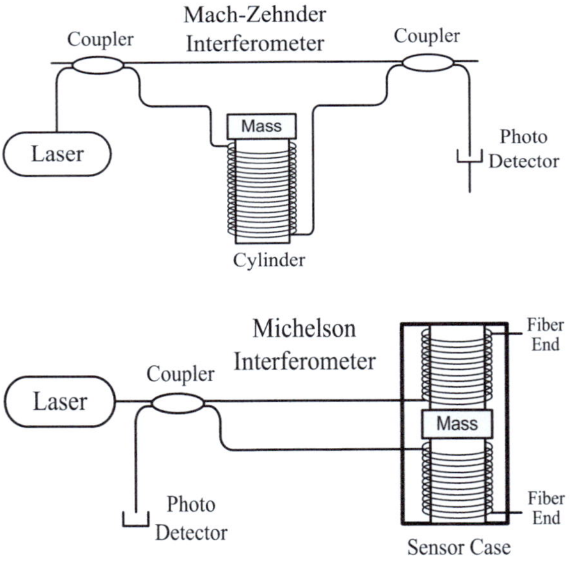

Fig. 5.6 MZ (left) and Michelson (right) fiber acceleration sensing system

There are numerous types of FP fiber acceleration sensors. Figure 5.7 illustrates a schematic diagram of a typical FP fiber acceleration sensing unit. The sensor operates on a mass-spring system with a reflector M_2 affixed to the end face of the mass block, creating an FP cavity in conjunction with the fiber end face M_1. When the acceleration signal is applied, a relative displacement occurs between the inertial mass and the fiber, altering the FP cavity length, which in turn changes the phase of the light wave. By measuring the phase change, the acceleration value to be measured is obtained. The dynamic measurement accuracy of the FP acceleration sensor can reach magnitude of 10^{-7} g/$\sqrt{\text{Hz}}$, and the measurement frequency band can reach the kHz.

Among the three types of high-sensitivity interferometric fiber acceleration sensors—compliant cylinder, diaphragm, and FP cavity—the displacement of the mass relative to the substrate directly affects the FP cavity. Consequently, there is no traditional transducer, but the sensitivity is constrained by the FP interference principle. In the other two types of interferometric acceleration sensors, the compliant

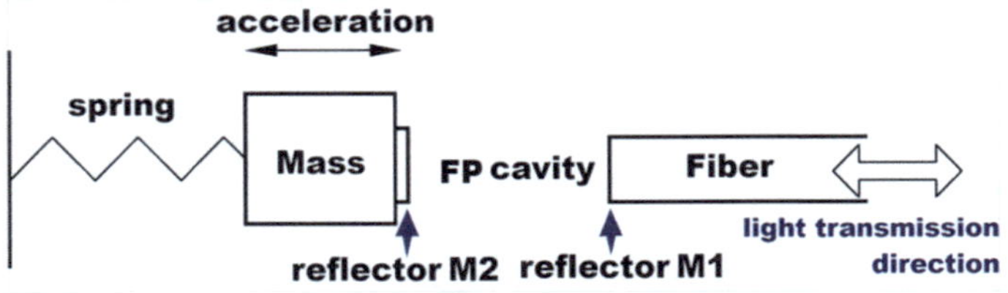

Fig. 5.7 Schematic diagram of the optical fiber FPI acceleration sensor probe

cylinder and the diaphragm function as transducers that convert the small displacement of the mass relative to the base into changes in optical path, thereby enhancing the sensor's phase shift sensitivity. Table 5.1 presents the major performance parameters of the interferometric fiber acceleration sensor. The sensitivity of the interferometric fiber acceleration sensor based on the compliant cylinder is ~10^4 rad/g, and the phase shift resolution of the system is 10 μrad/$\sqrt{\text{Hz}}$, thus the minimum detectable acceleration of the system is 1 ng/$\sqrt{\text{Hz}}$.

5.2.1 Membrane-Type Acceleration Sensor

A vector sensing unit necessitates a highly directional response. The structural design of the sensing unit incorporates popular vibration-sensitive materials, such as compliant rubber and polymer. A cylinder is the simplest structure; however, a special lateral vibration restraint structure, such as a composite compliant cylinder structure, must be designed. A diaphragm elastomer does not require any additional structure and has two inherent advantages:

(1) Since the area is much greater than the thickness, the diaphragm has an intrinsic directional response determined by its geometric feature. The strongest response occurs along the perpendicular direction to the diaphragm, whereas the response in the parallel direction is the weakest;
(2) With identical materials and external dimensions, the stiffness coefficient of the diaphragm is smaller than that of the cylinder, indicating that the diaphragm is more sensitive. Thus, the diaphragm structure is widely utilized in various vibration measurement instruments.

The membrane-type fiber acceleration sensing scheme employs a thin, sensitive diaphragm as the elastic element. Sensing optical fibers are attached to either side of the diaphragm, with a mass block secured at its center (Fig. 5.8). When the diaphragm experiences an external acceleration field and deflects, the deflection is transformed into phase changes via the sensing optical fibers. The magnitude of the external acceleration is determined by measuring these phase changes.

As depicted in Fig. 5.8, with the acceleration in the z-direction, the inertial mass moves relative to the diaphragm, causing a deflection of the elastic diaphragm. The deformation in the z-direction at various points on the diaphragm is referred to as deflection. According to the theory of elastic mechanics, the deflection distribution on the diaphragm is centrally symmetrical, and the diaphragm deflection at a distance r from the center is,

$$w(r) = w_0\left[1 - \frac{r^2}{R_{\text{out}}^2} + \frac{2(1+\mu_{\text{dia}})r^2}{(3+\mu_{\text{dia}})R_{\text{out}}^2}\ln\left(\frac{r}{R_{\text{out}}}\right)\right] \tag{5.13}$$

where, w_0 is the deflection at the center of the diaphragm, and μ_{dia} is the Poisson ratio of the diaphragm material. Excluding the influence of the sensing fiber on the mechanical characteristics of the diaphragm, the equivalent spring constant of the system is precisely that of the diaphragm for the mass inertial system composed of

Fig. 5.8 Structure of membrane sensing unit

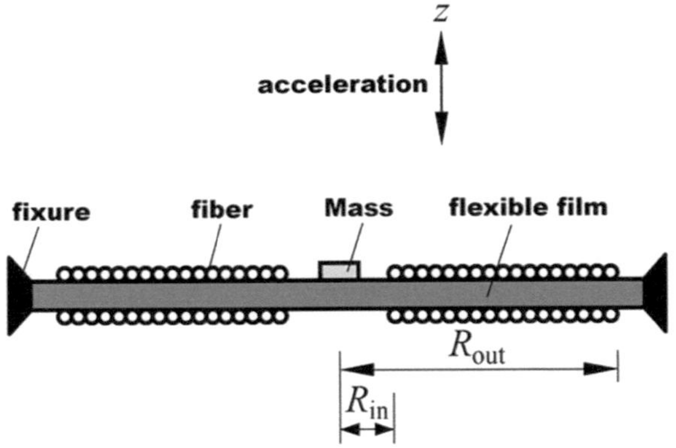

the diaphragm, and can be expressed as:

$$K_{dia} = \frac{ma}{w_0} = \frac{E_{dia}}{1-\mu_{dia}^2}\frac{4\pi H^3}{3C_R R_{out}^2} \tag{5.14}$$

where, a is the measured acceleration, E_{dia} is the Young's modulus of the diaphragm, H is the thickness of the diaphragm; $C_R = 4\left(\frac{1-r_{dia}^2}{4} - \frac{r_{dia}^2(\ln r_{dia})^2}{1-r_{dia}^2}\right)$. r_{dia} is the ratio of the inner diameter to the outer diameter of the diaphragm, where the inner and outer diameter refers to radius R_{in} and radius R_{out} of the elastic part of the diaphragm, due to the presence of the inertial mass. Assuming uniform contact between the diaphragm and the optical fiber, and that the strain distribution on the diaphragm surface matches that on the sensing fiber, the strain response of the diaphragm under vibration is non-uniform. This means that the strain modulation varies at different parts of the sensing fiber. Consequently, it is necessary to integrate along the fiber length to obtain the change in fiber length, Δl, and thus determine the phase change of the sensing fiber.

$$\Delta\varphi = 4nk_\lambda(1-p_c)\Delta l \tag{5.15}$$

p_c is the photo-elastic coefficient of optical fiber, and $p_c \approx 0.21$ for germanium doped fiber. When a fiber-optic acceleration sensor comprises a circular diaphragm with edge support, clamp fixation, and a central loading inertia mass, the relationship between the phase modulation of the optical fiber and the measured acceleration can be deduced (as detailed in the relevant literature) [3] according to a sensing system shown in (Fig. 5.9).

$$\Delta\varphi = \frac{3nk_\lambda C_R m R_{out}^2\left(1-\mu_{dia}^2\right)(1-p_c)}{D_S E_{dia} H^2(3+\mu_{dia})} \times\left[(\mu_{dia}-1)\left(1-r_{dia}^2\right) - 2(\mu_{dia}+1)r_{dia}^2 \ln r_{dia}\right]\cdot a \tag{5.16}$$

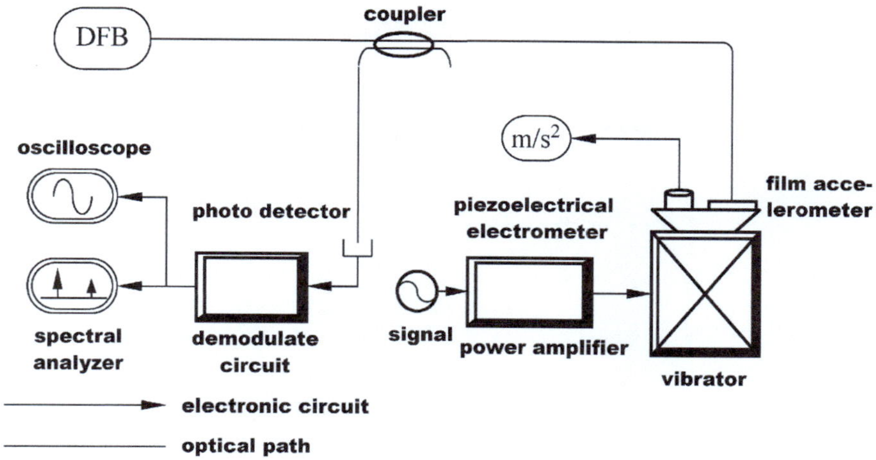

Fig. 5.9 Schematic diagram of membrane-type optical fiber accelerometer test system

It is important to note that the axial and transverse strains produced by the sensing fiber itself, due to physical quantities such as vibration, affect the change in the fiber's refractive index. However, the relative difference between the variation in the refractive index and the length of the fiber is quite small, and the signs are opposite, resulting in a small total phase change. This can be improved by using a specially designed sensitive material or structure within the sensing unit to increase the axial strain by several orders of magnitude while keeping the transverse strain constant. At this juncture, axial strain becomes the primary factor influencing the change in refractive index, which is smaller in magnitude than the relative change in length and opposite in sign. The total phase changes in the fiber are significantly increased after superposition, resulting in sensitization by the sensing unit structure.

Assuming the working wavelength $\lambda = 1550$ nm, and the optical fiber is subjected to uniform strain. For a fiber of unit length (1 m), a microstrain (10^{-6}) corresponds to a phase change of approximately 10 radians. It has been reported that the phase demodulation accuracy of an interferometer can reach 10^{-6} rad [8], so the corresponding strain of 10^{-13} can be obtained, indicating the high sensitivity of phase modulated fiber sensor.

The sensing performance of membrane fiber acceleration sensors is influenced by the properties of elastic materials (such as polymer diaphragms and stainless steel corrugated diaphragms), their shape, support modes (including center and edge support), and fixation methods (such as simple support fixation and clamp fixation). The comparison reveals that the membrane-type sensor exhibits good linearity, with a minimum detectable acceleration of better than 100 $\mu g/\sqrt{Hz}$. The effects of the four factors on the performance of the acceleration sensing system are illustrated in Table 5.2.

Table 5.2 Comparison of structural performance in diaphragm-type optical fiber acceleration sensing

	Membrane material		Membrane shape		Support mode		Fixed form	
	Polymer	Corrugation	Circular	Rectangle	Centrality	Border	Simply support	Grab
Sensitivity	>			>	>>		>>	
Working frequency band		>	>			>		>
Cross-desensitivity		>>	>>			>>		>>
Degree of linearity	=	=		>>		>>	=	=
Practicability	>>		>>		=	=	=	=

[*] *Note* The table uses ">" to represent factor 1 better than factor 2; ">>" to represent factor 1 far better than factor 2, and "=" to represent comparable performance under the two factors

5.2.2 *Interference Fiber Acceleration Sensor Based on Compliant Cylinder*

Compliant materials are highly sensitive to vibration signals and are typically fashioned into compliant cylinders for vibration sensing. As the sensing element of a fiber acceleration sensor, the sensing fiber is wound tightly around the compliant cylinder with a specific tension. When the vibration signal is present, the compliant cylinder is compressed or stretched, resulting in a corresponding strain on the optical fiber. Utilizing the sensing fiber as the signal arm of the interferometer, the strain change in the fiber can be determined by measuring the optical phase shift, thereby obtaining the acceleration signal.

For sensitivity improvement, a fiber acceleration sensing unit typically includes a mass block and two compliant cylinders that form the push–pull structure, as depicted in Fig. 5.6. The two optical fiber arms of the Michelson interferometer are wound around the upper and lower two cylinders under a certain tension. When the sensor is subjected to acceleration, an additional force proportional to the acceleration is applied to the end surface of the compliant cylinder. This causes one end to be compressed and the other to be stretched, forming a push–pull structure. As the compliant cylinder is compressed (or stretched) along the axial direction, it will inevitably expand (or shrink) along the diameter direction, causing the optical fiber to deform accordingly. Thus, the alteration in the outer diameter of the compliant column is converted into a variation in the OPD between the two arms of the interferometer. The push–pull mechanism enhances the sensitivity to twice the phase shift of a single compliant cylinder. Utilizing the reflective structure of the Michelson interferometer, light traverses the sensing arm twice, resulting in the sensor having four times the sensitivity of a single sensing arm at the resonance frequency. That is $\Delta\varphi = 4nk_\lambda l(\frac{\Delta l}{l} + \frac{\Delta n}{n})$. In a mass inertial system, the equivalent elastic coefficient of the system is $2K_{\text{eff}}$, where K_{eff} represents the equivalent elastic coefficient of a single wound fiber compliant cylinder.

To integrate the established mechanical model of the compliant cylinder with the stress analysis of optical fiber from Sect. 3.8, the changes in length and outer diameter of the compliant cylinder are deduced first, followed by the derivation of the equivalent elastic coefficient and the sensitivity of the sensor. The relationship between the change of the outer circumference Δc and the phase change $\Delta\phi$ of the interferometer is:

Where, n is the refractive index of the fiber core, λ is the light wavelength, μ_{f} is the Poisson ratio of fiber, and p_{11} and p_{22} are the strain coefficient of the fiber core. Then the expression of phase shift sensitivity of fiber acceleration sensor is then shown as:

$$\Delta\phi = 4 \times \frac{2\pi n}{\lambda} N \Delta c \left\{ 1 - \frac{1}{2} n^2 \left[(1 - \mu_f) p_{12} - \mu_f p_{11} \right] \right\} \tag{5.17}$$

$$\frac{\Delta\phi}{a}(\omega) = \left(\frac{\Delta\phi}{a} \right)_0 \times \frac{1}{\sqrt{(1 - \frac{\omega^2}{\omega_n^2})^2 + 4\xi^2 \frac{\omega^2}{\omega_n^2}}} \quad (A)$$

$$\left(\frac{\Delta\phi}{a}\right)_0 = \frac{8\pi^2 nb\mu N}{\lambda HX} \times \frac{m}{K_{\text{eff}}} \times \left\{1 - \frac{1}{2}n^2\left[\left(1-\mu_f\right)p_{12} - \mu_f p_{11}\right]\right\} (B) \quad (5.18)$$

where, b is the outer diameter of compliant cylinder, ω and ω_n are the resonance frequency; K_{eff} is the equivalent elastic coefficient of the compliant cylinder; the height of the compliant cylinder $H = N \cdot d_f$, d_f is fiber diameter; parameter X reflects the influence of the fiber wound around the compliant cylinder on it; μ is the Poisson ratio of the cylinder.

The push–pull structure significantly enhances the performance of the sensing system. When the acceleration signal influences the sensing unit, the two compliant cylinders that constitute the push–pull structure undergo compressive and tensile strain respectively. The phase changes in the two interference arms are of the same magnitude but opposite in sign, resulting in constructive interference. When the phase changes in the two arms caused by external noise, such as ambient temperature, are identical, this results in destructive interference and effectively suppresses the noise. Consequently, the push–pull structure significantly enhances the signal-to-noise ratio of the system compared to the conventional interference structure.

By multiplexing the three single-component sensing units, a three-component accelerator with sensitive axes perpendicular to each other is composed, as shown in Fig. 5.10. The three components of acceleration can be detected simultaneously, with the structure depicted in Fig. 5.11a. Here, S1, S2, and S3 represent the single-component sensing units, each with a sensitive axis perpendicular to the others.

Due to the comparable stiffness of the pure compliant cylinder in both the axial and cross directions, it is challenging for the sensing unit to suppress the vibration response in the cross direction, leading to a decline in single-component sensing performance. To enhance the difference in vibration sensitivity between the axial and crossover directions, a special design of a compound compliant cylinder is required.

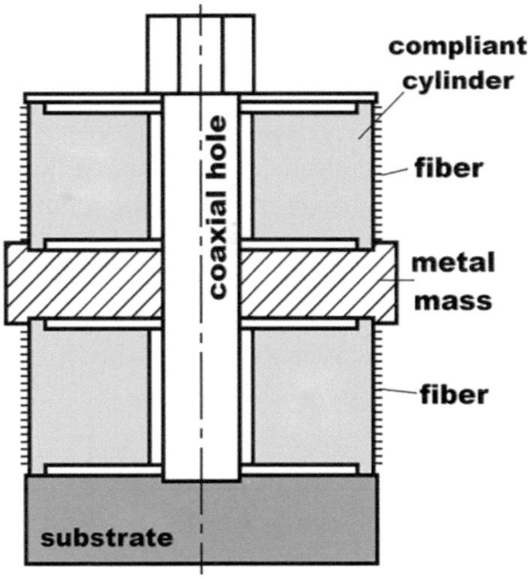

Fig. 5.10 Optical fiber acceleration sensor based on compliant cylinder

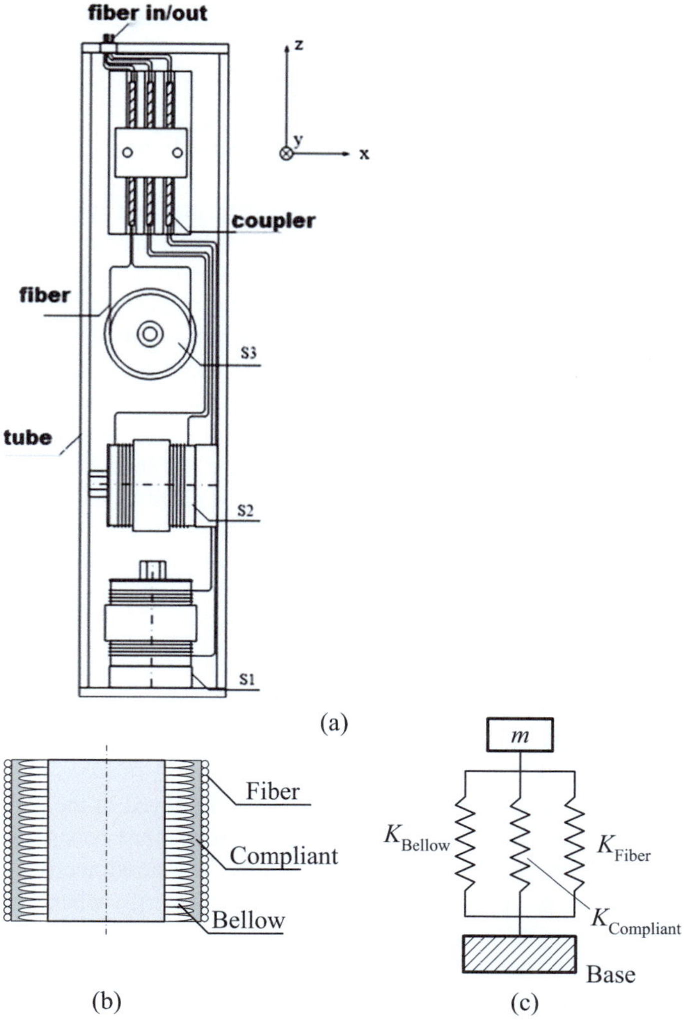

Fig. 5.11 Schematic diagram of the 3-component acceleration sensor structure **a** structure of the 3-component accelerator, **b** cross-section of a single-component compound sensing cylinder, **c** the effective vibrator model of the compound cylinder.

For instance, a compliant material is enveloped around a bellows structure to create a composite sensitive cylinder. The bellows, known for its superior axial elasticity and minimal radial elasticity, provides an effective unidirectional response to vibration signals. Thus, the transverse deformation of the cylinder, when combined with compliant materials, is effectively suppressed under vibration. This ensures that each sensor probe remains sensitive only to vibrations along the central axis. Figure 5.11b depicts a profile diagram of a single-component composite sensitive cylinder. The structure consists of an optical fiber layer, a compliant material layer, and a bellows, arranged from the outside in. Figure 5.11c depicts the effective vibrator model of the compound compliant cylinder.

Generation and Analysis of the Vectorality of Hydrophone

A traditional hydrophone, also referred to as an acoustic pressure hydrophone, detects the acoustic pressure signal $p(r, t)$ at a specific location within the sound field. A single acoustic pressure hydrophone processes information within a simple $\{p\}$ space, thereby neglecting vector information. Sound pressure hydrophones can also locate with an array. The diverse and complex positioning methods of acoustic pressure hydrophones have constituted an independent research field. This section begins with the most simple and commonly used phase positioning method, then delves into the focus of this section, vector hydrophones.

The fundamentals of the phase positioning method are shown in Fig. 5.12. The system consists of two matrix arrays separated by a distance of l. When a plane acoustic wave incidents at an angle of θ with the direction of the line connecting the two arrays, the phase difference between the signals received by the two arrays is:

$$\Delta\phi = \frac{l\cos\theta}{\lambda} \text{or } \cos\theta = \frac{\lambda\Delta\phi}{l} \tag{5.19}$$

where λ is the wavelength of the sound wave, and after measurement of the phase difference Δϕ, the one-dimensional direction angle θ of the sound wave can be calculated using Eq. (5.19).

The directionless acoustic pressure sensor array employed in the phase method has two limitations: (1) the distance l between the arrays must be significantly less than the distance between the sensor array and the measured acoustic excitation source to satisfy the plane wave approximation condition; (2) when l exceeds the acoustic half wavelength ($\lambda/2$), uncertainty in positioning arises because cosθ is a periodic function. As shown in the following Eq. (5.20), for all $n = 0,1,2,\ldots$, the solutions may be reasonable.

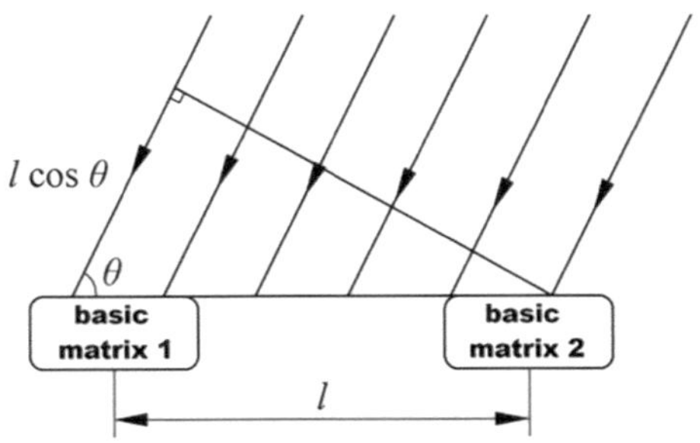

Fig. 5.12 Schematic of phase positioning

$$\cos\theta = \frac{\lambda(2n\pi + \Delta\phi)}{l} \tag{5.20}$$

For a practical phase positioning system, under the limiting conditions (1) and with determined signal noise, a higher positioning accuracy can be achieved with a greater l from Eq. (5.19). However, increasing l will introduce localization uncertainty under limiting conditions (2). This introduces the concept of matrix beam forming to solve this paradox.

The so-called matrix array beam refers to the directional single sensor matrix depicted in Fig. 5.12, which is no longer a single nondirectional sound pressure sensor. An example of a typical single beam and its basic parameters are shown in Fig. 5.13. The directionality of the sensor array is achieved through the signal processing technology of a directionless sensor array, known as beam-forming technology, which is widely utilized in radar positioning. The vector hydrophone employs the vector properties of a single sensor to implement the simplest cosine beam.

For an ordinary hydrophone, the essence of the measured sound pressure reveals the density of water at a specific point in the sound field, which is a scalar quantity. The definition of a vector sensor is an acoustic reception transducer that measures the vector of an underwater acoustic field, such as the acoustic pressure gradient (∇p), particle velocity (v), particle acceleration (a), or mass displacement (s). The vector hydrophone expands the one-dimensional scalar signal domain $\{p\}$ measured by a multidimensional signal domain encompassing $\{p, v, a, s, \nabla p\}$.

Traditional piezoelectric vector sensors can be categorized into three types: dual acoustic hydrophones, motionless shell types, and resonant sphere types. The first two types measure the sound pressure gradient, whereas the resonant sphere type measures particle velocity. Each type has its own unique characteristics. The resonant sphere type, for instance, is widely used due to its high sensitivity and ease of implementation. However, its installation is complex, and the working conditions are

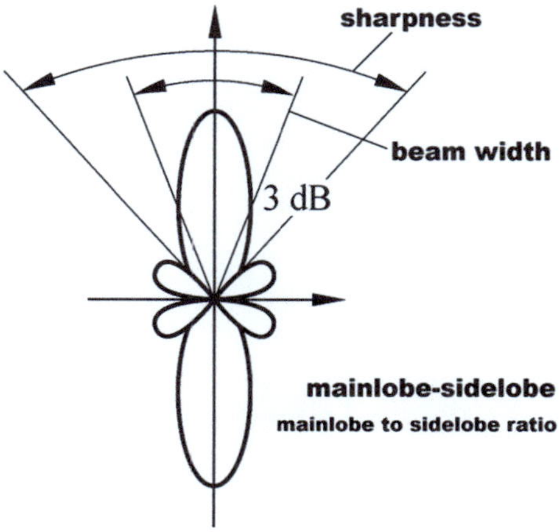

Fig. 5.13 A typical sensor beam

demanding. Interference fiber vector hydrophones, on the other hand, are primarily based on sound pressure gradient measurement.

5.2.3 The Fundamental Architecture of a Fiber Optic Vector Hydrophone (Acoustic Pressure Gradient Sensor)

Figure 5.14 illustrates the structure of a fiber vector hydroacoustic sensing unit. In this setup, both arms of the Michelson interferometer serve as sensing fibers, which are wound around the left and right compliant cylinders. These cylinders function as transducers, each with a diaphragm (acoustic receiver) affixed to their end surfaces.

This fiber optic vector hydrophone can be considered a combination of two fiber optic acoustic pressure sensors, as described in Sect. 5.2.2. Due to the coupler's effect, the optical path difference (OPD) between the left and right sensing fibers corresponds to the output difference of the left and right acoustic pressure sensors. Thus, this fiber vector hydrophone is of the sound pressure gradient type. Assuming the left and right structure of the sensor is symmetrical, with a diaphragm surface area of S and a distance of d between the two membranes, the force situation of the sensor in the plane wave sound field is depicted in Fig. 5.15.

Assuming the acoustic pressure amplitude of the plane wave sound field is p_0, the forces F_1 and F_2 on the left and right diaphragms are expressed as follows:

$$\begin{aligned} F_1 &= p_0 S \cos \omega t \\ F_2 &= p_0 S \cos(\omega t + 2\pi d/\lambda_a) \end{aligned} \tag{5.21}$$

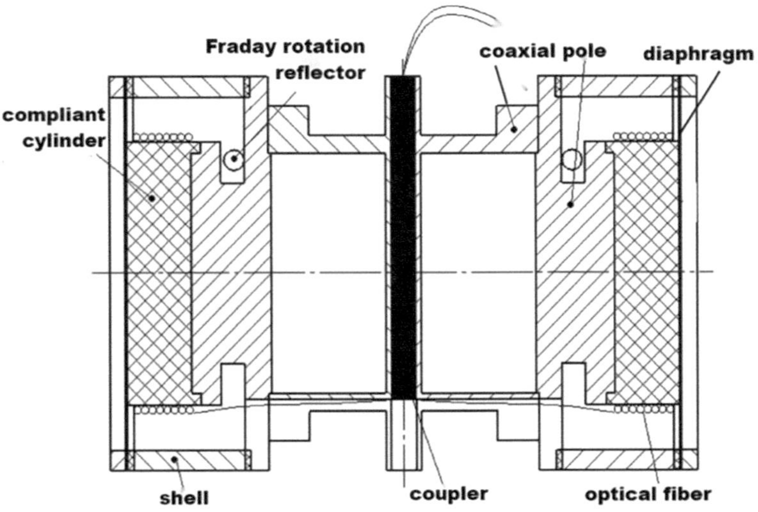

Fig. 5.14 Structure diagram of the sensing unit of the fiber vector hydrophone

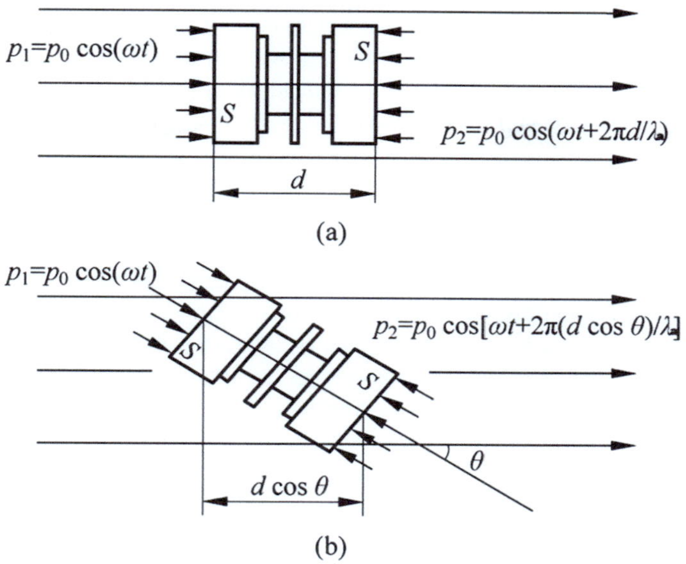

Fig. 5.15 Force analysis of vector hydrophone in plane acoustic wave field **a** parallel to **b** having an angle of θ with the hydrophone axis

Therefore, the output sensing signal is proportional to (F_1–F_2). When $d < < \lambda_a$ (acoustic wavelength), it can be expressed as:

$$\begin{aligned} F_1 - F_2 &= p_0 S \cos \omega t - p_0 S \cos(\omega t + 2\pi d/\lambda_a) \\ &= -2p_0 S \sin(\pi d/\lambda_a) \sin(\omega t + \pi d/\lambda_a) \\ &\approx \frac{-2p_0 S \pi d}{\lambda_a} \sin(\omega t + \pi d/\lambda_a) \end{aligned} \tag{5.22}$$

When the longitudinal main axis of the fiber vector hydrophone forms a certain angle θ with the propagation direction of the plane acoustic wave, the phase difference of the sound pressure signals on the stressed surfaces of the two diaphragms becomes $d\cos\theta/\lambda$. By substituting this into Eq. (5.22), the phase shift response of the sensor is found to be proportional to (F_1–F_2):

$$F_1 - F_2 = \frac{-2p_0 S \pi d(\cos\theta)}{\lambda_a} \sin(\omega t + \pi d/\lambda_a) \tag{5.23}$$

Equation (5.23) clearly illustrates that the cosine direction of the sensor response and the sensor's sensitivity to the acoustic pressure gradient diminish as the frequency of the acoustic wave signal rises. The measurand for this type of vector hydrophone is the sound pressure gradient, essentially a dual sound pressure sensor that can be divided into left and right components. The consistency between these two acoustic pressure sensors impacts the directional performance of the vector hydrophone.

Assuming the sensitivity of the left and right acoustic pressure sensing units is k_1 and k_2 respectively. If $k_1 \neq k_2$, and there is a scale coefficient $k = \frac{k_2}{k_1}$, then k will influence the directionality of the vector sensor. The phase shifts ϕ_1 and ϕ_2 of the two sensing units are as follows respectively:

$$\left.\begin{aligned} \phi_1 &= a_1 p_1 \\ \phi_2 &= a_2 p_2 = t a_1 p_2 \end{aligned}\right\} \Rightarrow \Delta\phi = \phi_1 - \phi_2 = a_1(p_1 - t p_2) \tag{5.24}$$

The performance of a vector hydrophone is typically described using the ratio Q, which is the quotient of the maximum sensitivity $\Delta\phi_{\max}$ and the minimum value $\Delta\phi_{\min}$, as a measurement parameter. Q can be expressed in the form of Eq. (5.25) and reflects the impact of the consistency k of the sound pressure sensing units on the directionality of the fiber vector hydrophone.

$$Q = \frac{\Delta\phi_{max}}{\Delta\phi_{min}} = \frac{(1-k)}{\sqrt{[1 - k\cos(2\pi l/\lambda)]^2 - k^2 \sin^2(2\pi l/\lambda)}} \tag{5.25}$$

5.3 Fiber-Optic Rotation Sensor (Gyro)

The fiber optic gyro, born in 1976, has evolved into a mainstream product with groundbreaking significance in the field of inertial navigation. The principles, processes, and key technology of the fiber optic gyro differ greatly from those of the traditional gyroscope. It boasts advantages such as high reliability, longevity, rapid response, and a large dynamic range, and holds an indispensable position in military, aerospace, and civilian applications.

Fiber optic gyroscopes are classified into two types: the interference fiber optic gyro (I-FOG) and the resonant fiber optic gyro (R-FOG). There have been reports on active ring resonators in the early stages, but due to the "dead zone" problem caused by mode locking, the passive ring interferometer has ultimately become the mainstream in fiber gyroscope technology.

5.3.1 Interferometric Fiber Optic Gyroscope

Principles of I-FOG

Utilizing a low-loss SM fiber, a multi-turn light path can effectively enhance the Sagnac effect and provide sufficient rotation rate sensitivity without the need for a resonator. The Sagnac phase shift of the I-FOG, caused by rotational speed Ω, is expressed as:

$$\Delta\phi_R = \frac{2\pi LD}{\lambda c}\Omega \tag{5.26}$$

where, λ is the light wavelength in vacuum, D is fiber ring diameter, $L = ND$ is the total length of fiber, and N is the number of turns of the fiber ring. Constant rate leads to a constant phase difference and the output light intensity is expressed as a cosine function according to the double-beam interference principle,

$$I = I_0[1 + \cos\Delta\phi_R] \tag{5.27}$$

There is a monotonic phase measurement interval of ± rad symmetric about the y-axis (as shown in Fig. 5.16), corresponding to the single-value working range of ± . The single value interval of the rotation speed is determined using the following equation,

$$\Omega_\pi = \frac{\lambda c}{2LD} \tag{5.28}$$

For a fiber optic gyro with total fiber length of $L = 1$ km, diameter $D = 10$ cm, working wavelength $\lambda = 1550$ nm, the measurement range of rotation speed $\Omega_\pi = 2.325$ rad/s $= 133°$/s, and phase difference $\Omega\mu = 0.152°$/h corresponding to the speed of 1 μrad, the dynamic range is about 130 dB.

The applications requiring large speed range should adopt fiber optic ring scheme with short fiber length and small diameter. For example, a fiber optic gyro with a total length of 100 m and a diameter of 3 cm has an angular velocity of $\Omega_\pi = 4440°$/s; and a deviation rate of $\Omega\mu = 5.07°$/h. Since the interferometric fiber optic gyro operates within several interference fringes near an OPD of zero, the coherence requirement of the light source is relatively low. Consequently, the use of a broadband light source can significantly reduce the parasitic effects in the system.

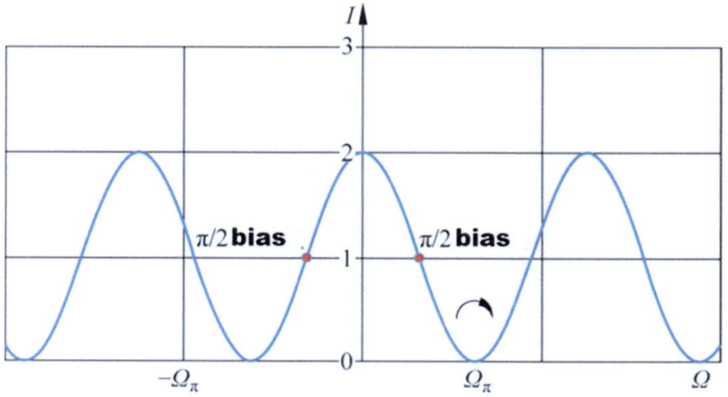

Fig. 5.16 Typical response curve of I-FOG

Theoretical Sensitivity and Optical Fiber Length

The theoretical sensitivity of passive systems is dictated by photon shot noise. In an I-FOG system, photon shot noise encompasses both the quantum efficiency of the detector and deviations in responsivity from the ideal. Substituting this deviation into the system calculates the noise equivalent phase difference $\sigma_{\Delta\varphi}/\sqrt{\Delta f_{\mathrm{bw}}} = 0.77 \times 10^{-9}/\sqrt{P}(rad)$ [1], where P represents power in units of watts. Suppose the working wavelength is 850 nm, the bias power is 1 μW, and the equivalent shot noise is 10^{-12} W/$\sqrt{\mathrm{Hz}}$, the equivalent phase difference is 1 μrad·$\sqrt{\mathrm{Hz}}$. Compared to the aforementioned values, it is evident that the sensitivity of fiber I-FOG is extremely high.

As photon energy decreases with increasing wavelength, the number of photons with the same power increases. Consequently, under the condition of a constant fiber ring diameter and light source power, the phase difference resulting from the rotation speed is inversely proportional to the wavelength. Simultaneously, the signal-to-noise ratio is proportional to the square root of the wavelength, as estimated by the $10^{-\alpha L/20}$, where α represents fiber attenuation. There exists an optimal fiber length, L_{op}, which equals $8.7/\alpha$. It is evident that this optimal fiber length is significantly larger than the actual length of the fiber used, primarily due to the trade-offs involving dynamic range, fiber ring volume, and sensitivity.

On the other hand, the length of the fiber used by FOG typically does not exceed 1 km, thus affecting the performance of the communication window from 850 to 1550 nm. The key considerations in selecting the working wavelength include component cost, reliability, and adherence to standards.

Noise, Drift, and Scaling Factors

The output signal of a Fiber Optic Gyroscope (FOG) at rest is defined as drift, which is a random function composed of white noise and a slow change function. White noise, also known as the photon shot noise limit, is represented by the square root of the standard deviation of the equivalent rotation speed or measurement bandwidth, expressed in units of °%/h/Hz. The equivalent noise power spectral density is represented by the square of the standard deviation, or $(°\%/\mathrm{h})^2$/Hz. The random walk unit is in hours (h), used solely to evaluate white noise and often confused with drift.

Drifts are utilized to assess the peak-to-peak range of long-term variations in the average output signal, typically measured in degrees per hour (°/h). In Fiber Optic Gyroscopes (FOG), the noise limit is defined by detector noise, which is influenced by the returned optical power; however, the drift resulting from residual non-reciprocity can theoretically be eliminated. In various applications, a fast response demands low noise, whereas navigation systems prioritize minimizing drift.

Scale factor: is 2005 released terminology of aerospace science and technology, which defines the ratio of the output change to the measured input change.

Bandwidth: The fastest response time of I-FOG, which is the transmission time of light in the fiber ring (approximately 1 μs for a 200 m fiber), corresponds to a theoretical bandwidth of 100 kHz. However, the subsequent signal processing speed diminishes the actual bandwidth to the order of kilohertz.

5.3.2 Resonant Fiber Optic Gyro

1. Fundamentals of an optical fiber ring cavity

R-FOG enhances the Sagnac effect through multibeam interference within a ring cavity. Compared to I-FOG, the theoretical sensitivity of R-FOG is $F/2$ times the fiber length, where F represents the fineness of the ring cavity. Thus, greater sensitivity can be attained by employing a shorter optical fiber. However, with a high coherence light source, the interference within the ring cavity cannot effectively diminish various parasitic effects, and the demodulation technology becomes more complex.

Referencing the discussion in Chap. 3, the operating principle of the fiber ring resonator is very similar to that of the FP cavity (see Fig. 5.17). Two low-loss fiber couplers substitute for the FP mirrors, replacing the reflected light of the FP cavity with multiple circulations of the transmission light beam in the ring cavity to enable multi-beam interference. Assuming the ring cavity has no transmission loss and uses two identical couplers, the finesse of the ring cavity is expressed as:

$$F = \frac{3}{C} \tag{5.29}$$

C represents the insertion loss of the coupler, analogous to the transmittance of an FP cavity. The output at the transmission end of the coupler can be expressed as:

$$T(k) = \frac{1}{1 + F\sin^2\left(\frac{nkL_r}{2}\right)} \tag{5.30}$$

The propagation constant $k = 2\pi/\lambda$, nL_r is the optical path of light traveling one around in the ring cavity, $F = 4(1-C)/C^2$. When the one round optical path nL_r

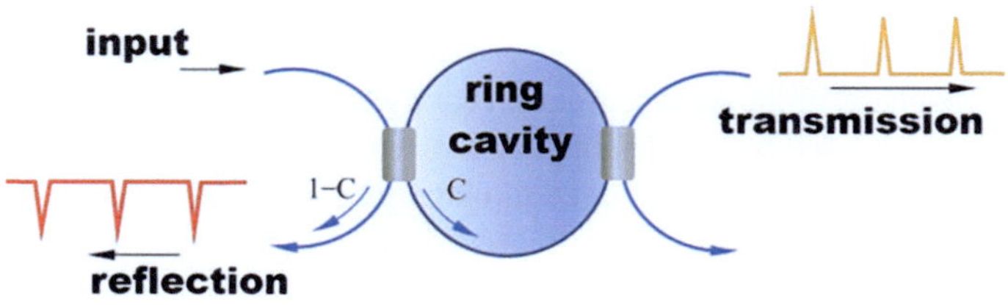

Fig. 5.17 Fiber-optic annular cavity structure

of the ring cavity is an integer multiple of the wavelength, the resonance occurs, and light is emits from the transmission end. The period of the resonant response corresponds to the free spectral range $\Delta\sigma_{\mathrm{Fr}} = \frac{1}{nL_r}\left(\mathrm{nm}^{-1}\right)$, and the bandwidth of the transmission peak is $\Delta\sigma_{\mathrm{FWHM}} = \frac{\Delta\sigma_{\mathrm{Fr}}}{F}$. If we assume that the frequency of the light source ($\sigma_0 = \frac{1}{\lambda_0}$) matches the resonance peak of the ring cavity, then the transmitted light power becomes a function of the rotational speed.

$$P(\Omega) = P_{max}\frac{1}{1 + F\sin^2\left(\pi\,\Delta L_R'(\Omega)\sigma_0\right)} \tag{5.31}$$

where $\Delta L_{R'}(\Omega)$ represents the change in cavity length due to rotation, with $\Delta L_{R'}(\Delta\Omega_{\mathrm{Fr}}) = \lambda_0$, also known as $\Delta\Omega_{\mathrm{Fr}} = \frac{2\lambda_0 c}{L_r}D$, and the Full Width at Half Maximum (FWHM) of the angular rate response $\Delta\Omega_{\mathrm{FWHM}}$ ($\pm\pi/2$ rad in equivalent double-beam interference) is:

$$\Delta\Omega_{\mathrm{FWHM}} = \frac{\Delta\Omega_{\mathrm{Fr}}}{F} = \frac{2\lambda_0 c}{FL_rD}, \; \Omega_\pi = \frac{\lambda_0 c}{2LD} \tag{5.32}$$

The spectral response curve of the reflection end output is precisely complementary to that of the transmission end. Achieving the ideal contrast requires that the coupling ratio be exactly the same for both couplers and that the resonator is without loss, a condition that is nearly impossible to achieve in practice. However, when employing a ring cavity with a coupler, it is necessary to satisfy only the condition that the coupling ratio equals the cavity loss, thus achieving both high precision and ideal contrast. Moreover, a resonance light path without fusion splicing can be achieved through the coupler preparation process. However, the high fineness of the R-FOG necessitates a light source with a large coherence length. For instance, if the finesse of an R-FOG with a fiber length of tens of meters is 100, the coherent length of the light source should be several kilometers, corresponding to a line width of 10 to 100 kHz.

2. Signal processing of the R-FOG

The signal processing of R-FOG typically involves two steps. First, the open-loop bias signal is obtained using modulation and demodulation techniques, which involves finding the derivative of the output response without modulation. Second, this signal is used as the input (error) signal for the closed-loop feedback circuit, enhancing the linearity and stability of the scaling factors. Unlike I-FOG, the output of R-FOG does not have an extremum corresponding to zero rotational speed; instead, it matches the resonance frequency of the light source and cavity length. Generally, the light source frequency is modulated by the driving current, and the cavity length is adjusted by PZT. The primary issue with this biased modulation and demodulation method is the presence of nonlinear and parasitic effects, as well as the high symmetry required for the two signals.

The signal processing system of the R-FOG is depicted in Fig. 5.18. After demodulating the two oppositely transmitted signals, namely the clockwise (CW) and counter-clockwise (CCW) signals, one of them is utilized as a bias error signal to ensure the system operates at the resonance wavelength, The other serves as a feedback signal corresponding to the measured angular speed, providing an additional frequency shift through the closed loop circuit, i.e., the difference between the resonance frequencies of the clockwise (CW) and counterclockwise (CCW) signals, denoted as Δf_{R}:

$$\Delta f_R = \frac{ND}{n\lambda} \cdot \Omega \tag{5.33}$$

where D is the diameter of the fiber ring, N is the number of turns, n is the refractive index and λ is the wavelength of the light source.

As illustrated in Fig. 5.18, despite the fiber length of R-FOG being shorter than that of I-FOG, the system employs a significantly greater number of components, such as couplers and modulators, compared to I-FOG. Consequently, its complexity considerably diminishes the advantage of its fiber length.

3. Reciprocity and parasitic effects of the ring cavity

The reciprocity of the ring resonator primarily originates from the birefringence of the optical fiber. Due to the presence of two orthogonal polarization modes in a non-ideal single-mode (SM) fiber, a SM fiber ring resonator exhibits two sets of resonances. This occurs because the refractive indices differ slightly, causing the two polarization modes to travel different optical paths per circle around the fiber ring. The two sets of resonances are identical in opposite propagation directions without rotation. In fact, there will always be energy coupling between two orthogonal polarization modes when light transmits through a single-mode (SM) fiber. These parasitic crossover modes reduce the symmetry of the resonance signal, and although the reciprocal

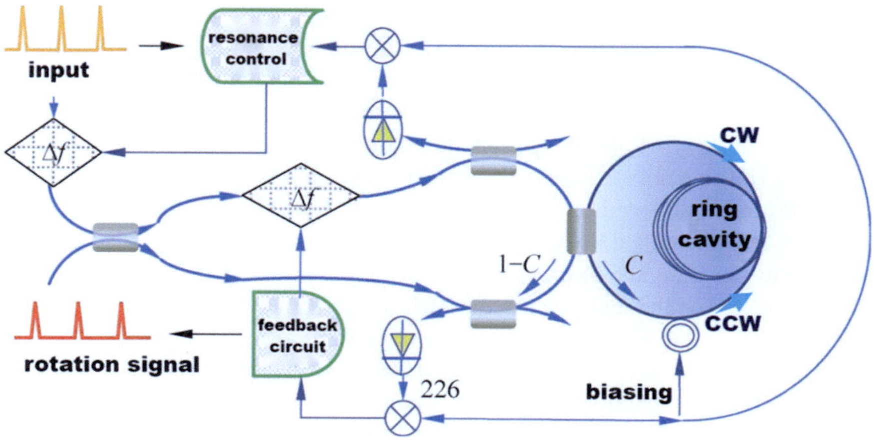

Fig. 5.18 R-FOG system structure

structure cannot completely eliminate the asymmetry, it ensures that the same level of asymmetry occurs in both opposite directions.

Moreover, due to the use of high coherence light sources, both polarization attenuation and polarization suppression result in parasitic intensity modulation that cannot be eliminated with broadband light sources. The most effective solution currently available is single polarization high birefringent fiber, including types such as panda and elliptical core fiber, which are primarily responsible for the popularity of PM fiber rings in R-FOG.

In a PM fiber resonator, some optical energy is still coupled into the cross-polarization mode, resulting in a small parasitic resonance peak that drifts with temperature. This, in turn, causes asymmetry and instability in the main resonance peak. For instance, consider a fiber resonator consisting of 10 m of PM fiber with a birefringence beat length of 1 mm. The drift of the parasitic resonance corresponding to a temperature change of 0.1°C can span the entire free spectral range of the cavity. As an effective countermeasure, the fiber spindle at the midpoint of the ring cavity can be fusion-spliced at a 90° angle.

In addition to the above issues related to reciprocity, Rayleigh backscattering and Kerr effects in R-FOG also become prominent because of the high coherence of the light source. The parasitic interference between the main frequency and the backscattering waves, along with the Kerr effect resulting from increased optical power in the resonator, are the primary causes of R-FOG drift. Other parasitic effects unrelated to coherence, such as the Faraday effect and transient effects, are similar to those observed in I-FOG.

5.4 Fiber-Optic Hydrogen Sensor

Based on the sensing principle, the reported fiber hydrogen sensors are categorized into interferometer, microlens, evanescent field, and fiber grating types. Among the interferometric fiber sensors, the MZ interference hydrogen sensors were first introduced in 1984, followed by the fiber Fizeau interferometric hydrogen sensor in 1994, as depicted in Fig. 5.19. Both types are based on the hydrogen-sensitive property of Palladium (Pd) membrane. The expansion effect produced by the hydrogen absorption of Pd membrane causes axial strain in the optical fiber, leading to the phase change of the transmitted light. The measurement of hydrogen concentration can be achieved by detecting the phase change in the output signal of the interferometer. A Pd film is plated on the sensing arm of the MZ interferometer plates, but on the outer casing surface of the Fizeau interferometer (Fig. 5.20).

The Fiber MZ interferometric hydrogen sensor struggles to eliminate the influence of ambient temperature drift and is significantly affected by polarization state fading. Although the Fizeau interferometer overcomes the inherent disadvantages of the MZ interferometer, its demodulation technology is complex and challenging. Consequently, a fiber broadband (white light) interferometric hydrogen sensor has emerged.

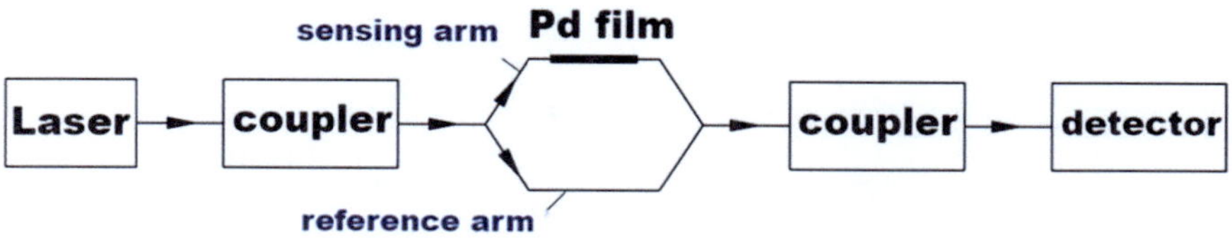

Fig. 5.19 Fiber MZ interferometric hydrogen sensor

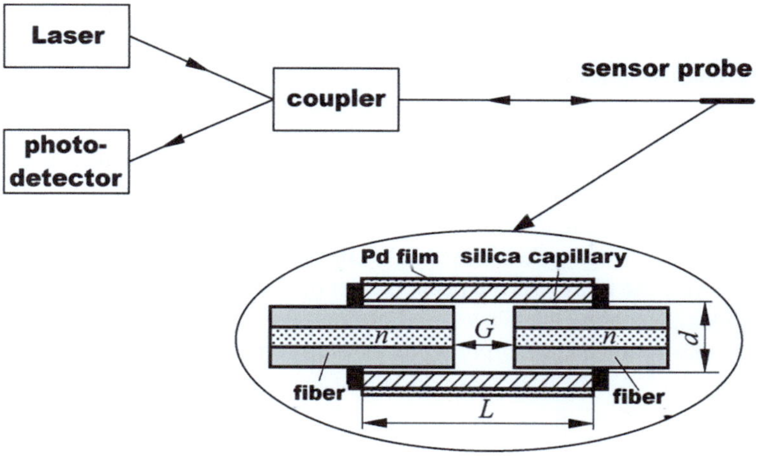

Fig. 5.20 Fiber Fizeau interferometric hydrogen sensor

5.4.1 Fiber Fizeau Broadband Interferometric Hydrogen Sensor

A fiber Fizeau broadband interference system primarily comprises three components: a broadband source, a sensor unit, and a demodulation unit (see Fig. 5.21). In comparison to a single-wavelength laser interferometric hydrogen sensing system, the broadband light interferometric hydrogen sensing system utilizes a low-coherence, wide-spectrum light source. The power fluctuations of the light source and optical fiber transmission losses have minimal impact on measurement accuracy. Additionally, the spectrum encompasses a combination of multiple single-wavelength information, which significantly exceeds the light intensity information of a single-wavelength system. Therefore, high-resolution measurement can be achieved.

As the concentration of the measured hydrogen gas varies, the hydrogen-sensitive film within the sensor undergoes deformation due to hydrogen absorption and desorption. This deformation leads to a change in the sensor cavity length, which in turn causes the interference spectral signal reflected to the detection end to alter accordingly. By calculating the change in cavity length using the interference spectrum data, the concentration of hydrogen is ultimately determined.

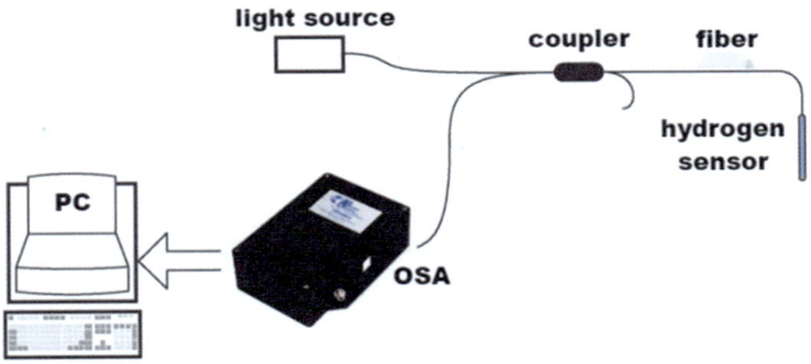

Fig. 5.21 Structure of a broadband Fizeau interferometric system

Pure palladium (Pd) film was initially used as a sensitive element. However, due to hydrogen embrittlement, the operating temperature of pure Pd film was typically above 300 °C, restricting its application in hydrogen measurement at room temperature and pressure. Doping Pd film with other metals, such as copper (Cu), silver (Ag), and gold (Au), can eliminate the phenomenon of hydrogen embrittlement, enhance its hydrogen detection capabilities, and prolong the stability of the Pd film during hydrogen absorption and desorption processes to some extent.

5.4.2 Cavity Length Demodulation and Hydrogen Sensing Based on Correlation Principles

Mutual Correlation and Demodulation Principle

Cross-correlation is typically used to describe the degree of similarity between two distinct signals. Consider two real signal sequences $x(a, n)$ and $y(b, n)$, with limited energy; their cross-correlation sequence $R_{xy}(a, b)$ is defined as:

$$R_{xy}(a, b) = \sum_{n=-\infty}^{\infty} x(a, n)y(b, n) \tag{5.34}$$

$R_{xy}(a, b)$ reaches its maximum value when $x(a, n)$ and $y(b, n)$ are precisely similar. Therefore, to calculate the variable a in the sequence $x(a, n)$, it is necessary to construct a suitable transformation function $y(b, n)$, which can be obtained through experimental test or theoretical deduction. The unknown variable a can be solved by adjusting parameter b untill reaching the maximum $R_{xy}(a, b)$ where $a = b$. This is the principle of mutual correlation demodulation.

In the fiber broadband interference hydrogen sensor, the reflective interference waveform $I_R(G, \lambda)$ solely depends on the cavity length G of the sensor, with a stable intensity output of the broadband light source. The interference stripes become denser as the cavity length increases. Consider a transformation function $I(g, \lambda)$ with length g as the variable, where the value of g includes all the possible cavity length of the interferometer, and the waveform of $I(g, \lambda)$ changes with g. When $g = G$, $I(g, \lambda)$ has the most similar waveform to that of $I_{SP}(G, \lambda)$, and the maximal integral value of the product of the two functions. Therefore, the cross-correlation functions of $I_{SP}(G, \lambda)$ and $I(g, \lambda)$ is:

$$R(g) = \int_{\lambda_{min}}^{\lambda_{max}} I_R(G, \lambda) \cdot I(g, \lambda) d\lambda \tag{5.35}$$

Among them, λ is the working wavelength. From Eq. (5.35), the cross-correlation function $R(g)$ is a function of the simulation cavity length g. The simulated cavity length of g is the real cavity length G to be measured when R(g) takes the maximum.

Assume LED to be the light source of the sensing system with an emission spectrum of Gaussian distribution, which can be described as:

$$I_0(\lambda) = I_0 \exp\left[-\frac{(\lambda - \lambda_0)^2}{(\Delta\lambda)^2}\right] \tag{5.36}$$

where λ_0 is the peak wavelength and I_0 is the peak intensity corresponding to λ_0, $\Delta\lambda$ is the spectral width of light source. For the fiber Fizeau interferometric sensor, the normalized interference spectrum is:

$$I_R(G, \lambda) = R\left[1 + \eta_F - 2\sqrt{\eta_F} \cos\left(\frac{4\pi G}{\lambda}\right)\right] \tag{5.37}$$

where F is the beam transmission coupling coefficient. After filtering out the DC term, one get

$$I_{\eta_F}(\lambda) = -2R\sqrt{\eta_F} \cos\left(\frac{4\pi G}{\lambda}\right) \tag{5.38}$$

Calculation of the Coupling Coefficient and Cavity Length Demodulation

For a broadband light interferometer, the contrast of interference fringes diminishes due to the divergence of the transmission beam within the air cavity and the aperture effect of the incident single-mode (SM) fiber. Treating the incident fiber as a point light source results in a theoretical deviation from the experimental outcomes. Using the end surface of the reflective fiber as the symmetry center, the surface of the incident fiber is mirrored. By solving the integrals of the two light fields of the returned Gaussian beam and the Gaussian beam transmitted in the fiber, a simple expression of the coupling coefficient can be obtained [7]:

$$\eta_F = \frac{\left|\int \psi_{\mathrm{e1}}(r,z)ds\right|}{\left|\int \psi_{\mathrm{e1}}(r,z)ds'\right|} = \frac{\left|\int_0^a \psi_{\mathrm{e1}}(r,z)rdr\right|}{\left|\int_0^\infty \psi_{\mathrm{e1}}(r,z)rdr\right|} \tag{5.39}$$

Under paraxial conditions, the Gaussian beam of the fundamental mode is freely transmitted through the optical fiber into the air cavity of the sensor,

$$\psi_{e1}(r,z) = A\frac{w_0}{w_1}\exp(-\frac{r^2}{w_1^2})\exp(-\frac{jk_0r^2}{2R} - j\varphi) \tag{5.40}$$

where, w_0 is the mode field radius of the Gaussian beam of the fundamental mode; $w_1(z) = w_0\sqrt{1+(z/z_R)^2}$ is the mode field radius of the Gaussian beam along the z axis (fiber axis) in free space, $z_R = \frac{\pi w_0^2}{\lambda}$ is the Rayleigh distance;$R(z) = \frac{z^2+z_R^2}{z}$ is the radius of the equal phase plane, $k_0 = \frac{2\pi}{\lambda}$ is propagation constant;$\phi(z) = -\arctan\left(\frac{z}{z_R}\right)$ is additional phase difference.

Given the significant computational demands of the relevant algorithms, an initial cavity length can be established during demodulation. The precise cavity length can then be determined through cross-correlation demodulation within a specific region around the initial cavity length (e.g., 3 μm). The estimation of the initial cavity length can be achieved using straightforward, low-accuracy methods, such as the fringe counting method.

The length of the Fizeau cavity is firstly adjusted to the preset minimum value, and the maximum of the cross-correlation function at $g = 0$ is obtained and stored. The expression is:

$$R_{max}(0) = 2(v_2 - v_1)R^2 \tag{5.41}$$

Normalize the maximum value of the obtained cross-correlation function $R_{MAX}(G)$ with respect to R_{max}, and the theoretical cavity length can be derived from Eq. (5.35). Figure 5.22 illustrates the comparison of the cavity lengths obtained using the coupling coefficient method and the demodulation algorithm. It is evident that the difference in the sensor cavity lengths obtained by these two methods is less than 3 μm, thereby eliminating the possibility of a gross error.

Measurement of Hydrogen Gas Concentration

The hydrogen sensing system monitors online the long-term trace change of hydrogen content in a sealed tank. As a static measurement, the static accuracy indicators can be selected in corresponding sensor design. The sensing unit adopts EFPI structure as shown in Fig. 5.23. The Pd–Ag alloy film thickness is 300 nm, the sensor pitch $L = 3$cm, E_{PA} is the Young modulus of Pd–Ag alloy, α_{PA} is the thermal expansion coefficient of Pd–Ag alloy,α_S and E_s are the thermal expansion coefficient and the Young modulus of quartz capillary.

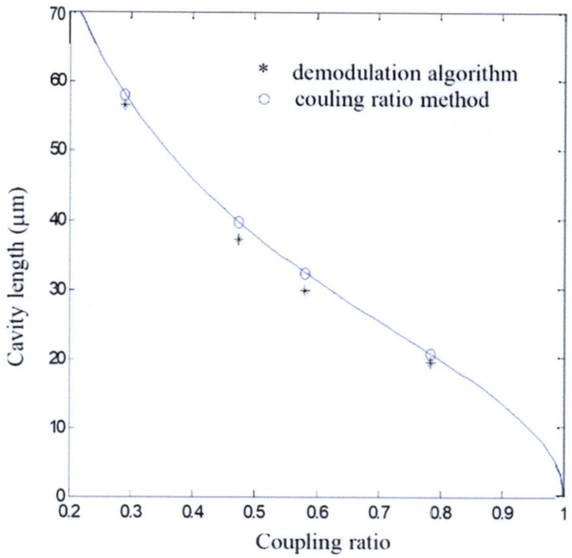

Fig. 5.22 Comparison of the sensor cavity length obtained by coupling coefficient method and the demodulation algorithm

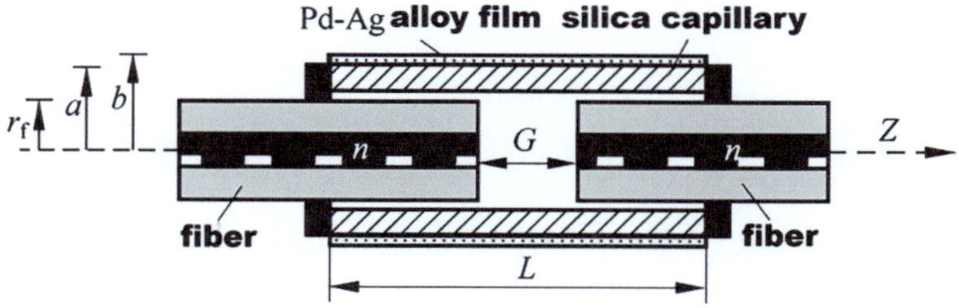

Fig. 5.23 Schematic structure of sensing EFPI

When the sensing unit is placed in a hydrogen environment, the relationship between G, the change in cavity length, and C, the hydrogen concentration, as expressed in the previous section, can be obtained as follows:

$$\Delta G = 0.12\alpha\frac{\sqrt{C}}{k}\cdot\frac{E_{\text{PA}}(b^2-a^2)L}{(b^2-a^2)E_{\text{PA}}+\left(a^2-r_f^2\right)E_s} \tag{5.42}$$

Among them, E_{PA} takes the empirical value 150 Gpa, E_s=73 Gpa, inner diameter $r_{\text{f}} = 62.5\ \mu m$, outer diameter $a = 150\ \mu m$, hydrogen concentration $C = 4\%$, Sievert coefficient $k = 350\ \sqrt{\text{Torr}}(1\text{Torr} = 133.322\ \text{Pa})$, and let the correction coefficient of theoretical amplitude $\alpha = 0.95$. The available calculated response curve of the

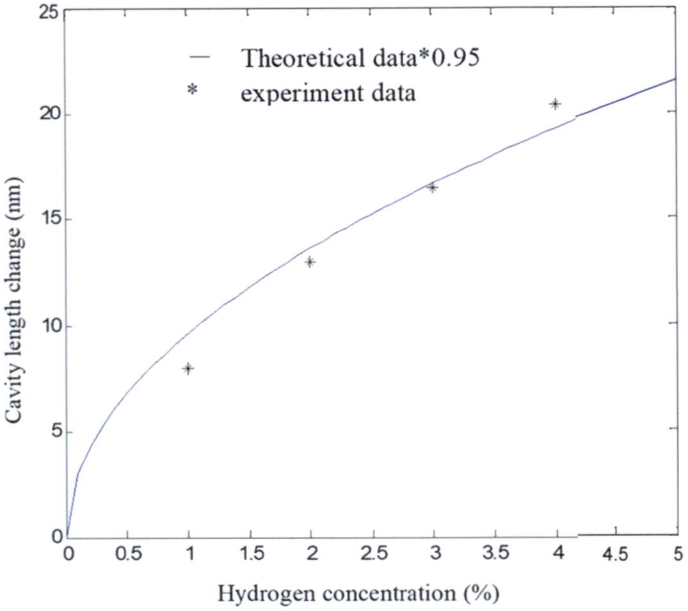

Fig. 5.24 The relationship curve of cavity length and hydrogen concentration

sensor to hydrogen concentration is shown in Fig. 5.24, from which the hydrogen concentration is measured by wavelength drift.

5.5 Long-Period Grating MZ Interferometric Refractive Index Sensor

5.5.1 Principle of the Long-Period Grating MZ Interferometer

Long period gratings (LPGs) are created by periodically modulating the refractive index of the fiber core at a scale of approximately 100 μm using ultraviolet radiation. According to the coupled mode theory, coupling occurs between the core-guided mode and the co-propagating cladding mode in LPGs, resulting in a transmission spectrum that forms an absorption peak at the resonance wavelength $\lambda_{R,j}$, which satisfies the phase matching condition.

$$\lambda_{R,j} = \left(n_{\text{eff}}^{\text{co}} - n_{\text{eff},j}^{\text{cl}}\right)\Lambda \tag{5.43}$$

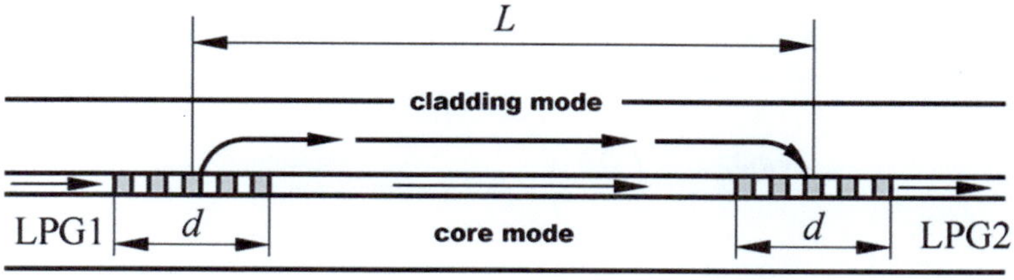

Fig. 5.25 Schematic diagram of LPG-MZ interferometer

where $\wedge$ is the period of LPG; $n_{\mathrm{eff}}^{\mathrm{co}}$ and $n_{\mathrm{eff},j}^{\mathrm{cl}}$ are the effective refractive index of the fiber core guide mode and the j-th order envelope mode, respectively. According to the dispersion of optical fiber, the effective refractive index of the optical fiber core mode is determined by the refractive index of the optical fiber material (fiber core and cladding) and the fiber core diameter. Additionally, the effective refractive index of the cladding mode $n_{\mathrm{eff},j}^{\mathrm{cl}}$ is also related to the external refractive index.

Two LPGs with a transmission loss of 3 dB are cascaded to form a LPG-MZ interferometer, as depicted in Fig. 5.25. The light wave coupled from the first grating to the cladding mode is coupled back to the core guide mode at the second grating and interferes with the light that passes directly through the fiber core, creating interference fringes within the original stopband of the LPG. The transmission output of the LPG-MZ interferometer can be derived from coupled mode theory and the transmission matrix [9]:

$$T = t^2 + r^2 \exp(-2\alpha L) - 2tr \exp(-\alpha L) \cos \psi \tag{5.44}$$

where α is the loss coefficient of the cladding mode, $\alpha = 1$ represents lossless, $\alpha = 0$ indicates complete loss; L is the distance between the centers of the two gratings, d is the length of the grating, t and r are defined as:

$$t = \cos^2 sd + \left(\frac{\Delta\beta}{2s}\right)^2 \sin^2 sd \tag{5.45}$$

$$r = \left(\frac{\kappa\kappa^*}{s^2}\right)^2 \sin^2 sd \tag{5.46}$$

κ is coupling ratio between modes; ψ is the relative phase difference between the core and cladding modes after passing through a pair of LPGs, and can be expressed as:

$$\psi = \arctan\left(-\frac{\Delta\beta}{2s} \tan sd\right) + \Delta\beta d - (\beta_{\mathrm{co}} - \beta_{\mathrm{cl}})L \tag{5.47}$$

where β_{co}, β_{cl} are the propagation constants of the core and cladding modes, respectively. s and $\Delta\beta$ are defined as

$$s^2 \equiv \kappa\kappa^* + \left(\frac{\Delta\beta}{2}\right)^2, \Delta\beta \equiv \beta_{co} - \beta_{cl} - \frac{2\pi}{\Lambda} \tag{5.48}$$

For a small coupling coefficient of the grating, $\Delta\beta$ can be considered 0 near the resonance wavelength, and the relative phase difference ψ can be approximated as:

$$\psi \approx -(\beta_{co} - \beta_{cl})L = -\frac{2\pi}{\lambda}\Delta n_{eff}L \tag{5.49}$$

where Δn_{eff} is the effective refractive index difference between the core mode and the cladding mode, which is defined as:

$$\Delta n_{eff} \equiv n_{eff}^{co} - n_{eff,j}^{cl} \tag{5.50}$$

At that time $\psi = 2m\pi$, the interference fringes of the transmission spectrum exhibited a minimum at the peak wavelength λ_p:

$$\lambda_p = -\frac{\left(n_{eff}^{co} - n_{eff,j}^{cl}\right)L}{m} \text{(m is a negative integer)} \tag{5.51}$$

5.5.2 *LPG-MZ Interferometric Refractive Index Sensor*

Any perturbation that causes a change in the relative phase difference affecting one arm of the interferometer modulates the interference fringes. The optical path of the cladding mode varies with the refractive index surrounding the LPG-MZ interferometer, resulting in the movement of the interference fringes. Consequently, the external refractive index can be measured by observing the movement of the interference fringes.

The effective refractive index of the fiber cladding mode varies with the refractive index surrounding the LPG-MZ interferometer. However, the change in the external refractive index has little effect on the effective refractive index of the core mode, rendering the change in the effective refractive index of the core mode negligible. Consequently, the relative phase difference of the LPG-MZ interferometer can be expressed as:

$$\psi = -\frac{2\pi}{\lambda}\left[\left(n_{eff}^{co} - n_{eff,j}^{cl}\right)L - \Delta n_{eff,j}^{cl}L\right] \tag{5.52}$$

where $\Delta n_{\text{eff},j}^{\text{cl}}$ represents the change in the effective refractive index as a function of the external refractive index. Equation (5.52) illustrates that the alteration in the relative phase difference, induced by fluctuations in the external refractive index, is associated with the cladding mode and the distance L between the centers of the two gratings: as L increases, so does the relative phase difference change due to variations in the external refractive index. To maintain a constant phase difference of $2m\pi$ at the peak wavelength, the corresponding shift in the interference peak wavelength $\Delta\lambda_p$ is:

$$\Delta\lambda_p = \frac{\Delta n_{\text{eff},j}^{\text{cl}} L}{m}, \text{ (m is a negative integer)} \tag{5.53}$$

According to Eqs. (5.51), (5.53) can be expressed as:

$$\Delta\lambda_p = -\lambda_p \frac{\Delta n_{\text{eff},j}^{\text{cl}}}{\Delta n_{\text{eff}}} \tag{5.54}$$

From Eq. (5.54), when the external refractive index increases, $n_{\text{eff},j}^{\text{cl}}$ also becomes larger, causing the blue shift of the interference fringes. Moreover λ_p, the movement of the interference fringes depends solely on the order of the cladding mode and the interference peak wavelength, not on the distance L between the two gratings' centers.

It is important to note that the previously mentioned interference fringe shift is solely dependent on the change in relative phase difference. This refers to the shift of interference fringes within the stopband, specifically, the shift of interference fringes relative to the stopband of the LPG. In reality, the entire LPG-MZ interferometer functions as a refractive index sensor for the measurand sample. Consequently, the stopband of the LPG itself will shift in response to changes in the external refractive index, and the extent of this shift is:

$$\Delta\lambda_{R,j} = -\Delta n_{\text{eff},j}^{\text{cl}} \Lambda \tag{5.55}$$

As evident from Eq. (5.55), the stopband of the LPG also undergoes a blue shift with an increase in the external refractive index. Consequently, the absolute wavelength shift $\Delta\lambda_{\text{Dip}}$ of the LPG-MZ interferometer is the aggregate of the LPG stopband shift and the internal interference fringe shifts within the stopband, namely:

$$\Delta\lambda_{\text{Dip}} = \Delta\lambda_{R,j} + \Delta\lambda_p \tag{5.56}$$

Since the interference fringe of the LPG-MZ is produced exclusively within the stopband, and by tracing the interference peak near the center of the stopband, the assumption of $\lambda_p \approx \lambda_{R,j}$ is made. From Eqs. (5.43), (5.51), and (5.54) to (5.56), $\Delta\lambda_{R,j} \approx \lambda_p$ and $\Delta\lambda_{Dip} \approx 2\Delta\lambda_{R,j}$ can be deduced. It is evident that the sensitivity of the LPG-MZ interferometer is twice that of a single LPG.

To improve the refractive index sensitivity of LPG-MZ interferometer, $\Delta n_{\mathrm{eff},j}^{\mathrm{cl}}$ should be increased. There are two methods:

(1) LPG-MZ interferometer with corroded cladding;
(2) Double-cone LPG-MZ interferometer.

In the corroded cladding LPG-MZ interferometer, a decrease in the cladding diameter alters the waveguide structure, thereby strengthening the evanescent field of the cladding mode in the surrounding medium. This, in turn, enhances the sensitivity of the refractive index measurement for the LPG-MZ interferometer. Experimental results indicate that the interference fringes in the transmission spectrum experience a red shift as the cladding diameter decreases. Moreover, the higher the order of the cladding mode, the greater the fringe shift observed. Simultaneously, the interference intensity also increases.

From the above discussion, it is evident that the LPG-MZ is essentially a novel MZ interferometer constructed from a single fiber.

References

1. Liwei W (2007) Investigation of fiber optic hydrogen array and key technology. Tsinghua University, BeiJing
2. Qingping S (2011) Research on the key technology of Interferometric fiber optic detector in seismic towing cable. Tsinghua university, Beijing
3. Kai Y (2008) Study of core shaft interference fiber hydrophone element and demodulation technology. Tsinghua university, Beijing
4. Zhang H (2011) Time division multiplexing technology of fiber hydrophone based on matching interferometric structure. Tsinghua university, Beijing
5. Nan Z (2005) Study of some key issues of fiber optic acceleration sensor. Tsinghua university, Beijing
6. Lefevre HC (2002) Fiber optic gyro. National Defense Industry Press, Beijing
7. Zhen Y (2011) Design and implementation of fiber optic hydrogen sensing system based on white light interference. Tsinghua university, Beijing
8. Jinfei D (2006) Study of fiber optic grating refractive index sensing technology. Zhejiang university, Hangzhou
9. Dandridge A, Tveten AB, Giallorenzi TG (1982) Homodyne demodulation scheme for fiber-optic sensors using phase generated carrier. IEEE J Quaut Electron QE-18:1647–1653

MIX
Papier aus verantwortungsvollen Quellen
Paper from responsible sources
FSC® C105338

FSC
www.fsc.org

If you have any concerns about our products,
you can contact us on
ProductSafety@springernature.com

In case Publisher is established outside the EU,
the EU authorized representative is:
Springer Nature Customer Service Center GmbH
Europaplatz 3, 69115 Heidelberg, Germany

Printed by Libri Plureos GmbH
in Hamburg, Germany